Principles of Cell Energetics

BOOKS IN THE BIOTOL SERIES

The Molecular Fabric of Cells
Infrastructure and Activities of Cells

Techniques used in Bioproduct Analysis
Analysis of Amino Acids, Proteins and Nucleic Acids
Analysis of Carbohydrates and Lipids

Principles of Cell Energetics
Energy Sources for Cells
Biosynthesis and the Integration of Cell Metabolism

Genome Management in Prokaryotes
Genome Management in Eukaryotes

Crop Physiology
Crop Productivity

Functional Physiology
Cellular Interactions and Immunobiology
Defence Mechanisms

Bioprocess Technology: Modelling and Transport Phenomena
Operational Modes of Bioreactors

In vitro Cultivation of Micro-organisms
In vitro Cultivation of Plant Cells
In vitro Cultivation of Animal Cells

Bioreactor Design and Product Yield
Product Recovery in Bioprocess Technology

Techniques for Engineering Genes
Strategies for Engineering Organisms

Principles of Enzymology for Technological Applications
Technological Applications of Biocatalysts
Technological Applications of Immunochemicals

Biotechnological Innovations in Health Care

Biotechnological Innovations in Crop Improvement
Biotechnological Innovations in Animal Productivity

Biotechnological Innovations in Energy and Environmental Management

Biotechnological Innovations in Chemical Synthesis

Biotechnological Innovations in Food Processing

Biotechnology Source Book: Safety, Good Practice and Regulatory Affairs

BIOTECHNOLOGY BY OPEN LEARNING

Principles of Cell Energetics

PUBLISHED ON BEHALF OF :

Open universiteit and **Thames Polytechnic**

Valkenburgerweg 167
6401 DL Heerlen
Nederland

Avery Hill Road
Eltham, London SE9 2HB
United Kingdom

Butterworth-Heinemann Ltd
Linacre House, Jordan Hill, Oxford OX2 8DP

 PART OF REED INTERNATIONAL BOOKS

OXFORD LONDON BOSTON
MUNICH NEW DELHI SINGAPORE SYDNEY
TOKYO TORONTO WELLINGTON

First published 1992

British Library Cataloguing in Publication Data
A catalogue record for this book is
available from the British Library

Library of Congress Cataloguing in Publication Data
A catalogue record for this book is
available from the Library of Congress

ISBN 0 7506 15044

Composition by Thames Polytechnic
Printed and Bound in Great Britain by
Thomson Litho, East Kilbride, Scotland

The Biotol Project

The BIOTOL team

OPEN UNIVERSITEIT, THE NETHERLANDS
Prof M. C. E. van Dam-Mieras
Prof W. H. de Jeu
Prof J. de Vries

THAMES POLYTECHNIC, UK
Prof B. R. Currell
Dr J. W. James
Dr C. K. Leach
Mr R. A. Patmore

This series of books has been developed through a collaboration between the Open universiteit of the Netherlands and Thames Polytechnic to provide a whole library of advanced level flexible learning materials including books, computer and video programmes. The series will be of particular value to those working in the chemical, pharmaceutical, health care, food and drinks, agriculture, and environmental, manufacturing and service industries. These industries will be increasingly faced with training problems as the use of biologically based techniques replaces or enhances chemical ones or indeed allows the development of products previously impossible.

The BIOTOL books may be studied privately, but specifically they provide a cost-effective major resource for in-house company training and are the basis for a wider range of courses (open, distance or traditional) from universities which, with practical and tutorial support, lead to recognised qualifications. There is a developing network of institutions throughout Europe to offer tutorial and practical support and courses based on BIOTOL both for those newly entering the field of biotechnology and for graduates looking for more advanced training. BIOTOL is for any one wishing to know about and use the principles and techniques of modern biotechnology whether they are technicians needing further education, new graduates wishing to extend their knowledge, mature staff faced with changing work or a new career, managers unfamiliar with the new technology or those returning to work after a career break.

Our learning texts, written in an informal and friendly style, embody the best characteristics of both open and distance learning to provide a flexible resource for individuals, training organisations, polytechnics and universities, and professional bodies. The content of each book has been carefully worked out between teachers and industry to lead students through a programme of work so that they may achieve clearly stated learning objectives. There are activities and exercises throughout the books, and self assessment questions that allow students to check their own progress and receive any necessary remedial help.

The books, within the series, are modular allowing students to select their own entry point depending on their knowledge and previous experience. These texts therefore remove the necessity for students to attend institution based lectures at specific times and places, bringing a new freedom to study their chosen subject at the time they need and a pace and place to suit them. This same freedom is highly beneficial to industry since staff can receive training without spending significant periods away from the workplace attending lectures and courses, and without altering work patterns.

Contributors

AUTHORS

Dr R.D.J. Barker, Leicester Polytechnic, Leicester, UK

Dr T.G. Cartledge, Nottingham Polytechnic, Nottingham, UK

Dr F. Dewhurst, University of Leicester, Leicester, UK

Dr R.O. Jenkins, Leicester Polytechnic, Leicester, UK

EDITOR

Dr G. D. Weston, Leicester Polytechnic, Leicester, UK

SCIENTIFIC AND COURSE ADVISORS

Professor M. C. E. van Dam-Mieras, Open universiteit, Heerlen, The Netherlands

Dr C. K. Leach, Leicester Polytechnic, Leicester, UK

ACKNOWLEDGEMENTS

Grateful thanks are extended, not only to the authors, editors and course advisors, but to all those who have contributed to the development and production of this book. They include Mrs N.A. Cartledge, Mrs S. Connor, Dr M. de Kok, Ms H. Leather, Ms J. Skelton, and Professor R. Spier. Special thanks go to the Open universiteit of The Netherlands for providing access to relevant teaching materials.

The development of this BIOTOL text has been funded by **COMETT, The European Community Action Programme for Education and Training for Technology**. Additional support was received from the Open universiteit of The Netherlands and by Thames Polytechnic.

Project Manager: Dr J. W. James

Contents

How to use an open learning text

An open learning text presents to you a very carefully thought out programme of study to achieve stated learning objectives, just as a lecturer does. Rather than just listening to a lecture once, and trying to make notes at the same time, you can with a BIOTOL text study it at your own pace, go back over bits you are unsure about and study wherever you choose. Of great importance are the self assessment questions (SAQs) which challenge your understanding and progress and the responses which provide some help if you have had difficulty. These SAQs are carefully thought out to check that you are indeed achieving the set objectives and therefore are a very important part of your study. Every so often in the text you will find the symbol Π, our open door to learning, which indicates an activity for you to do. You will probably find that this participation is a great help to learning so it is important not to skip it.

Whilst you can, as an open learner, study where and when you want, do try to find a place where you can work without disturbance. Most students aim to study a certain number of hours each day or each weekend. If you decide to study for several hours at once, take short breaks of five to ten minutes regularly as it helps to maintain a higher level of overall concentration.

Before you begin a detailed reading of the text, familiarise yourself with the general layout of the material. Have a look at the contents of the various chapters and flip through the pages to get a general impression of the way the subject is dealt with. Forget the old taboo of not writing in books. There is room for your comments, notes and answers; use it and make the book your own personal study record for future revision and reference.

At intervals you will find a summary and list of objectives. The summary will emphasise the important points covered by the material that you have read and the objectives will give you a check list of the things you should then be able to achieve. There are notes in the left hand margin, to help orientate you and emphasise new and important messages.

BIOTOL will be used by universities, polytechnics and colleges as well as industrial training organisations and professional bodies. The texts will form a basis for flexible courses of all types leading to certificates, diplomas and degrees often through credit accumulation and transfer arrangements. In future there will be additional resources available including videos and computer based training programmes.

Preface

The outstanding features of all living systems are their abilities to bring about chemical changes and to reproduce. These two features are intimately linked. A pre-requisite for cell reproduction is the capability to produce the many and varied compounds which make up cells. With a few exceptions, the chemicals available to cells from their environment are different in both kind and quantity to those cells require for the production of cellular structures. Cells therefore must carry out many different chemical conversions to produce the right chemicals in the right proportions and at the right time. It would however be wrong to consider metabolism merely as a collection of chemical reactions.

Maintenance and propagation of living systems are energy-demanding and much of cellular metabolism is concerned with generating usable forms of energy. This energy is used to carry out a wide range of cellular tasks such as maintaining an osmotic balance, repairing or replacing damaged cell components, taking up nutrients, driving biosynthesis, mobility and so on.

Central to understanding metabolism is the need to understand energetics. It is for this reason that the three BIOTOL texts which examine metabolism, ('Principles of Cell Energetics', Energy Sources for Cells' and 'Biosynthesis and the Integration of Cell Metabolism'), use energy as a recurrent and underpinning theme.

This text begins with a brief, orientating chapter which leads immediately into a study of the thermodynamic principles necessary to understand cellular metabolism. Also important in understanding metabolism is the study of enzymes, the mediators of metabolism, and the cofactors which enable enzymes to carry out their functions. These topics are examined prior to discussion of the central energy-generating mechanisms encountered in heterotrophic systems.

These central pathways include the primary breakdown of carbohydrates and their subsequent oxidation via the tricarboxylic acid (Krebs) cycle. The mechanisms of energy generation in anaerobic systems are also examined. These aspects of metabolism underpin many areas of biology, applied biology and biotechnology ranging from physiology through to fermentation technology. Opportunity has been taken to stress the metabolism of micro-organisms because of the diversity they display and because of their importance to a wide range of human activities.

This text provides readers with the prospect of understanding the central issues of metabolism which they will be able to develop further either by studying partner BIOTOL texts ('Energy Sources for Cells' and 'Biosynthesis and the Integration of Cell Metabolism') or by some other route.

Written by experienced teachers, the author:editor team has produced a logically developed discussion of the central and key issues of metabolism which will enable readers to build up their knowledge in a step-by-step manner.

Our thanks to all who have contributed to the production of a text that will enable others to experience the excitement of understanding this import aspect of biology.

Scientific and Course Advisors: Professor M.C.E. van Dam-Mieras
Dr C.K. Leach

Introduction

Introduction

1.1 General introduction

Organisms are composed of lifeless molecules which individually obey chemical and physical laws though collectively in a living cell they display extraordinary additional properties. The cell is the fundamental unit of living matter - a discrete entity which is bounded by a lipid bilayer, the cell membrane.

All living cells are complicated, highly organised structures in which each structure within every cell has specific functions and each molecule within that structure has a particular role. Perhaps the most incredible property of cells is their power to self replicate with usually-perfect accuracy and the ability to package the genetic message, the deoxyribonucleic acid, into a tiny part of their interiors.

Cells are extremely complex containing mainly proteins, carbohydrates, lipids and nucleic acids in their organic structures. Even *Escherichia coli*, a simple bacterium, has approximately three thousand proteins and one thousand nucleic acid types within its cell. Cells require energy just to maintain the organisation of these complex molecular mixtures and they also need to import compounds from which the biomolecules can be made. Cells therefore have the power to feed in the sense of attaining both energy and the required nutrients from the environment. Within such cells a vast array of chemical processes take place. This collection of processes we usually call metabolism.

three BIOTOL texts

Three BIOTOL texts have been produced to examine the whole breadth of metabolism. These are,

- 'Principles of Cell Energetics';
- 'Energy Sources for Cells';
- 'Biosynthesis and the Integration of Cell Metabolism'.

This brief introductory chapter is designed to provide a general introduction to all three texts and to give an overview of the various facets of cell metabolism. It also outlines the structure of this text and explains the assumptions about the knowledge that readers have prior to commencing the study of cell metabolism. The primary objective of this chapter is to ensure that you gain maximum benefit from the opportunities this BIOTOL text offers you to learn about cell metabolism.

1.2. Assumed knowledge

In producing any text, assumptions have to be made about the previous experience of readers. BIOTOL texts are no exception. Thus, although BIOTOL texts are carefully designed to facilitate the process of learning, some assumptions have had to be made about the knowledge base of readers. This text has been prepared on the basis that readers will be familiar, in broad terms, with the cellular nature of organisms and with the subcellular organisation of cells. It is also assumed that readers will be familiar with the major molecular species found within cells. If you are not familiar with these topics, we recommend the BIOTOL texts, 'The Molecular Fabric of Cells', and 'Infrastructure and Activities of Cells'.

1.3 BIOTOL texts and the division of metabolism

Cell metabolism is about the processes involved in using the substances cells procure from their environment to produce both the energy and cellular materials they need to grow and reproduce. We can subdivide these processes into three major groups of activities. These are:

- the generation of a usable form of energy;
- the generation of simple molecules to act as precursors of cell constituents;
- the synthesis of new cell constituents.

In dealing with the complexities of metabolism, it is eminently sensible to divide the topic into groups of activities. These include:

- the development of knowledge of the underpinning thermodynamic principles needed to understand metabolism;
- the principles of the common mechanisms encountered in metabolism (properties and modes of action of enzymes and coenzymes, electron transports coupled to phosphorylation, substrate level phosphorylation);
- the common and central metabolic pathways encountered in catabolism with particular emphasis on the catabolism of carbohydrates.

These topics are covered in this text. The other two BIOTOL texts dealing with metabolism examine the following aspects:

- 'Energy Sources for Cells' - builds on the knowledge gained from this text and examines the diversity of energy sources and generating systems found amongst organisms;
- 'Biosynthesis and the Integration of Cell Metabolism' - explores how cells make new cell constituents and how metabolism as a whole is co-ordinated.

In order to put these three texts into context, we need to briefly consider the range of starting materials for metabolism used by cells. These starting materials are, of course, the nutrients that cells absorb from their environment. In the next sections we will examine the diversity of the nutrients used by cells which will enable us to identify a variety of nutritional types. Do not be overawed by this diversity. Much of the metabolism of these various types is based upon common principles and concepts. Throughout this text we will mainly concentrate on these common mechanisms and metabolic pathways.

We have placed special emphasis on the metabolic activities of micro-organisms. This is because of the diversity of the metabolic activities they display and of their importance in biotechnology. Nevertheless, many of the central pathways described are directly relevant to understanding the metabolism of plant and animal systems. You should, however, realise that with these multicellular systems, we need, not only to consider metabolism at a cellular level but also to consider it in terms of specialised cell and organ function associated with differentiation and morphogenesis. These latter features extend metabolism into the domains of plant and animal physiology. The spatial organisation and control of metabolism in plant and animal systems are described in the BIOTOL texts, 'Crop Physiology' and 'Functional Physiology'. Here we confine our discussion to metabolic processes at the cellular level.

1.4 Nutritional requirements of living systems

To succeed, that is to grow and reproduce, living systems must be able to procure from their environment all of the substances they require to produce both energy and cellular materials. Such substances are their food or nutrients and must be obtained by, or be supplied to, each organism according to its needs. As we shall see, such needs vary enormously depending on the type of living system. Before looking at the diversity of these different nutrients we can make some generalisations concerning the essentials required for successful growth.

composition of cells

Most living systems consist of 70 to 90% by weight of water, a compound which has always to be available for life. The remaining 10 to 30%, called the dry weight, contains surprisingly few elements; there are about ninety natural elements in the Earth's crust but less than thirty are found in, and required by, living systems.

Π Make a list of as many of these thirty or so elements required by living systems as you can. Try to put those required in greatest quantity to the front of your list. (Do this before reading on).

Your answer should include:

carbon	50% of the dry weight
oxygen	20% of the dry weight
nitrogen	14% of the dry weight
hydrogen	8% of the dry weight
phosphorus	3% of the dry weight
sulphur) potassium) sodium)	each approximately 1% of the dry weight

Other elements include magnesium, calcium, iron, manganese, chlorine together with some additional metals and halogens.

The first six elements in the list above constitute around 96% of the dry weight of the cell and can generally be supplied in one or a variety of forms. Some organisms may accept many alternative sources of each element but, as we shall see, others are fastidious in their requirements. It must be remembered that elements which are only required in tiny or trace amounts are nevertheless vital to the well-being of the cell; for example only a small fraction of a percentage of the dry weight of an organism is copper, but it is a required cofactor for some enzymes. Interestingly copper at higher concentrations is toxic, particularly to algae and fungi. In addition to elements needed albeit at relatively low levels, some elements are occasionally required in substantial amounts by specific cells; for example diatoms require large amounts of silica for their silica-based exoskeletons.

In the laboratory we must supply all of the necessary ingredients for successful growth of the organism. The nature of the carbon source is very important, particularly whether it is in the form of an organic or inorganic substrate. The carbon source often acts both as an energy source and as a substrate for growth. Nitrogen is generally supplied in reduced form as the ammonium ion though photosynthetic bacteria prefer nitrogen in nitrate form. Sulphur is generally supplied to organisms as sulphate but some bacteria prefer reduced sulphur compounds or elemental sulphur.

The majority of metals can be supplied individually or it may be possible for an organism to obtain a sufficient quantity from impurities in the medium.

anaerobes, aerobes and facultative anaerobes

The availability of hydrogen and elemental oxygen is not usually a problem, being provided as water or as constituents of organic compounds. We must, however, mention molecular oxygen separately from elemental oxygen. The latter, as we have just indicated, is required by all living systems. Molecular oxygen, however, is only required by organisms growing aerobically - termed aerobes. Conversely anaerobes grow in the total absence of molecular oxygen and in fact this molecule is often toxic to them. There are some organisms which are able to grow both in the presence and in the absence of molecular oxygen and these are referred to as facultative anaerobes. Eukaryotic cells are, almost without exception, strictly aerobic; a few fungi such as yeasts of the genus *Saccharomyces* are facultative anaerobes and a few protozoa and possibly one or two fungi and algae are strict anaerobes. Although the majority of bacteria are aerobic there are many facultative and anaerobic species.

growth factors

Let us introduce another term - growth factor. In biology, this term is used in two different ways. It is a term which is applied to organic compounds which cannot be synthesised by an organism but are required for growth. For example a yeast which could not synthesise the amino acid valine would need to obtain this compound from its environment. Without it the yeast would not be able to grow since this amino acid is needed for protein synthesis.

The term growth factor is however also used in another way especially in discussing the growth of plant and animal cells. In multicellular systems it is important that the growth and development of different parts of the organism occur in a controlled and well regulated way. For example the organs of our bodies (heart, liver, kidneys etc) need each to be of an appropriate size. The control of growth of different cell types is achieved by a series of 'messenger' (hormone-like) molecules. These regulators are often referred to as growth factors. For example the growth of epidermal cells of mammals is regulated by, amongst other things, a specific epidermal growth factor. Such growth factors are not nutrients, but are essential for co-ordinated growth.

In this text, we predominantly use the term growth factor to denote nutrients which particular cells cannot synthesise but which are essential for cell metabolism and growth. You should perhaps think of these growth factors as accessory nutrients.

Π Make a list of the compounds which may be regarded as accessory nutrients.

Vitamins are the group of compounds usually quoted but some living systems also require one or more amino acids and nitrogenous bases such as purines or pyrimidines.

Within the world of microbes some organisms, for example *Escherichia coli*, grow on a simple salts medium containing metal ions, ammonium sulphate, phosphate and glucose. Thus it can synthesise all of its necessary organic compounds from glucose. Other organisms however are more nutritionally demanding, an extreme example being *Leuconostoc mesenteroides* which must be supplied with glucose, ammonium chloride, six minerals, acetate, nineteen amino acids, three purines, a pyrimidine and ten vitamins. One is tempted to ask how it survives in the 'wild'.

We have now considered the chemical nature of the medium composition. However for successful growth, cells require a suitable environment. Principally we look for a suitable temperature for growth, protection from extremes of pH, correct molecular oxygen tension and protection from harmful radiation. It may be necessary to provide light. Selection of a suitable environment is not always a straightforward task due to the diversity of cells, especially amongst micro-organisms. For instance some bacteria have an optimum temperature for growth of approximately 15°C and do not grow above 25°C. Others grow optimally at 55°C and will not grow below 44°C.

We have seen that each cell requires a variety of elements and nutrients for successful life. Much of this unit will be devoted to how cells utilise these nutrients to their advantage, converting some resource to a whole host of complex structural and functional components and degrading some nutrients during the production of energy. The overall process is incredibly complex and is generally termed intermediary metabolism.

We have so far noted the diversity of nutritional requirements exhibited by the range of living systems and shortly we shall investigate the flexibility exhibited by individual organisms. Before this, however, we shall study the nutritional classifications of living organisms, that is the way in which organisms may be grouped according to their required carbon source and energy source. This type of grouping is particularly important when studying micro-organisms.

1.5 Nutritional classification of living systems

descriptive terms of nutritional categories

Until around 100 years ago living systems were divided into two categories; autotrophs - for example plants - and heterotrophs -for example animals. Autotrophs were considered to be organisms which could exist in a totally inorganic environment whereas heterotrophs required organic compound(s) for growth. An alternative name for heterotrophs was organotrophs. As the knowledge of the diversity of micro-organisms widened it became apparent that the two categories were inadequate in that many organisms could not simply and absolutely be assigned to one of the two categories, autotroph and heterotroph.

The first attempt at a more detailed classification came when scientists distinguished organisms on the basis of the principal carbon source together with the principal energy source. Carbon sources could be either organic or inorganic and the energy could be derived from light or from chemicals. Thus four categories were possible as follows:

principal carbon source	organic	heterotroph
principal carbon source	inorganic	autotroph
principal energy source	chemicals	chemotroph
principal energy source	light	phototroph

The term 'troph' is from the Greek 'to feed'; thus an autotroph is an organism which 'self-feeds'; in this case uses inorganic nutrients as a carbon source.

The first terms shown above are relatively straight forward but the terminology now starts to become more complicated. The next step was to put the terms together, for example an organism which utilises organic compounds as principal carbon source and chemicals for energy is termed as chemoheterotroph.

Π Can you name the other three combinations and define the nutritional requirements of such organisms?

The three other possibilities are: chemoautotroph; photoheterotroph; photoautotroph.

A chemoautotroph is an organism which utilises inorganic compounds as principal carbon source and chemicals as an energy source.

A photoheterotroph is an organism which utilises organic compounds as carbon source and light as an energy source.

Finally, a photoautotroph is an organism which utilises inorganic compounds as carbon source and light as an energy source. In summary:

energy source	carbon source	name
light	organic compounds	photoheterotroph
light	inorganic compounds	photoautotroph
chemical	organic compounds	chemoheterotroph
chemical	inorganic compounds	chemoautotroph

lithotrophs

mixotrophs

Examination of the summary table now raises the question - as carbon source is separated into organic and inorganic sources, why not energy sources in chemotrophs? In fact it has been known for many years that some organisms can grow in a totally inorganic environment meaning that they must use inorganic chemicals as an energy source. Organisms using inorganic chemicals are termed lithotrophs (the Greek 'litho' meaning rock). Organisms therefore using inorganic chemicals as both carbon and energy source are called chemolithotrophs - these being a type of chemoautotroph. One other group occurs which uses inorganic compounds for energy but organic compounds as a carbon source, the lithotrophic heterotrophs or mixotrophs. These organisms, restricted to some bacterial species produce energy from the oxidation of inorganic compounds but their reducing power for biosynthesis and carbon for assimilation are both obtained from organic compounds.

We have now distinguished the major nutritional groups (photoheterotrophs, photoautotrophs, chemoheterotrophs and chemoautotrophs).

Π Can you estimate how extensive each group may be across the whole of living systems and give general examples of types of organisms fitting into each group?

Higher plants use light as an energy source and are therefore phototrophs. They utilise carbon dioxide as carbon source and are therefore photoautotrophs. This group also includes green algae, blue green algae (the cyanobacteria) and some photosynthetic bacteria; overall a very large group.

Photoheterotrophs are a very restricted group containing some members of the purple and green photosynthetic bacteria.

The largest single group are the chemoheterotrophs, a group containing all animals, protozoa, fungi and the majority of bacteria.

Chemoautotrophs (chemolithotrophs) are restricted to bacteria but form a large group which is very important in nature as it includes bacteria involved in the nitrogen, sulphur and iron cycles.

Some micro-organisms can apparently change their nutritional status. For example some algae grow photoautotrophically when in the light for several hours per day. However if they are incubated continuously in the dark they readily grow as chemoheterotrophs. Such organisms are said to be facultative autotrophs; those which will only grow in an autotrophic mode are said to be obligate autotrophs.

This section has been somewhat lengthy and has probably yielded several unfamiliar terms. However we shall be meeting these regularly throughout the chapters of this unit so it is worthwhile making sure that you understand them at this stage; regular exposure to them later will make them seem almost second nature to you.

1.6 Intermediary metabolism - the processes involved

This section will largely be used to define the terms employed during our studies of intermediary metabolism. It is wise not to try to learn these definitions word-perfectly but better to explore and understand them.

Intermediary metabolism is often defined as the sum total of all of the processes which occur within a cell. The term is apt because metabolism proceeds in a stepwise fashion via intermediates, which are often referred to as metabolites. In one sense the definition of intermediary metabolism is correct but it is perhaps only part of the real meaning of the process. It does not indicate the incredible complexity of metabolic sequences, their near perfect accuracy and efficiency, superb co-ordination and the overall degree of integration.

The processes of intermediary metabolism can be divided into several areas which are intimately related. Firstly the cell has to use nutrients or sunlight as an energy source; secondly it has to acquire and/or convert nutrients into building blocks for macro-molecular biosynthesis; thirdly the macro-molecules are synthesised using the energy and building blocks provided. Finally the cell has to organise these processes, monitor concentrations and control turnover of macro-molecules. Overall - a daunting task indeed.

catabolism and anabolism

Metabolism has for many years been separated into two areas, catabolism and anabolism. Catabolism is the process by which intra- or extracellular molecules are degraded to yield smaller molecules which are either waste products or building blocks for biosynthesis. During catabolism there is generation of reducing power such as NADH + H^+, $FADH_2$ and NADPH + H^+. (Do not be too concerned if you are not familiar with these molecules, we will be dealing with them in detail later). The subsequent reoxidation of these cofactors, principally NADH + H^+, releases energy which the cell converts to ATP, the major chemical energy currency. NADPH + H^+ is largely used as a source of reducing power for anabolic reactions.

Anabolism is the building up or biosynthesis of complex molecules such as proteins and nucleic acids, from simple intra- or extracellular precursors. This is an energy requiring process.

Throughout living systems catabolic processes are known to be extremely diverse. Anabolic processes, however, are thought to be remarkably similar in all living systems; a phenomenon first called the 'unity of biochemistry' by Kluyver in 1926 and still held as correct today.

amphibolic pathways

Catabolism and anabolism are obviously very closely linked for metabolism to be successful. It was found many years ago that some individual pathways can in fact be both catabolic and anabolic. Such pathways are termed amphibolic (the Greek 'amphi' means both). They are catabolic in that they are able to degrade molecules to yield energy but are anabolic in that they produce building blocks for biosynthesis.

Let us now try to get an overview of anabolism and catabolism taking into account the types of energy sources used by organisms (as discussed in Section 1.5 of this chapter).

SAQ 1.1

Consider Figure 1.1. There are several words omitted from the diagram, indicated by question marks. Try to fill in the missing words and spend a few minutes studying the completed diagram.

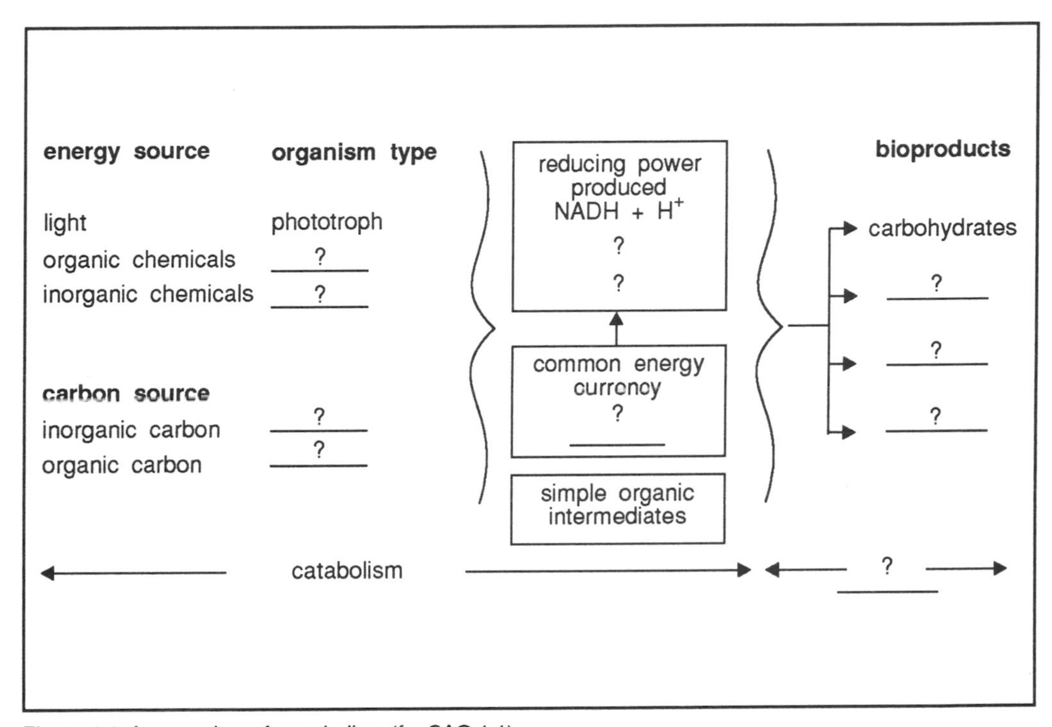

Figure 1.1 An overview of metabolism (for SAQ 1.1).

The completed version of Figure 1.1 serves to demonstrate the diversity of nutritional sources (on the left-hand side). But particularly note that whatever the source of energy and carbon, the same types of intermediates (reducing power in the form of NADH, NADPH, $FADH_2$ and energy in the form of ATP) are involved. Note also that these common intermediates are the starting point for the production of cell constituents (on the right-hand side of Figure 1.1).

We noted in Section 1.5 that organisms can be divided into three groups depending on their requirements for molecular oxygen; namely aerobes, anaerobes and facultative anaerobes. We have also examined a very superficial overview of metabolism (Figure 1.1). Thus we can now attempt to define and understand the term respiration and subdivide it into aerobic and anaerobic respiration. The conventional understanding of respiration when applied to ourselves is breathing, that is the intake of oxygen and the output of carbon dioxide. But this does not explain respiration at the molecular level which, in the case of higher animals, is essentially the oxidation of substrates to carbon dioxide and the reduction of oxygen to water. As we shall see, there are other types of respiration.

aerobic respiration

We will be examining the processes of oxidation and reduction in detail in Chapter 3. For now we will simply define oxidation as the removal of electrons from a compound or element and reduction as the addition of electrons to a compound or element. The oxidation of substrates during catabolism therefore involves the removal of electrons from the substrate. This implies that there is a need for an electron acceptor. In aerobic cells the electron acceptor is always molecular oxygen. During the transfer of electrons from these substrates to oxygen, energy is released. We can represent this process by:

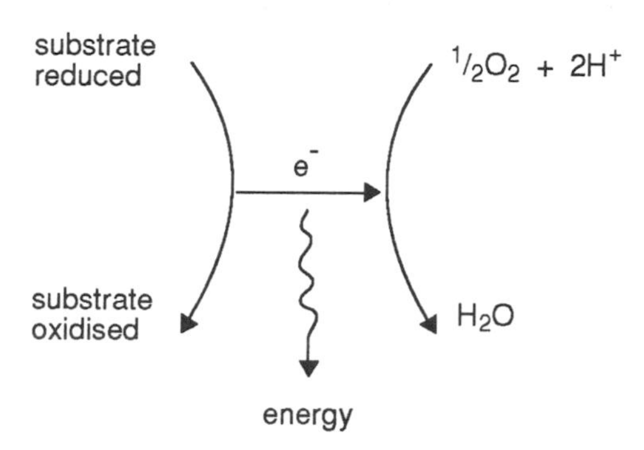

The energy released during the oxidation is harnessed by the cells to produce ATP by means of a process called oxidative phosphorylation and involves an electron transport chain. A major part of this text examines the processes involved in the oxidation of substrates, the nature of the electron transport chain and the mechanism by which the transport of electrons along the chain is coupled to oxidative phosphorylation. In aerobic respiration the oxidation of substrates results in the reduction of molecular oxygen to water.

anaerobic respiration

Anaerobic respiration, on the other hand, never involves molecular oxygen; it is an ATP yielding process in which inorganic compounds such as nitrate or sulphate serve as terminal electron acceptors. Organisms may carry out anaerobic respiration in a totally anaerobic environment or in air but molecular oxygen is not involved. Again we will examine this process in detail in a later chapter.

fermentation

A few micro-organisms, notably some bacteria and yeasts of the genus *Saccharomyces*, can carry out a process called fermentation. Fermentation is an anaerobic, energy-yielding process in which organic substrate(s) act as electron donor and also as electron acceptor. There are usually a variety of carbon end-products, some more oxidised and some more reduced than the starting material. We will see that in fermentation the net oxidation state of the carbon products is identical to the oxidation state of the substrate(s). Essentially what is happening is a re-arrangement of the atoms of the substrate. Note that the term 'fermentation' is a precisely defined, anaerobic process. It is often misused to describe industrial processes such as 'penicillin fermentation', however this process is an aerobic one and cannot therefore be a fermentation. Fermentation is a process we will examine in the final chapter of this text.

photosynthesis

The fourth way in which some organisms can derive chemical energy is from light by the mechanism of photosynthesis. Photosynthesis is the process by which certain organisms convert solar energy to chemical (ATP) energy. It is a two stage process; the first stage or light reaction is the production of ATP and reduction of $NADP^+$ to $NADPH + H^+$ using light energy. In the second stage the ATP and reducing power are used to reduce carbon dioxide to cellular products.

chemolithotrophs

Finally some organisms (the chemolithotrophs) use the oxidation of inorganic substrate to produce a usable source of cellular energy. Photosynthesis and chemolithotrophy are topics covered in the BIOTOL text, 'Energy Sources for Cells'.

1.7 Catabolism - an introduction

the three phases of metabolism

Let us study an introductory, very simple diagram of catabolic processes as shown in Figure 1.2. As you can see the diagram is divided horizontally into three phases, I, II and III. Many of the compounds or groups should be familiar to you, for example the starting materials - polysaccharides, proteins and lipids. The overall diagram should give the impression of diversity of starting materials, almost like a funnel channelling compounds to a relatively restricted number of central metabolites. Many such diagrams in the literature, particularly if describing catabolism in higher organisms, would not include the fourth starting point termed 'other compounds'. After identifying examples of these other compounds we shall consider why we have included them in this diagram.

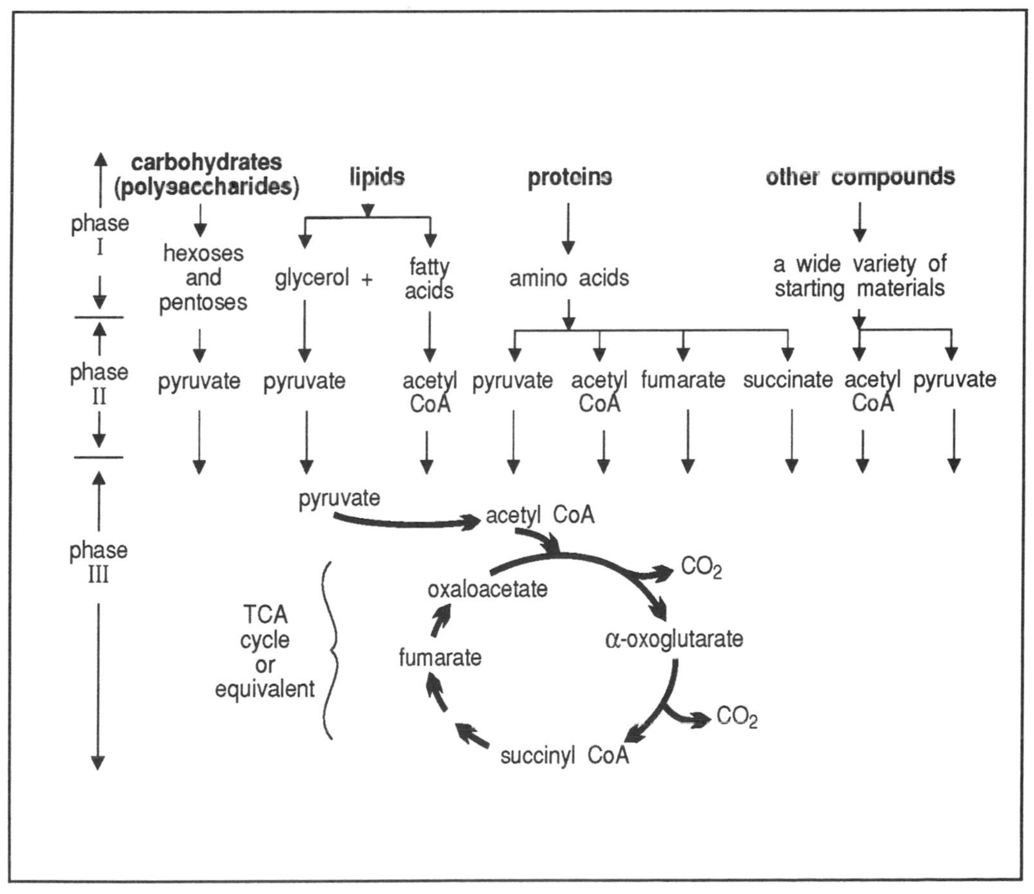

Figure 1.2 The overall pathway of heterotrophic metabolism. TCA = tricarboxylic acid cycle.

Π Write down any general classes of compounds which you consider may be metabolised other than polysaccharides, proteins or lipids?

Such classes will include nucleic acids, aromatic compounds and hydrocarbons.

Π Can you make any generalisation as to which compounds are acceptable carbon and energy sources for micro-organisms compared to compounds acceptable to higher organisms?

In fact higher animals including mammals are very restricted in the types of compounds which they are able to utilise as the main sources of carbon and energy. Such organisms will utilise proteins which contain conventional amino acids, many lipids - particularly neutral lipids and some polysaccharides but relatively few other organic compounds.

Even the use of polysaccharides by higher animals is restricted; for example alpha-linked glucose storage polysaccharides such as glycogen and starch are ideal carbon and energy sources but the β-linked glucose polysaccharides such as cellulose cannot be digested by higher organisms.

microbial infallibility

Virtually any organic compound known can be utilised by some micro-organism (especially bacteria). This is a surprising, sweeping statement due to the extremely large number of compounds known but it is seemingly correct. This ability of micro-organisms was first suggested by Gale in 1952 and referred to as the 'principle of microbial infallibility'. Individual bacteria, particularly of the family *Pseudomonadaceae* show tremendous versatility being able to degrade many hundreds of different organic compounds. It must be remembered, however, that there are many bacterial species which are of restricted ability, some even more restricted than higher animals in the number of compounds which they can utilise.

degradation of complex compounds

Thus phase I (Figure 1.2) of metabolism is characterised by the degradation of complex compounds to relatively simple compounds recognised as the entry points into the central pathways of metabolism. The simple compounds at the end of phase I are the same in all living systems; it is only the number of starting materials which varies. Energy changes in terms of ATP production or utilisation are low or zero in this phase.

breakdown into common intermediates

Phase II of metabolism is virtually identical in all living systems. From common entry precursors all living systems use well defined pathways to produce central metabolites such as pyruvate, acetyl CoA and TCA cycle intermediates. You may not be familiar with the TCA (tricarboxylic acid) or Krebs cycle as it is sometimes called. If not, do not worry, we will be describing this later. Essentially it is an arrangement of reactions which operate as a cycle. During phase II, reducing power and a small amount of energy are generally produced by substrate level phosphorylation, that is direct production of ATP (we shall learn details of this in later chapters). The term 'small amount' is used here when comparing the ATP yield with that possible in phase III. As we shall see later, however, this ATP yield is important - vital indeed to certain micro-organisms as it may be their only source of ATP.

central metabolic pathways

Phase III is shown in the diagram as 'TCA cycle or equivalent'. We again can divide organisms into two main groups: aerobes and anaerobes. Aerobes (from higher animals through to simple bacteria) have a complete TCA cycle. In these systems, the cycle represents the final oxidation of carbon nutrients. It represents a major energy generating system. In anaerobes the TCA cycle is incomplete and the cycle is not used for energy generation.

1.8 Anabolism - an introduction

Any studies on individual cells or whole organisms, however superficial, will very quickly give an idea of the complexity of such systems both in terms of overall structure and function and in terms of the numbers of biological compounds which they contain. All of these biological compounds have to be obtained from the medium or synthesised from suitable nutrient sources. Such biosyntheses proceed in an orderly, stepwise manner until all the cellular requirements are made enabling the cell at least to survive and at most to grow and divide. Some organisms such as photosynthetic organisms can obtain all of their carbon compounds from carbon dioxide but non-photosynthetic organisms generally need an organic carbon source, the compound often serving as an energy source as well. The production of new biomolecules invariably requires energy and often involves the reduction of carbon compounds. Anabolism is therefore characterised as requiring energy and reducing power. Whatever the starting point(s), the early stages in anabolism, that is the production of simple molecules such as amino acids, fatty acids and carbohydrates, are very similar in all living systems. It is generally agreed that all living cells have to obtain or synthesise twelve simple molecules, termed the precursor metabolites, from which all other biomolecules can be produced.

precursor metabolites

The precursor metabolites are:

glucose-6-phosphate; fructose-6-phosphate;

glyceraldehyde-3-phosphate; 3-phosphoglycerate;

phosphoenol pyruvate; pyruvate;

acetyl CoA; oxaloacetate;

succinyl CoA; α-oxoglutarate;

ribose-5-phosphate; erythrose-4-phosphate.

If at present you are not familiar with intermediary metabolism this may appear as a complicated list of chemicals. However they are all relatively simple compounds and are metabolically closely related. You will become very familiar with them through usage. Do not attempt to learn them just yet. By the time you have completed this text you will be familiar with them and also know how they are produced.

Π From our studies to date we have implied that some forms of catabolic processes must precede anabolism. Very briefly, what would you consider to be the major requirements for catabolic and anabolic processes and what are the products of each?

Catabolism requires a source of nutrients (or light) from which it can obtain reducing power in the form of, for example NADH + H^+, and also ATP. In addition it yields a variety of simple molecules which must include the twelve precursor metabolites.

Anabolism requires reducing power, precursor metabolites and ATP and will, in several stages, produce further precursors (for example amino acids) and subsequently all of the cellular biopolymers.

products of catabolism used in anabolism

The links between these two processes then are ATP, precursor metabolites and reducing power - NADH + H^+, NADPH + H^+ and $FADH_2$. (Look back at Figure 1.1, it shows the links very clearly). All are products of catabolism and requirements of anabolism. One of the problems which is common to all living cells and which we must bear in mind at all times when studying metabolic strategy is that cells do not have the capacity for directly storing large amounts of ATP or reducing power. They do not have the ability to operate an equivalent of a bank vault in times of plenty in which to deposit ATP. On the other hand there is always a demand for at least a small amount of ATP to keep cells alive. Thus catabolic and anabolic processes are always occurring and must be constantly adjusted to meet the needs of each process.

We are now in a position to explain the layout of the text.

1.9 Arrangement of chapters

From our discussion so far it should be self-evident that the production and use of energy and reducing power by cells are central issues of metabolism. The study of energy relationships is called thermodynamics and it is the principles and concepts of thermodynamics which we will deal with in Chapter 2. Reducing potential and compartmentalisation are closely interrelated with thermodynamic considerations and these are the topics which are the focus of Chapter 3. Chapters 2 and 3 therefore seek to develop the knowledge of energy and oxidation/reduction processes necessary to understand metabolism.

The other important aspect of metabolism is of course the many chemical changes involved in using nutrients to generate a usable form of cellular energy and reducing power and to produce new cell constituents. From your knowledge of the diversity of nutrients and the chemical complexity of cells it should be self evident that there is an enormous array of such chemical changes. Such chemical changes normally would occur at extremely slow rates. Biological systems are, however, equipped with special catalysts, the enzymes, which speed up these reactions. Thus to fully appreciate the processes of metabolism, we must also be familiar with the properties of enzymes. Although this text has been written on the assumption that you are aware of the existence of enzymes and have knowledge of their general properties, Chapter 4 provides some helpful reminders of these general properties. It does, however, extend into the kinetic properties of enzymes, the regulation of their activities and the role of cofactors and coenzymes; all essential to an understanding of the processes and regulation of metabolism.

The remainder of the text focuses on the three principal stages of catabolism (Figure 1.2), with particular emphasis on the catabolism of carbohydrates. Catabolism of other major substrates (lipids and proteins) and the generation of energy, reducing power and biosynthetic precursors in photo- and chemoautotrophs are covered in the BIOTOL text, 'Energy Sources for Cells'.

Anabolism is the major topic of the BIOTOL text, 'Biosynthesis and the Integration of Cell Metabolism'.

A key feature of living systems is that the various catabolic and anabolic processes are not conducted in a haphazard way. They are carefully controlled. Throughout the text, we will draw your attention to aspects of the regulation of the various processes we have described. The full extent of metabolic regulation can, however, only be appreciated from a knowledge of both catabolism and anabolism. Therefore we have chosen to include the major discussion of the regulation of metabolism in the final text of this unit, 'Biosynthesis and the Integration of Cell Metabolism'.

Summary and objectives

This chapter serves as an introduction for the BIOTOL unit on metabolism and also as an introduction to this text, which is the first of the three that make up the unit.

We have examined the nutritional requirements of cells and shown that living systems can be grouped according to the nature of their energy sources and the nature of their carbon source. We explained that intermediary metabolism is complex involving two basic, but closely co-ordinated processes, catabolism and anabolism.

Catabolism involves the production of reducing power, energy (in the form of ATP) and twelve essential precursor metabolites. Anabolism is the process by which energy, reducing power and precursor metabolites are used to produce new cell constituents.

Now that you have completed this chapter you should be able to:

- list the nutritional requirements of living cells;
- describe the roles of nutrients;
- describe a nutritional classification of organisms based on their carbon and energy requirements;
- define and discuss the terms metabolism, catabolism and anabolism;
- discuss heterotrophic catabolism and, with the aid of a flow diagram, indicate the phases of generation of reducing power, of energy and of precursor metabolites;
- show, using a diagram, how catabolism and anabolism are intimately related.

Energy and equilibria: an introduction to biochemical thermodynamics

Energy and equilibria: an introduction to biochemical thermodynamics

Thermodynamics are important in many areas of biology. Many texts in this field, however, offer relationships between various thermodynamic entities without explaining fully what those entities are nor how the relationships are derived. The reader is often left with little real understanding. In this text we have elected to give a fuller discussion of the principles underpinning thermodynamics. This we believe will prove to be a greater value to the reader's subsequent studies and work than a collection of half understood truisms. As a consequence this chapter is rather a long one. Do not attempt to study it all in one sitting. Study each section and make sure you have understood it before moving on to the next.

In the first part of the chapter we explain why the study of thermodynamics is important and describe the fundamental approaches to the study of thermodynamics. This ground work is built upon by consideration of heat, work and energy interconversions. Much of the chapter examines the laws of thermodynamics and their application to biochemical problems. The thermodynamic concepts you will learn by studying this chapter have wide application in biology, chemistry, physical sciences and in technology.

2.1 Why study thermodynamics?

The value of thermodynamics to the biotechnologist lies in the type of information it can provide.

This may be concisely summarised as follows:

1) Relationships between quantities of heat and work in biological systems may be obtained from the application of the first law of thermodynamics.

   These relationships can be applied to both non-equilibrium systems and to systems at equilibrium. The information we derive from these relationships is vital to understanding metabolism and for the design and operation of bioreactors. It forms a basis for predicting the influence of physical and chemical parameters on various processes.

2) For systems at equilibrium the effect of changes in temperature, composition etc on a variety of physio-chemical and biological phenomena can be predicted by applying the first and second laws of thermodynamics.

   The ability to predict the position of equilibrium is fundamental to the manipulation of biological systems to optimise yields of desired products in biotechnology. It is also basic to the understanding of biochemical processes.

3) Non-equilibrium processes involving temperature, electromotive force (emf), or concentration gradients and the flow (or flux) of heat, mass and electricity can be analysed by non-equilibrium or irreversible thermodynamics. This extension of classical thermodynamics gives a unified method of treating the transport processes which lie at the heart of all living organisms and all bioreactors and other chemical engineering plants.

2.1.1 Macroscopic and microscopic perspectives

macroscopic and microscopic approaches

Two contrasting but ultimately complementary approaches have characterised the development of the science of biology. One approach involves the study of large populations whilst the other is concerned with the behaviour of individual entities, organisms, cells or even atoms and molecules. We may distinguish these two approaches as the macroscopic and microscopic perspectives.

definition of a system

In all sciences, analysis begins with conceptually (or even physically) separating a defined region of space or a finite portion of matter, termed a system, from its surroundings. The surroundings contain everything outside the system which has a direct bearing upon the properties of the system. The system is then described using parameters related to the behaviour of the system, and to its interactions with its surroundings.

macroscopic properties

The macroscopic description of a system involves the specification of a small number of fundamental measurable properties.

Volume, temperature, pressure and composition are examples of the macroscopic properties which may be used to characterise a physical or chemical system.

Macroscopic properties useful for the scientific description of systems are characterised by:

- being few in number;
- involving no special assumptions concerning the structure of matter;
- generally being measurable directly.

Pressure is a typical macroscopic property and was exploited for the successful description and understanding of physical systems long before the development of the atomic, molecular and kinetic theory concepts upon which our modern understanding of pressure is based.

microscopic properties

The microscopic description of a system in contrast requires assumptions to be made about the structure of matter. The existence of atoms and molecules is assumed for example and postulates are developed relating to their motion, structure, energy states and interactions. Characteristically it is necessary to specify a number of quantities which can not be directly measured and which are not immediately suggested by our sense perceptions.

From a microscopic viewpoint, a system could be regarded as being an assemblage of a huge number N of molecules which individually can exist in a large number of discrete states of energies, E_1, E_2 .... E_n. Interactions between molecules involve collisions and electrostatic and other field forces and for their description probability concepts are applied. The equilibrium state of a system is taken as being the state of highest probability with each energy state containing a definite numerical population of molecules under a given set of conditions.

macroscopic properties are time averages of microscopic properties

The two view points are not incompatible because macroscopic properties are simply time averages of microscopic characteristics. Our perception of macroscopic characteristics tends to remain unchanged as it is, after all, the perception of reality through our senses. In contrast the microscopic viewpoint is constantly being changed in the light of advances in our theoretical understanding. The basis for the justification of our theoretical assumptions lies in the comparison of the conclusions reached from the microscopic viewpoint with the deductions made from the macroscopic perspective.

Π Which of the following statements characterise quantities used to give a macroscopic description of a system?

1) they are few in number;

2) they involve assumptions concerning the structure of matter;

3) they are suggested by our sense perceptions;

4) they cannot be measured.

The answer is 1) and 3). The fact that they are directly measurable and that there is no need for assumptions on the structure of matter is the strength of this approach.

2.1.2 Thermodynamic approaches

Max Planck in his classic 'Treatise on Thermodynamics' noted that the subject developed as a result of three distinct methods of investigation.

thermodynamic energy (U) entropy (S)

He wrote that the most fruitful approach starts from a few very general empirical facts, mainly the two fundamental principles (laws) of thermodynamics. This approach gave us two key properties needed to describe systems, the thermodynamic energy (U) and the entropy (S). This macroscopic approach exploiting U, S and a few other fundamental properties such as mass, volume etc is the basis of classical thermodynamics and has been widely applied in chemistry, engineering and biology as we will see in this and the next chapter.

Helmholtz Clausius, Maxwell and Boltzmann statistical mechanics and statistical thermo-dynamics

The two other approaches to thermodynamics outlined by Planck are of much less general application. They are the approach of Helmholtz which confines itself to the initial hypothesis that heat is due to motion and the detailed kinetic theory treatments of Clausius, Maxwell and Boltzmann. The latter approach treats heat as being due to the definite motions of molecules and considers atoms as discrete masses. This has been exploited in statistical mechanics or statistical thermodynamics.

2.2 Basic concepts

Thermodynamics are concerned with systems and in particular with the energetics of systems. They are concerned with the bulk properties of matter rather than with the molecular properties.

The study of thermodynamics evolved from attempts to understand steam engines but the interrelationships between useful work, energy and equilibria hold for all systems including biological ones.

2.2.1 Types of systems

boundary

A system is any specified part of the universe such as an organism, a fermenter or a test tube. It is separated from its surroundings, the rest of the universe, by a boundary the characteristics of which define the type of system.

isolated system/adiabatic process

An isolated system (Figure 2.1) has a boundary which is impermeable to both matter and all forms of energy so that it exchanges neither heat nor matter with its surroundings. Any process which occurs in such a system is termed an adiabatic process. Perfectly isolated systems do not exist but the concept is useful and a reaction in a Dewar (Vacuum) flask is a close enough approximation for many purposes.

closed system

isothermal system

In a closed system (Figure 2.1) the boundary is permeable to energy but there is no transfer of matter between the system and its environment. If the temperature is kept constant we have an isothermal system.

open system

An open system is one which can exchange both energy and matter with its surroundings. These types of systems are represented in Figure 2.1.

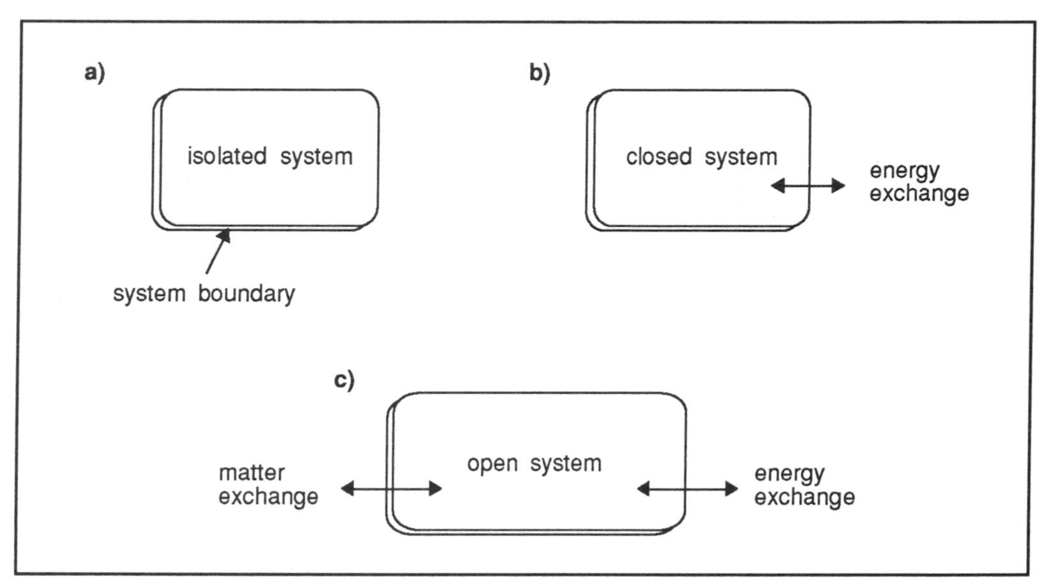

Figure 2.1 .Various systems as described in the text: a) isolated system; b) closed system; c) open system. Note that changes in the composition of matter in an isolated or a closed system can occur although there is no net gain or loss of matter. This change in composition can arise through chemical reactions within the system.

SAQ 2.1

Which of the following statements apply to:

1) an isolated system; 2) a closed system; 3) an open system.

a) There are no barriers to the exchange of matter and energy with the surroundings.

b) There is a surrounding barrier present which is permeable to energy but not matter.

c) There is a surrounding barrier present which is impermeable to both matter and energy.

d) All processes taking place in the system must be adiabatic. (Adiabatic is a word derived from the Greek adiabatos = 'impassable (to heat)'

State, with a brief explanation what type of system applies to a living organism.

Thermodynamics are used to describe energy conversions. Energy can take many forms (mechanical, electrical, thermal, chemical). In the next sections we will explore the various forms of energy and their interconversions.

2.2.2 Work and energy interconversion

definition of energy

Energy is defined as the capacity to do work. When a system does work its energy content falls. When work is done upon a system its energy content is increased.

definition of work

Work is organised motion and mechanical work is defined as force times distance. Although in a biochemical reaction mechanical work may be done against the pressure of the atmosphere if there is an increase in volume due to gas evolution, we normally think of work in a different way in relation to biological processes. Work can take many forms in biological systems and involve displacements against electrical, mechanical or chemical potential forces. These various forms of energy are interconvertible.

If F is a generalised force and dy is a generalised displacement then the work done on the system is:

$$dW = Fdy$$

We can use this generalised equation to apply it to different types of work and forces.

We must stress here that dW has a negative value when the energy content of the system decreases (= work is done by the system on the environment) while dW has a positive value when the energy content of the system increases (= work done by the environment on the system).

Table 2.1 shows the forces and displacements for several types of work.

type of work	work done on the system dW
volume change (mechanical work)	P x dV where P is the external pressure and dV the volume change
electrical work	E x dZ where dZ is the electrical charge transferred and E the potential
work of contraction of a fibre	F x dl where F is the force exerted and dl the change in length

Table 2.1 The relationship between work done on a system and different forms of pressure and displacements.

Work and energy are interconvertible and heat is a measure of the kinetic energy of the constituents of a system. The random movements of the atoms and molecules and their internal vibrations increase when heat is transferred to the system. Work is done when the motion induced by the transfer of energy is organised and not chaotic. For example the movement of ions or electrons in an organised way across a membrane constitutes work. In contrast if there is merely an increase in random motion no work is being done.

2.2.3 Types of work and energy

intensive and extensive factors

It is necessary to distinguish between intensive factors and extensive factors. Intensive, or potential factors as they are also called, are independent of the size of the system. Extensive factors (also called capacity factors) do depend on the size of the system; they are doubled when the size of the system is doubled.

Π Which of the following are intensive factors and which are extensive: temperature; mass; pressure; volume?

You should have decided that temperature and pressure are intensive factors - by doubling the size of an item you do not double the temperature! The other items (mass, volume) are extensive factors - doubling the quantity of matter will double both its mass and its volume.

Energy is the product of pairs of associated intensive and extensive factors a number of which are given in Table 2.2 below.

type of energy	intensive (potential) factor	extensive (capacity) factor
mechanical (PV)	pressure (P)	volume (V)
electrical (EZ)	electrical potential (E)	electrical charge (Z)
thermal (TS)	temperature (T)	entropy (S)
chemical ($M_i\mu$)	chemical potential (μ)	mass (M) or number of moles (N) of the ith species)

Table 2.2 Pairs of energy factors.

Π Note the term entropy, it is an extensive (capacity) factor. We will discuss this factor later.

The direction of change when two systems are brought into contact, with no barrier between them, is determined by the difference in potential factors. The systems will change in such a way as to equalise the magnitude of the potential factors throughout the combined system. A system reaches a state of complete equilibrium when no localised differences in any potential factors exist within the system. Equilibrium is thus a time - invariant state and a system in equilibrium will show no tendency to change.

Let us consider what happens when two vessels filled with gas at different pressures are connected to each other. Pressure is an intensive (potential) factor and will equalise until it is the same in both vessels. If the vessels are rigid the pressure only will change but if we have joined together two balloons the volumes will alter to equalise the pressures. In this second case the extensive (capacity) factor volume has changed but it was the potential factor gradient which determined the direction of change and equilibrium was attained when the potential gradient had been eliminated.

Now let us turn our attention to thermal energy.

Π Given that the thermal energy of a system is TS, what happens when two solid bodies of differing temperatures are brought into contact so that heat exchange between them can take place?

The potential factor here is temperature. In the system described there is a gradient of the potential factor. Thus heat will flow from the hotter to the colder body until both are at the same temperature.

Π When the temperature of the two bodies is the same, will they have the same thermal energy?

The answer is no, unless the bodies are of exactly the same shape, size and composition. The thermal energy is the product of two factors, temperature (T) and entropy (S) (see Table 2.2). Thus the thermal energy of body (1) will be T_1S_1 and the thermal energy of body (2) will be T_2S_2. Thus the thermal energy of the two bodies will only be the same when $T_1 = T_2$ if S_1 and S_2 are also equal.

The chemical potential μ is the intensive factor for chemical energy and the extensive (capacity) factor is the amount of the chemical species present which can be expressed as number of particles of the particular species, the number of moles (N) or the mass (M). Chemical potential determines the direction of chemical and physical processes which tend to continue until the chemical potential is equalised throughout the system. In both types of change there is a spontaneous movement from higher to lower chemical potential and systems are in equilibrium with each other when their chemical potentials are identical.

Let us check your understanding of which intensive factors go with which extensive factors before proceeding further.

SAQ 2.2

1) Fill in, from the list provided, the gaps in the following table. (Do this without looking at Table 2.2).

type of energy	intensive factor	extensive factor
mechanical	?	?
electrical	?	?
thermal	?	?
chemical	?	?

chemical potential; entropy; mass; volume; temperature; electrical charge; entropy; electrical potential, pressure.

2) Decide whether this statement is true or false:

Two systems are said to be in equilibrium when the total energy in each system is the same.

Living systems can never be in a state of true and complete equilibrium but a state of apparent or dynamic equilibrium may be transiently attained for example during steady state metabolism or growth in a fermenter. Classical equilibrium thermodynamics can fortunately be applied without serious problems to many processes in living systems but irreversible thermodynamics must sometimes be used.

Finally in this section we need to introduce the concepts of state functions and path functions.

2.2.4 The state and path functions

If we consider the intensive property temperature, we are considering a property which is part of the description of a particular state. It is independent of the path by which the system reached that particular state. For this reason it may be termed a 'state function' in contrast to a 'path function'. The work which has gone into the attaining of a particular state and which depends upon the path taken to reach that state is an example of a path function.

Now that we have introduced some of the terms used in thermodynamics, we can examine the laws of thermodynamics and their application to biochemical systems.

2.3 The beginnings of biochemical thermodynamics

The foundations of modern biochemical thermodynamics can be traced back to 12 minutes past eight on the morning of February 3rd 1783 when a Guinea pig was placed in an ice calorimeter. This was in effect an adiabatic container surrounding a closed vessel containing the Guinea pig and the amount of heat transferred was measured by the amount of ice melted. These studies by Lavoisier and Laplace linked heat production with oxygen consumption and carbon dioxide formation. It was subsequently established that the heat evolved by a small mammal metabolising carbohydrate to carbon dioxide and water was, within the limits of experimental error, the same as the heat evolved when the same substance was completely burnt in air or oxygen.

Biological oxidation was perceived as being simply equivalent to a slow form of combustion. It would appear, therefore, that biological processes follow the same set of thermodynamic rules (laws) as normal chemical processes. Thus in order to apply thermodynamic principles to biological systems, we need to consider the laws of thermodynamics.

2.3.1 The zeroth law of thermodynamics

Definition:

> If two systems A and B are each in thermal equilibrium with a third system C, then A and B are also in equilibrium with each other

The temperature of a system gives an indication of the potential a system has to transfer heat to another system. Experience teaches us that when two closed systems A and B, able to exchange energy but not matter with each other, are in thermal equilibrium with a third system C they are also in thermal equilibrium with each other.

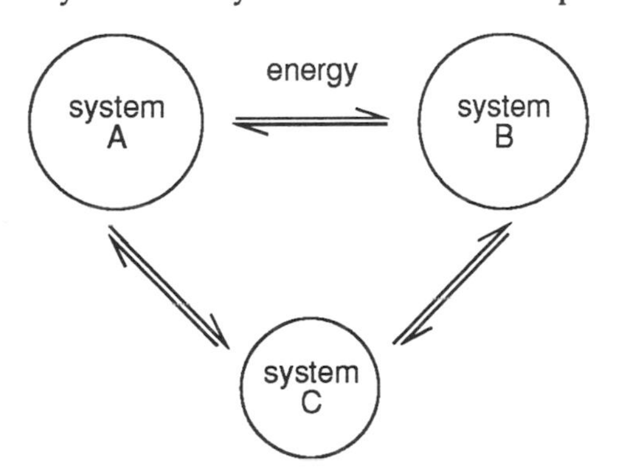

If A is in equilibrium with C and B is in equilibrium with C, then A and B are in equilibrium with each other

Zeroth Law of thermodynamics

If there is no net flow of heat from A to C or from B to C then there will be no net flow of heat between A and B. This empirical fact is the zeroth law of thermodynamics. Its odd name indicates that logically it should precede the other laws of thermodynamics.

Like all the laws of thermodynamics it rests on observation, experiment and experience and not upon theory.

This law defines the concept of temperature, a thermodynamic state function. For a discussion of temperature scales the reader should consult Atkins P.W, Physical Chemistry, 3rd Edition Oxford, Chapters 1 and 7.

2.3.2 The first law of thermodynamics

Definition:

> Changes in a body inside an adiabatic enclosure from a given initial state to a given final state involves the same amount of work independently of how the change is brought about.

The experimental studies of Joule in particular established that mechanical work could be transformed into heat and that when the same amount of work was performed the same rise in temperature was produced in a given quantity of water in an adiabatic enclosure. The adiabatic enclosure prevented exchange of heat or matter with the surroundings but work could be performed on the system. It was observed that all forms of work produced an equivalent amount of heat no matter what the form of the work was.

first law of thermo-dynamics

This led to the formulation of the first law of thermodynamics which states that the change of a body inside an adiabatic enclosure from a given initial state to a given final state involves the same amount of work independently of how the change is brought about.

Since energy is defined as the capacity to do work, we might consider the first law of thermodynamics in another way. The amount of energy released or adsorbed by a body to change from a given initial state to a given final state is the same irrespective of how the change is brought about.

Π Is the amount of energy released during the complete oxidation of a given quantity of a fatty acid by oxygen in a micro-organism greater or smaller than the amount of energy released by burning the same amount fatty acid in oxygen? (Assume that these systems are in adiabatic enclosures).

The answer is that they are the same providing both processes start with the same amount of fatty acid in the same state and produce the same products (CO_2 and H_2O) in the same state.

There are a lot of important consequences that arise from the first law of thermodynamics so let us investigate these in more detail.

2.3.3 Internal energy

A system is taken from state A to state B by the performance of a quantity of work W_{ad} on the system. The process is adiabatic. The first law implies that W_{ad} is independent of the path. W_{ad} depends on the states of the system only. We can therefore consider that the system has a property, a state function, U such that:

$$W_{ad} = U_B - U_A \qquad \text{(E - 2.1)}$$

The function U is termed the internal energy. The change in internal energy ΔU when the system goes from state A to state B under adiabatic conditions is:

$$\Delta U = U_B - U_A$$

If the system changes under non-adiabatic conditions so that it is in thermal contact with its surroundings the work involved in the process of changing from state A to state B may not be the same as it was under adiabatic conditions. The change in internal energy

ΔU will however be the same as this is a state function and independent of the path. Therefore where ΔQ is the heat absorbed by the system from its surroundings during the process and W is the work done on the system by its surroundings we may write:

$$\Delta U = \Delta Q + W \qquad (E - 2.2)$$

The sum ΔQ + W represents the mechanical equivalent of all the thermal and mechanical operations of the surroundings upon the system. ΔQ is of course dependent upon there being a temperature difference between the system and its surroundings. If there is no temperature difference then $\Delta Q = 0$.

principle of conservation of energy

In other words the change in the internal energy is the algebraic sum of the mechanical equivalents of the effects produced within the environment when a system is transformed from state A to state B. This, in fact, is a statement of the Principle of Conservation of Energy. The first law of thermodynamics is nothing more than the principle of conservation of energy applied to phenomena involving the production or absorption of heat.

This is of course an important concept. We will begin to realise its relevance to biological systems in general and cells in particular.

Now let us broaden Equation (E - 2.2) to apply it to other forms of work.

To remind you of the various types of work re-read Table 2.1.

Π Now substitute these expressions into Equation (E - 2.2) by filling in the blank boxes in Table 2.3. In each case the internal energy U is a function of any two of three functions.

	type of system	form of the first law	U is a function of
1)	Hydrostatic (mechanical)	$\Delta U =$ ☐	☐
2)	Contracting/relaxing fibre	$\Delta U = \Delta Q + Fdl$	TFl
3)	Electrical cell	$\Delta U =$ ☐	☐

Table 2.3 Application of the first law of thermodynamics to different types of systems.

You should have come to the following conclusions:

1) Hydrostatic (mechanical work) $\Delta U = \Delta Q - PdV$ and U is a function of P, V and T (since ΔQ is a function of temperature).

2) Contracting/relaxing fibre $\Delta U = \Delta Q + Fdl$ and U is therefore a function of T, F and l.

3) Electrical cell $\Delta U = \Delta Q + EdZ$ and U is therefore a function of T, E and Z.

All three equations have the same general form. Note however the signs (+ and -), these are important. You should remember these, or better still, understand why these are positive in some cases and negative in others.

Let us consider case 1) first. When the system expands (that is dV is positive), then work has been done on the surroundings in the expansion. Therefore PdV (the work done) is lost from the system and PdV has a negative value. In case 2), work has to be done to stretch a fibre. Thus in increasing the length of a fibre (that is dl is positive), all work (Fdl) is done on the fibre and has therefore a positive value.

2.3.4 Sign conventions

Thermodynamics consider energy changes from the point of view of the system. Energy expended by the system in doing work on its surroundings is lost from the system and therefore is given a minus sign.

When W = +8 kJ this means 8 kJ of energy have been supplied to the system by doing work on it. When $\Delta Q = -8$ kJ this means that 8 kJ of energy have left the system as heat and when W = -8 kJ this means that the system has done 8 kJ of work on its surroundings.

SAQ 2.3

The heat evolved during a biochemical reaction in solution is 152.48 kJ. One mole of gas is evolved but no gaseous reactants are used during the reaction which takes place at 18°C and 1 atmosphere pressure.

Calculate the total change in the thermal energy (ΔU) of the system.

Remember PV = nRT (the gas law). The value of R, the gas constant is, 8.314 JK^{-1} mol^{-1}.

2.3.5 Enthalpy

enthalpy is a state function

In biology we are interested not in the actual internal energy change ΔU but in how much of this energy is available for biological purposes such as growth, biosynthesis etc. Most biological reactions occur at constant pressure and under these conditions work done associated with volume changes such as gas absorption or evolution is not available for biological purposes. It is therefore convenient to identify another function of state (H) the enthalpy. Enthalpy is also called the heat content or heat of formation.

The change in enthalpy (heat content) is given by:

$$\Delta H = \Delta U + P\Delta V \qquad \text{(E - 2.3)}$$

or if the increase in volume is due to the formation of n moles of gas:

$$\Delta H = \Delta U + nRT$$

Π Can you say why H must be a function of state?

The enthalpy, H, is a function of state because it involves only U, P and V which are themselves functions of state.

The enthalpy change, ΔH is the quantity of heat absorbed by a closed isothermal system at constant pressure when it changes from one thermodynamic state to another without performing any work other than that associated with the volume change.

Π Volume changes during reactions in solution are negligible. What then is the relationship between ΔH and ΔU for an isothermal biochemical reaction in solution in which no gas is evolved?

If you think about this carefully when $\Delta V = 0$ then substituting in to (E - 2.3) gives:

$$\Delta H = \Delta U$$

Experimentally ΔU and ΔH are obtained by measuring the heat of reaction using an adiabatic calorimeter in which the temperature change of the calorimeter during a reaction can be measured and this enables us to calculate ΔQ and thus ΔU. If there is no volume or pressure change then $\Delta Q = \Delta U$.

Two major types of calorimeters are used. The constant volume adiabatic calorimeter uses a constant volume reaction vessel surrounded by a water bath. The calorimeter is insulated to prevent heat loss from the system. The temperature change ΔT produced by the reaction initiated in the vessel enables the heat of reaction at constant volume ΔQ_v, to be calculated from a knowledge of the heat capacity C of the calorimeter.

$$\Delta Q_v = C\Delta T$$

The second type of calorimeter is the constant pressure calorimeter which is similar but operates at a constant pressure.

2.3.6 Terminology

exothermic reaction

An exothermic reaction is one in which the system gives out heat to its surroundings so that the enthalpy of the system falls (ΔH is negative).

endothermic reaction

An endothermic reaction is one in which the system absorbs heat from its surroundings (ΔH is positive). For biochemical reactions the main contribution to ΔH comes from energy changes associated with the formation or breaking of chemical bonds.

For carrying out energy calculations, a standard reference value is chosen and the energy of all other states can be calculated from this reference value (this is analogous to using sea level as a reference value in geography). For thermodynamic calculations, the chosen standard conditions are: 298.15 K and 1 atm and all substances in their standard states (pure substance as it exists at 298.15 K and 1 atm). According to this convention, the enthalpy of the elements at 298.15 K and 1 atm = 0. The energy of compounds can be calculated from this reference value by taking into account the energy effects during their formation. For compounds in solution of course both formation of the compound and its dissolution have to be taken into account.

The standard enthalpy of formation is the enthalpy change associated with the formation of a compound from its elements in their standard states at 298.15 K.

Π If you know that acetic acid is a liquid at 298.15 K and it has the formula $C_2H_4O_2$ and that oxygen and hydrogen are gases at this temperature and carbon is a solid in the form of graphite write down an equation for the formation of acetic acid to which the term standard enthalpy of formation applies.

The equation is:

$$2C_{(graphite)} + 2H_{2(g)} + O_{2(g)} \rightarrow C_2H_4O_{(l)}$$

(note that subscripts g = gas, l = liquid).

Thus the standard enthalpy of formation of acetic acid refers to the enthalpy change associated with converting two moles of graphite, two moles of gaseous hydrogen, two moles of gaseous oxygen to one mole of liquid acetic acid at 298.15 K and 1 atm.

Of course, if we use non-standard conditions, or we change the physical state of the reactants or products (for example use a different allotrope of carbon), this will influence the enthalpy change. In this case we would have to define the conditions in describing the enthalpy change.

2.3.7 Hess's Law

Definition:

Whether a reaction is performed in several stages or in a single step the overall heat change will be the same.

Hess's law of constant heat summation

The first law of thermodynamics is often applied in the form of Hess's law of constant heat summation. This states that whether a reaction is performed in several stages or in a single step the overall heat change will be the same, ie the heat change of the single step will equal the algebraic sum (taking account of +/- signs for heat absorbed or evolved) of all the heat changes for the individual stages.

We can now consider the example of a bacterium which uses ethanol as a source of energy for growth.

We have available the following information. The heat of combustion (enthalpy change for complete oxidation) at 293 K and standard atmospheric pressure, for ethanol is - 1371 kJ mol^{-1} and the corresponding values are - 1168 kJ mol^{-1} and - 876 kJ mol^{-1} for acetaldehyde (CH_3CHO) and for acetic acid (CH_3COOH) respectively. Thus we can write:

for alcohol oxidation:

1) $C_2H_5OH_{(l)} + 3\ O_{2(g)} \rightarrow 2\ CO_{2(g)} + 3H_2O_{(l)}$ $\Delta H = -1371$ kJ mol^{-1}

for acetaldehyde oxidation:

2) $CH_3CHO_{(l)} + 2\frac{1}{2}\ O_{2(g)} \rightarrow 2\ CO_{2(g)} + 2H_2O_{(l)}$ $\Delta H = -1168$ kJ mol^{-1}

and for acetic acid oxidation:

3) $CH_3COOH_{(l)} + 2\ O_{2(g)} \rightarrow 2\ CO_{2(g)} + 2H_2O_{(l)}$ $\Delta H = -876$ kJ mol^{-1}

Π Calculate the value of ΔH at 293 K and standard atmospheric pressure if the ethanol is oxidised to acetaldehyde. Also calculate how much extra energy would be available if the oxidation proceeded to acetic acid (ie the further reaction $CH_3CHO + \frac{1}{2}\ O_2 \rightarrow CH_3COOH$ took place). You might like to have an attempt at this before reading on.

Subtracting 2) from 1) gives us:

$CH_2H_5OH_{(l)} - CH_3CHO_{(l)} + \frac{1}{2}\ O_{2(g)} \rightarrow H_2O_{(l)}$ $\Delta H = (-1371 - (-1168))$ kJ mol^{-1}

which gives us:

$C_2H_5OH_{(l)} + \frac{1}{2}\ O_{2(g)} \rightarrow CH_3CHO_{(l)} + H_2O_{(l)}$ $\Delta H = -203$ kJ mol^{-1}

Thus for the oxidation of alcohol to acetaldehyde: $\Delta H = -203$ kJ mol^{-1}.

Subtracting 3) from 2) we get:

$CH_3CHO_{(l)} + \frac{1}{2}\ O_{2(g)} \rightarrow CH_3\ COOH_{(l)}$ $\Delta H = (-1168 - (-876))$ k J $mol^{-1} = -292$ kJ mol^{-1}

Thus for the oxidation of acetaldehyde to acetic acid: $\Delta H = -292$kJ mol^{-1}.

These values of ΔH apply only for the stated conditions. This example shows how we can predict the heat output from a simple bioreaction using thermochemical data.

If we consider the oxidation of amino acids derived from protein during nutrition we will see how Hess's law can be applied to a rather more complex biological problem.

We will use alanine as an example. Alanine may be metabolised to urea and carbon dioxide by a process of biological oxidation. Starting from protein consumed during the digestive process we may regard solid alanine as our starting point for the purposes of this calculation.

The equation for the complete oxidation of alanine is:

4) $CH_3CHNH_2COOH_{(s)} + 3\,^3/_4\,O_{2(g)} \rightarrow 3\,CO_{2(g)} + 3\frac{1}{2}\,(H_2O)_l + \frac{1}{2}N_{2(g)}$

$\Delta H^\circ = -1620\ kJ\ mol^{-1}$

Here we have used the superscript ° to represent standard enthalpy (that is enthalpy changes under standard conditions). This is a useful shorthand.

For the complete combustion of urea:

5) $CO\,(NH_2)_{2(s)} + 1\frac{1}{2}\,O_{2(g)} \rightarrow CO_{2(g)} + 2H_2O_{(l)} + N_{2(g)}$ $\Delta H^\circ = -632\ kJ\ mol^{-1}$

Every two alanine molecules will give rise to one urea so we may obtain the equation for the partial oxidation of alanine to carbon dioxide and urea by subtracting (0.5) x Equation 5) from Equation 4). This will give us Equation 6). Try this on a piece of paper for yourself.

(Note each of the two urea nitrogens will have come from an alanine molecule which only contains one nitrogen atom).

6) $CH_3CHNH_2COOH_{(s)} + 3\,O_{2(g)} \rightarrow 2\frac{1}{2}\,CO_2 + 2\frac{1}{2}H_2O_{(l)} + \frac{1}{2}CO(NH_2)_{2(S)}$

$\Delta H^\circ = (-1620 - (\frac{1}{2}) \times (-632)\ kJ\ mol^{-1} = -1304\ kJ^{-1}$

However the urea will not be in the form of solid urea and allowance must be made for the fact that it will be in solution.

The equation for the dissolution of urea in water to give for example a concentration of 0.1 mol l^{-1} is:

7) $CO(NH_2)_{2(S)} \rightarrow CO(NH_2)_2\ (aq\ 0.1\ mol^{-1})\ \Delta H^\circ = -15.0\ kJ\ mol^{-1}$

so to convert $\frac{1}{2}CO(NH_2)_{2(S)}$ to $\frac{1}{2}CO(NH_2)_2$ (aq 0.1 mol^{-1}) then the energy change = -7.5 kJ mol.

Thus if we add this to Equation 6) we can then write:

$CH_3CHNH_2COOH_{(S)} + 3O_{2(g)} \rightarrow 2\frac{1}{2}CO_2 + 2\frac{1}{2}H_2O_{(l)} + \frac{1}{2}CO(NH_2)(aq\ 0.1\ mol^{-1})$

$\Delta H^\circ = -1304 + (-7.5)\ kJ\ mol = -1311.5\ kJ\ mol^{-1}$

To have ignored the heat of solution would have lead to a 0.6% error in ΔH for the biological oxidation of alanine, for a smaller amino acid such as glycine the error would be greater (1.2%) as glycine contains a higher proportion of nitrogen.

A similar calculation can be made for ammonia as the end-product.

It is important to remember that when we are considering the energetics of chemical reactions we must not only remember enthalpy changes arising from the chemical changes but also take into account the energy changes resulting from the change in physical states.

SAQ 2.4

At 310 K the values of ΔU, for complete oxidation to carbon dioxide and water, for glucose ($C_6H_{12}O_6$) and palmitic acid ($C_{16}H_{32}O_2$) are -2810 kJ mol^{-1} and -10012 kJ mol^{-1} respectively.

Calculate ΔH for the two processes and discuss which of these two substances would be potentially the most useful as an energy store for an organism. Remember to allow for any uptake or evolution of gases. The reactions are:

glucose oxidation:

$$C_6H_{12}O_{6(s)} + 6O_{2(g)} \rightarrow 6CO_{2(g)} + 6H_2O_{(l)}$$

palmitic acid oxidation:

$$C_{16}H_{32}O_{2(s)} + 23O_{2(g)} \rightarrow 16CO_{2(g)} + 16H_2O_{(l)}$$

Take $R = 8.314 \text{ JK}^{-1} \text{ mol}^{-1}$

2.4 Entropy and the second law of thermodynamics

It is a well established empirical observation that there is a natural direction of change: heat flows from a hot body into its cooler surroundings, a solute will diffuse from a concentrated solution to a less concentrated one, hydrogen and oxygen react together to form water.

These are all examples of thermodynamically 'spontaneous' processes and they can only be reversed if work is done on the system. In controlled circumstances spontaneous change can be made to perform useful work.

Π Do 'spontaneous' processes occur rapidly and immediately?

spontaneous does not mean fast

A moment's thought will have enabled you to answer 'not necessarily'; consider the reaction between hydrogen and oxygen. When the two gases are mixed any change is at an undetectable slow rate in the absence of a catalyst or a stimulus such as a spark. Similarly glucose does not rapidly become oxidised to CO_2 and H_2O in the presence of oxygen. 'Spontaneous' refers to the direction not the rate of a process.

The first law of thermodynamics states that the total energy is conserved in any process. It follows from this that total energy does not determine the direction of spontaneous change. Any loss of energy by a system will be balanced by an increase in the energy of its surroundings. Yet spontaneous change occurs only in one direction.

2.4.1 The second law of thermodynamics

Butler's definition

Clausius, in the 1860s, recognised the existence of a further function of state which relates to the capacity of a system to undergo transformation and which gives a measure of this capacity. We can get an understanding of what this function is by considering the second law of thermodynamics. This is an empirical generalisation like the first law and has been stated in a number of ways, many of which are more useful to engineers and physicists than to biochemists. J.A.V Butler (Chemical Thermodynamics, 4th Edition,

1946, MacMillan, London, p32) has stated the second law in the form, 'spontaneous processes (those which may occur of their own accord) are those which when carried out under proper conditions can be made to do work'. He also added that if carried out reversibly they will yield a maximum amount of work; in the natural irreversible way the maximum work is never obtained. We will explore exactly what he meant in the following section.

It is always convenient to remember a simple definition of a law. We will use the definition:

> Spontaneous processes are those which when carried out under proper conditions can be made to do work.

2.4.2 Thermodynamic reversibility

biochemical reversibility

This term must not be confused with physical or biochemical reversibility which simply means that a process or transformation can be reversed or that a reaction proceeds at a measurable rate in both directions.

The term 'thermodynamically reversible' refers to a change so infinitesimally small that both the system and its immediate surroundings can be restored to their initial states without producing changes in the rest of the universe. Any process which was thermodynamically reversible would be infinitely slow and always very close to equilibrium.

All natural processes are in fact thermodynamically irreversible. Table 2.4 gives some examples.

mechanical	thermal	chemical
snapping of a stretched fibre; free expansion of a gas (into a space at lower pressure); collapse of a surface film	conduction or radiation of heat to cooler surroundings	chemical reactions; diffusion and mixing; osmosis; dissolution

Table 2.4 Some thermodynamically irreversible processes in biochemistry and biology.

Π The examples shown in Table 2.4 are described as thermodynamically irreversible. Can we in fact reverse them?

The answer is yes, but we would have to do some work on them. For example the expansion of a gas could be reversed by applying a higher pressure. The pressure would have to be generated by, for example, a pump. In other words, we would have to do some work on the gas. The energy to drive the pump would have to be generated elsewhere in the universe. Thus, according to our definition this would be a thermodynamically irreversible process. It would however be a physically reversible process.

The first law of thermodynamics tells us how much energy there is in 'total' in a system and if we can establish how much of this is unavailable we can determine how much is left to do useful work. A thermodynamically reversible process is one which is carried out in such a way as to develop the maximum work.

Reactions will only proceed spontaneously if they can perform work and hence this maximum work gives us a true measure of the tendency for a reaction to occur. The ability to perform work does not however, tell us anything about the rate at which a reaction will proceed.

2.4.3 Entropy and change

If we consider the melting of ice to give water at 0°C (273.15 K) there is a great increase in the number of microscopic states available to the system when the highly ordered solid is replaced by the more disordered liquid at the same temperature. The capacity for transformation is thus increased. In an ice-water mixture at 273.15 K the two phases are in equilibrium and reversible melting and thawing take place. If an infinitesimal amount of heat (dQ) is absorbed by an infinitesimal amount of ice, the ice will melt but the temperature will not significantly change. The change in thermal energy for the ice will be temperature multiplied by the capacity factor for thermal energy (TdS). The capacity factor for thermal energy is called entropy (S) (see Table 2.2).

We may thus write:

$$TdS = dQ$$

rearranging gives:

$$dS = \frac{dQ}{T} \qquad \text{(E - 2.4)}$$

for the entropy change for the ice.

The water will however have lost an infinitesimal amount of heat dQ and will have undergone an equivalent entropy change.

If we regard the ice as the system and the water as its surroundings we see that for an infinitesimal, reversible change:

$$\Delta S_{surroundings} + \Delta S_{system} = 0 \qquad \text{(E - 2.5)}$$

$$\Delta S_{total} = 0 \qquad \text{(E -2.6)}$$

In irreversible changes, the total entropy of the system and the surroundings increases (in other words there is a tendency to increase disorder). We may, therefore, write:

$$\Delta S_{surroundings} + \Delta S_{system} > 0 \qquad \text{(E - 2.7)}$$

ie has a positive value.

We may in fact write for all changes:

$$\Delta S_{total} \geq 0 \qquad \text{(E - 2.8)}$$

and indeed:

$$\Delta S_{universe} \geq 0$$

It is impossible to diminish the entropy of a system of bodies without thereby leaving behind changes in other bodies. Every process in nature takes place in such a way as to increase the sum of the entropies of all the bodies taking part in the process. In the limiting case of reversible processes the sum of the entropies remains the same.

Earlier we gave a definition of the second law of thermodynamics in terms of 'spontaneous processes are those which when carried out under proper conditions can be made to do work.' We can now describe the second law of thermodynamics in another way.

The entropy of a system increases in a thermodynamically irreversible process and remains unchanged in a reversible process. It can never decrease.

Thus either ΔS_{system} or $\Delta S_{surroundings}$ can be a negative quantity for a particular process, but their sum can never be less than zero.

SAQ 2.5

Fill in the blanks in the statements below using a selection of the expressions given in the box below. The second law of thermodynamics may be stated in the following way.

1) Entropy is a function of state which is defined for a reversible transformation by: dS = [ ].

2) In all thermodynamically reversible transformations the entropy change of the system plus its surroundings [ ].

3) All irreversible changes are accompanied by [ ] in entropy.

   At equilibrium the entropy of a system and its surroundings is at a maximum.

4) It follows from the above that total entropy may remain constant or increase but it can never [ ]. This is the basis for statements such as 'the entropy of the universe is increasing'.

$\frac{dQ}{T}$; $\frac{dU}{T}$; nRT; an increase; a decrease; an increase; a decrease; is zero, is positive; is negative.

2.4.4 Entropy at the molecular level

Entropy, as described above is the capacity factor for thermal energy. It is a function of state. Entropy, as a function of state, is a product of the macroscopic viewpoint. The microscopic viewpoint enables us to see entropy as a consequence of the number of microstates in which the system could find itself due to the arrangement of the atoms into groups and molecules and the way in which energy is distributed into kinetic motion, vibrations and rotations. It is in effect a function of the degree of disorder in the system.

A system with a high degree of disorder has a high S value. Likewise a system with a low degree of disorder has a low S value.

Boltzmann has related the thermodynamic probability of a particular microstate (Ω) to entropy in the following way:

$$S = k \ln \Omega \qquad \text{(E - 2.9)}$$

where k is the Boltzmann constant.

entropy can be regarded as a state of disorder or of the amount of information known about a system

The thermodynamic probability describes the number and relative importance of the possible microstates of a microscopically specified system. Ω is a measure of the disorder of a system and Equation (E - 2.9) above gives us a simple relationship between entropy and disorder. From information theory, the information increases as the number of choices for the arrangement of a system decreases (that is as the degree of disorder decreases). Thus we may regard entropy either as being a measure of the degree of disorder or as a measure of the lack of information about the exact state of a system.

SAQ 2.6

What would be the effect on the entropy of a system of 1) decreasing the volume and 2) increasing the temperature.

Care is needed when thinking about order in biological systems. For example the assembly of viruses from their subunits in aqueous solution might be regarded as an increase in order (that is a decrease in entropy). This seems to be the reverse of our previous argument that in *all* natural processes there is an increase in entropy. The explanation for this apparent paradox is as follows. Clearly the assembly of the virus itself involves an increase in order but the component subunits would be solvated if assembly did not occur. As the subunits associate to form the viruses, then water molecules are released leading to an increase in the entropy of the solution. The net result is an increase in the entropy of the system as a whole.

2.4.5 The Gibbs and Helmholtz functions

Chemical and biochemical reactions move towards an equilibrium with both reactants and products being present. At equilibrium there is no tendency for any net change to occur.

Consider a system, not in chemical equilibrium, undergoing an infinitesimal irreversible process involving transport of matter between phases in a heterogeneous system, or a chemical reaction in a homogeneous system, or both. These systems transfer heat dQ to or from a reservoir at temperature T.

If dS is the entropy change of the system and dS_R the entropy change of the reservoir then by the second law of thermodynamics:

$$dS_R + dS > 0$$

Therefore the change in entropy could be thought of a being produced by the transfer of heat (dQ) from a reservoir at temperature T to the system at the same temperature (see Equation (E - 2.4)). In this situation, the reservoir would lose energy and the system would gain it, thus we could write:

$$-\frac{dQ}{T} + dS > 0$$

or multiplying by T and changing the signs:

$$dQ - TdS < 0$$

(Notice the change from >0 to <0 because we have changed the signs).

As a result of the energy (heat) transfer to the system, the internal energy of the system will have changed by dU and if there has been any volume change the amount of work performed will be PdV. This may be expressed by:

$$dQ = dU + PdV$$

so application of the second law of thermodynamics to a system and its reservoir (surroundings) ultimately leads to the expression for the system:

$$dU + PdV - TdS < 0$$

For a system at constant temperature and volume, PdV = 0 and so the inequality becomes:

$$dU - TdS < 0$$

or:

$$d(U - TS) < 0$$

As U, T and S are all functions of state we have a new function of state (U - TS) which is called the Helmholtz function, A.

We may write:

$$dA < 0$$

A in systems of constant temperature and volume has its minimum value at the final equilibrium state.

If we take T and P as constant the inequality becomes:

$$d(U + PV - TS) < 0$$

But earlier (Equation (E - 2.3) we showed that $\Delta H = \Delta U + P\Delta V$ or:

$$dH = dU + PdV$$

thus:

$$d(H - TS) < 0$$

Again H, T and S are all functions of state so we can write:

$$dG < 0.$$

Gibbs function

where G is another new function of state called the Gibbs function. For a system at constant temperature and pressure, G decreases during an irreversible process and has a minimum value at the final equilibrium point.

We can write for a system changing from one thermodynamic state to another:

$$\Delta G = \Delta H - T\Delta S \quad \text{(E - 2.10)}$$

where ΔG is the Gibbs function change for the transition characterised by an enthalpy change ΔH and an entropy change ΔS.

The Gibbs function change (ΔG) for any chemical reaction provides a measure of the capacity to do work. In other words, it is the energy which is made available to perform useful work during a reaction. For this reason the Gibbs function is also referred to as the Gibbs free energy. (Note that in biological texts ΔG is simply referred to as the free energy of a reaction). It is important that you realise that the Gibbs function applies to processes which occur at constant temperature and pressure.

By analogy, the Helmholtz function (A) refers to processes which operate at constant temperature and volume.

SAQ 2.7

Which do you think is likely to be of most value in biology and biochemistry, the Gibbs function G or the Helmholtz function A?

2.4.6 Gibbs function and equilibrium

We have so far learnt that the Gibbs function change (ΔG) for any chemical reaction provides a measure of the capacity to do work. We have also learnt from Equation (E - 2.10) that the change in Gibbs function arises from a change in enthalpy (ΔH) and the thermal energy ($T\Delta S$) in the system. In this section we will explain the relationship between product and reactant concentration and ΔG.

Π What will be the value of ΔG if the system is in equilibrium?

At equilibrium there is of course no net forward or backward reactions and there will be no net change in enthalpy or entropy. Thus at equilibrium $\Delta G = 0$.

When a reaction proceeds spontaneously there is of course a change in both enthalpy and entropy and thus a change in the Gibbs function. As the reaction approaches equilibrium, ΔG decreases. We therefore visualise a situation in which the further a system is away from equilibrium, the greater its ΔG value and, therefore, the greater its capacity to do work. In other words the value of ΔG is related to the actual concentrations of reactants and products and their distance from the equilibrium position.

The standard Gibbs function change which we will give the symbol ΔG° can be defined as the Gibbs function change when all reactants and products at standard concentrations (1.0 mol l^{-1}) are allowed to reach equilibrium under standard conditions (298.15° K and 1 atm). Clearly the size of ΔG° will be dependent upon the final equilibrium position. The relationship between ΔG° and the equilibrium position can be written as:

$$\Delta G^\circ = -RT\ln K_{eq} \qquad \text{(E - 2.11)}$$

Remember that ΔG° refers only to standard conditions (that is, all reactants and products are at standard concentrations, temperature and pressure).

ΔG° as such is of limited direct use in biological systems because these systems usually operate at the physiological pH = 7. pH 7 means that the hydrogen ion concentration is 10^{-7} mol l^{-1}, therefore in biological systems often the term $\Delta G^{\circ'}$ is used. $\Delta G^{\circ'}$ is the standard Gibbs function change at pH 7. Thus $\Delta G^{\circ'}$ signifies the change in Gibbs function when reactants and products at a concentration of 1.0 mol l^{-1} are allowed to proceed to equilibrium at pH 7.0.

Note: strictly speaking we should use activities not concentrations. Activities are concentrations corrected for deviations from ideal behaviour. Ideal behaviour is only found in very dilute solution and thus in dilute solutions activity = concentration. In practice, however, because of the degree of uncertainty over physiological concentrations due to experimental errors and because of the lack of knowledge of activity coefficients for substances in the intracellular environment, it is normal to use concentrations rather than activities. The error in the assumption is usually not large and falls within experimental error. By convention the activity of water in aqueous solutions is generally taken as unity.

Thus we can write:

$$\Delta G^{o'} = -RT\ln K_{eq} \quad \text{(E - 2.12)}$$

where $\Delta G^{o'}$ is the standard Gibbs function change (standard free energy change) at pH7.

Converting to logarithms to the base ten we can write:

$$\Delta G^{o'} = -2.303\ RT\log_{10} K_{eq} \quad \text{(E - 2.13a)}$$

or:

$$\log_{10} K_{eq} = -\frac{\Delta G^{o'}}{2.303RT} \quad \text{(E - 2.13b)}$$

At 25° (298K) and R 8.314 JK^{-1} mol

then 2.303 RT = 2.303 x 8.314 x 298 = 5706 J mol^{-1} = 5.706 kJ mol^{-1}

Thus for 25°C we can re-write Equation (E - 2.13b) as:

$$\log_{10} K_{eq} = \frac{-\Delta G^{o'}}{5.7}$$

Now let us turn our attention to non-standard conditions.

The relationship between product and reactant concentrations and ΔG may be written as:

$$\Delta G = \Delta G^{o'} + RT\ln \frac{[\text{product of initial product concentrations}]}{[\text{product of initial reactant concentrations}]} \quad \text{(E - 2.14a)}$$

or at pH 7:

$$\Delta G = \Delta G^{o'} + RT\ln \frac{[\text{product of initial product concentrations}]}{[\text{product of initial reactant concentrations}]} \quad \text{(E - 2.14b)}$$

or:

$$\Delta G = \Delta G^{o'} + 2.303\ RT\log_{10} \frac{[\text{product of initial product concentrations}]}{[\text{product of initial reactant concentrations}]}$$

Thus for the reaction $A + B \rightleftarrows C + D$ we can write:

$$\Delta G = \Delta G^{o'} + 2.303\ RT\log_{10} \frac{[C]\,[D]}{[A]\,[B]} \text{ at pH7.} \quad \text{(E - 2.15)}$$

where [A], [B], [C] and [D] are the initial concentrations of reactants and products.

We will use Equations (E - 2.13) and (E - 2.15) in some examples.

SAQ 2.8

Use the above relationship to fill in the blanks in the following table. The relationships between K_{eq} and the sign, and magnitude of $\Delta G^{o'}$ should be noted.

K_{eq}	$\Delta G^{o'}$ kJ mol^{-1}
0.00001	[ ]
0.001	17.1
0.1	[ ]
1.0	0
10.0	-5.7
1000.0	[ ]
100000.0	[ ]

Table 2.5 The relationship between $\Delta G^{o'}$ and K_{eq} at 25°C.

SAQ 2.9

The enzyme phosphoglucomutase at pH7.0 and 298 K catalyses the following reaction in the presence of magnesium ions.

$$\text{glucose-6-phosphate} \rightleftarrows \text{glucose-1-phosphate}$$

An initial concentration of 0.20 mol l^{-1} glucose-6-phosphate falls to 0.19 mol l^{-1} at equilibrium.

Calculate:

1) K_{eq} for this reaction;

2) $\Delta G^{o'}$ for the conversion of glucose-6-phosphate to glucose-1-phosphate;

3) $\Delta G^{o'}$ for the reverse reaction when glucose-1-phosphate is converted to glucose-6-phosphate.

In answering this remember that:

$$[\text{G-1-P}] + [\text{G-6-P}] = 0.20 \text{ mol}^{-1}$$

The change in Gibbs function for forward and back reactions differs only in sign.

$$\Delta G_{\text{forward reaction}} = -\Delta G_{\text{back reaction}}$$

(Re-examine your answer to SAQ 2.9 to convince yourself of this).

As the Gibbs function is a function of state the values of $\Delta_f G$ etc are additive, where subscript $_f$ signifies 'formation'. Thus $\Delta_f G$ is the standard free energy of formation.

It follows that for a chemical reaction:

$$\Delta G^{o} = \Sigma \Delta_f G^{o} \text{ products} - \Sigma \Delta_f G^{o} \text{ reactants}$$

(The symbol Σ means 'the sum of the values of').

We remind you that when considering reactions in aqueous solution, water may be regarded as:

- a pure liquid in which case its standard state is unit activity;
- an aqueous solution of pure water has a concentration of 55.5 mol kg^{-1}. Note that this is not quite the same as 55.5 mol l^{-1}. It is conventional in biochemistry to use an activity of 1 for water in dilute aqueous solutions.

You should note that strictly speaking, standard Gibbs function change only applies to standard conditions (298.15° K, 1 atm). However, we often have to consider energy changes at temperatures other than 298.15 K (most commonly 310 K = 37°C). Just as we noted above that in biochemistry we often consider standard Gibbs function changes at pHs other than at standard pH (= pH 0), we can also consider Gibbs function changes at temperatures other than at 298.15° K. Just as with pH, we must, however, state the temperature. Here we will use the convention $\Delta G'$ to symbolise changes in Gibbs function when the reaction takes place at a stated temperature, pH and pressure other than standard contitions, but with reactants and products in their standard state.

We can calculate the Gibbs function change for the synthesis of the dipeptide DL leucylglycine from glycine and DL leucine at physiological temperatures and pH conditions from the data in Table 2.6.

substance	$\Delta_f G'$ kJ mol^{-1}
glycine	-367
DL leucine	-330
DL leucylglycine	-448
water	-235

Table 2.6 Gibbs function of formation values in aqueous solution at pH7.0 at 37°C.

The reaction is:

$$\text{glycine} + \text{DL leucine} \rightleftarrows H_2O + \text{DL leucylglycine}$$

$$\Delta G' = -448 + (-235) - [(-367) + (-330)] \text{ kJ mol}^{-1}$$

$$= -683 + 697 \text{ kJ mol}^{-1}$$

$$= +14 \text{ kJ mol}^{-1} \text{ at pH 7.0 and 37°C}$$

It follows from the additive nature of ΔG values that Gibbs function data referring to substances in their pure standard states (gas, solid, liquid) can be related to ΔG data for the substances in solution and in ionised or non-ionised states.

ΔG° of solution refers to the Gibbs function change associated with the concentration change to unit activity.

$\Delta_f G^\circ$ in solution = $\Delta_f G^\circ$ in standard state + ΔG° of solution; where $\Delta_f G^\circ$ in solution and ΔG° of solution refer to unit activity in aqueous solution.

Likewise we can write $\Delta_f G^\circ$ of ionised form in aqueous solution = $\Delta_f G^\circ$ of non-ionised form in aqueous solution + ΔG° of ionisation.

2.4.7 The effect of pH and temperature upon Gibbs function and equilibria.

Changes in pH may alter the degree of ionisation of substrates and if a specific form of the molecule (for example the unionised state) is involved as a product or reactant, the equilibrium constant and ΔG value will vary with pH. If $[H^+]$ is a reactant or product of the reaction in question then there will be a change in pH as the reaction proceeds which for an enzyme catalysed reaction will often mean a change in enzyme activity. It is therefore normal in biochemistry to carry out reactions in buffered solution.

If we consider a reaction:

$$A + B \rightleftarrows C + D + H^+$$

then:

$$\Delta G = \Delta G^\circ + RT \ln \frac{[C][D][H^+]}{[A][B]}$$

If we take all the reactants and products as being in their standard states under standard state conditions (see Section 2.4.6) with the exception of $[H^+]$ we may write:

$$\Delta G^{\circ'} = \Delta G^\circ + RT \ln [H^+]$$

($[A] = [B] = [C] = [D] = 1$)

$$= \Delta G^\circ + 2.303\ RT \log [H^+]$$

we may write:

$$\Delta G^{\circ'} = \Delta G^\circ - 2.303\ RT\ pH$$

We can see from this relationship that the change in Gibbs function is dependent upon pH.

Temperature may also be expected to influence the equilibrium position of a reaction.

Let us consider an exothermic reaction (one which generates heat). We could write the equation as:

$$A + B \rightleftarrows C + D + \text{heat}$$

We might anticipate that by raising the temperature we would push the equilibrium to the left. Thus we might anticipate that the value of $\Delta G'$ will be influenced by temperature. This is indeed so.

The relationship between temperature and $\Delta G'$ is described by the Gibbs-Helmholtz equation which may be conveniently expressed in its integrated form as:

$$\frac{\Delta G_2'}{T_2} = \frac{\Delta G_1'}{T_1} - \Delta H^\circ \left(\frac{T_2 - T_1}{T_1 T_2} \right)$$

The assumption has been made in deriving this equation that over a modest temperature range when you change the temperature from T_1 to T_2 the value of ΔH° does not change significantly. Experience shows this would be a reasonable assumption.

SAQ 2.10

The Gibbs function change $\Delta G'$ for the hydrolysis of ATP, under the conditions used to culture a micro-organism at 18°C was found to be -30.3 kJ mol^{-1} and $\Delta H^{\circ'}$ under the same conditions was found to be -20.1 kJ mol^{-1}.

The organism is a thermophile and may be cultured at 58°C. Calculate the value of $\Delta G'$ under these conditions. Comment upon the implications for physiological processes involving ATP.

The ionic strength of a solution can also effect $\Delta G'$ values.

Figure 2.2 and Table 2.7 below illustrate the influence of pH, ionic strength, temperature and magnesium ion concentration upon ATP hydrolysis. The effect of magnesium ions should not be surprising as more than 99% of the ATP in a cell is present as the Mg^{++} complex.

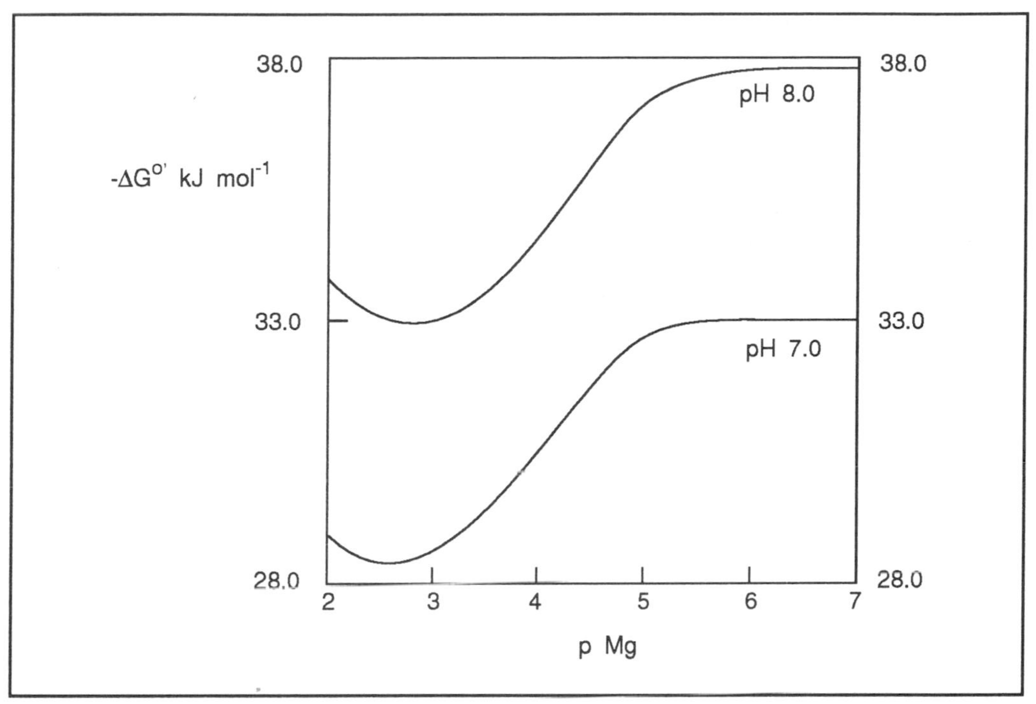

Figure 2.2 The pH and Mg^{2+} ion concentration dependence of $\Delta G^{o'}$ for the hydrolysis of ATP at 25°C and ionic strength I = 0.2. (pMg is minus the log of the magnesium ion concentration by analogy to pH). Based on Rosing J and Slater E.C. Biochimica et Biophysica Acta 267 (1972) p289.

25°C		37°C	
ionic strength (I)	$-\Delta G'$ (kJ mol^{-1})	ionic strength (I)	$-\Delta G'$ (kJ mol^{-1})
0.00	33.51	0.00	34.00
0.10	32.72	0.10	33.15
0.15	32.91	0.15	33.28
0.20	33.17	0.20	33.59

Table 2.7 The effect of temperature and ionic, strength on the value of $\Delta G^{o'}$ ($\Delta G'$) for ATP hydrolysis at pH7.0 25°C, 37°C and variable [Mg^{2+}].

The data demonstrated the need to specify conditions closely for biochemical reactions as a change of one pH unit or a 100 fold change in magnesium ion concentration can produce an almost 10 fold change in the equilibrium constant for ATP hydrolysis.

2.5 Nomenclature, units, symbols and functions

We have introduced a lot of different terms used in thermodynamics. Here we attempt to summarise what you have learnt by examining some of the nomenclature, units symbols and functions used in our discussions.

2.5.1 Nomenclature

Correct use of symbols and terminology is the key to avoiding confusion in thermodynamics. Read this section carefully.

Do remember also that in older literature in particular symbols, units and nomenclature often differ from modern practice.

Π It would be helpful to write out a sheet summarising the next section and pin it up where you can see it.

2.5.2 Units, symbols and conventions

SI units relationship between atmospheres, bar and Pascals

SI units are used (eg g, m) but for reasons of convenience one atmosphere is retained as the standard value for pressure rather than expressing the standard state value in Pascals (1 atm = 101.325 k Pa).

The bar (10^5Pa) has been recommended for reporting standard state data but is not universally used. Most, if not all, biochemical data tabulations refer to 1 atm and for solutions and solids (ie most biological cases) the differences are small enough to neglect for most purposes.

Physical states are represented as follows:

g - gaseous; l - liquid; s - solid; c - crystalline; aq - dissolved in water.

The subscript f and the term formation are used for the formation of a substance from its elements. The physical form (allotrope) of solids must be specified as must be the physical conditions under which the substance was formed.Thus $\Delta_f H$ would be an enthalpy of formation under the specified conditions ($\Delta_f H$ and not ΔH_f is used because the f modifies the operator Δ not the H, which is the symbol for a physical quantity - see McGlashan M.L, Physico - Chemical Quantities and Units, Royal Society of Chemistry Monographs for teachers No15, P.71).

2.5.3 Thermodynamic standard functions

These refer to standard reaction conditions which are defined as:

- Temperature = 298.15 K;
- Pressure, 1 atmosphere (101.325 k Pa);
- Composition, all components in their standard states.

The standard state of a substance is the pure substance (not in solution!) as it exists at 298.15 K and standard atmospheric pressure eg. gaseous carbon dioxide, solid glucose and liquid water.

For substances in solution the standard state of a solute or solvent is normally defined as unit activity.

significance of ^{o} and Δ

In the literature a superscript '$\ominus$' or 'o' is used to indicate thermodynamic functions relating to reactions under standard conditions and Δ to signify the difference between the values of a function for two states of a system. For convenience, we have adopted what has become the more common practice of using superscript 'o'.

G is the Gibbs function and ΔG^o is the change in Gibbs function for a reaction taking place under thermodynamic standard conditions (298.15 K, a pressure of one atmosphere and all the components in their standard states).

$\Delta G^{o'}$ is used to specify the charge in Gibbs function for a reaction in solution under standard conditions with all components in their standard states with the exception of hydrogen ion activity. Hydrogen ion concentration is usually 10^{-7} mol l^{-1} (pH7) unless otherwise specified.

$\Delta G'$ is the change in Gibbs function when the reaction takes place at a stated temperature, pH and pressure other than standard conditions but with reactants and products in their standard states.

ΔG is the change in Gibbs function when the reaction takes place at constant pressure and temperature under arbitrary but stated conditions (often corresponding to physiological or 'test tube' concentrations). In principal we can use similar conventions for other thermodynamic functions, for example enthalpy (H).

We point out that in the literature you may find the use of 'f' in thermodynamic functions for formation located after the physical quantity symbol, as in ΔH_f (= $\Delta_f H$). Non SI units are frequently to be found in older literature, particularly the use of the thermochemical calorie (1 calorie = 4.184 J) instead of the Joule to express energy.

2.5.4 Symbols for functions

There has also been a lack of consistency in the terminology and symbols used for thermodynamic functions. In this text we have used the following symbols:

thermodynamic energy (U)

U is the thermodynamic energy (internal energy, frequently given the symbol E) in Joules (J).

enthalpy (H)

H is the enthalpy (heat of formation or heat content) in Joules (J).

entropy (S)

S is the entropy in J K^{-1}.

Gibbs function (G)

G is the Gibbs function (free energy, net work-function or Gibbs free energy, frequently given the symbol F in older literature) in Joules (J).

Helmholtz function (A)

A is the Helmholtz function (maximum work function, available energy or Helmholtz free energy, also sometimes given the symbol F) in Joules (J).

heat (Q)

Q is heat absorbed in Joules (J).

temperature (T)

T is the thermodynamic (absolute) temperature in K.

Π As a form of revision, it might be sensible to write these symbols down and work through the previous sections to write down the relationships between them. In other words make a summary sheet.

2.5.5 Equilibrium constants

Strictly speaking, K_{eq} is the thermodynamic or true equilibrium constant calculated from the activities of the components at equilibrium. Activities are concentrations corrected for deviations from ideal behaviour, which is only found in very dilute solutions where:

$$\text{activity} \rightarrow \text{concentration}$$

K_c is the equilibrium constant calculated from the stoichiometric concentrations of the components at equilibrium.

Because of the degree of uncertainty over physiological concentrations due to experimental error etc and because of the lack of knowledge of activity coefficients for substances in the intracellular environment it is normal to use stoichiometric concentrations and to assume K_{eq} = approximately K_c in biochemical thermodynamic calculations. The error in the assumption is not usually large.

For our purposes it is usually sufficient to assume $K_{eq} = K_c$ and we use the common biochemical notation of K_{eq} to represent the equilibrium constant calculated from the stoichiometric concentrations of the components at equilibrium.

2.5.6 State functions and the Δ notation

We have used the symbol Δ to represent a change in particular functions. Remember however that changes in state functions are dependent only on the initial and final states. We should therefore signify the state. For example the state may be solid or liquid or solution. Thus we should use a notation to indicate this. Thus $X_{(s)}$ signifies the value of a state function when the system is a solid. Likewise the $X_{(l)}$ signifies the value of a state function when the system is a liquid, and $\Delta_{(s-l)}X = (X_{(s)}-X_{(l)})$ the change in X on melting.

We can use the notation $\Delta_{l \rightarrow g}X$ to signify a change in state function when the state is changed from a liquid to a gas.

This is however rather cumbersome. In most circumstances it is usually self evident which states a ΔX refers to so it is customary to simply write ΔX.

2.6 Applications of the Gibbs function and entropy to biological systems.

We may now move on to look at some of the ways in which thermodynamic data may be used to shed light upon biochemical processes. In particular we will look at some applications of the Gibbs function and of entropy.

2.6.1 Steady state equilibria and the identification of regulatory enzymes.

So far we have considered individual reactions in isolation. *In vivo* reactions are more usually part of a metabolic pathway.

A metabolic sequence such as a biochemical pathway $A \rightarrow B \rightarrow C \rightarrow D \rightarrow E$ may achieve a steady state flux through the pathway so that the concentration of substrates, for example [C] and [D] remain constant.

This leads to a dynamic equilibrium for which the mass action ratio γ of products over reactants ($\frac{[D]}{[C]}$) is an 'apparent equilibrium constant'. This mass action ratio is not a 'true' thermodynamic equilibrium which corresponds to a 'static' equilibrium. A 'static' equilibrium would result from mixing a fixed amount of substrate with enzyme and allowing the reaction to proceed until no further increase in product concentration was observed, (see Figure 2.3).

static and dynamic equilibria

A system of the static type of equilibrium corresponds more to *in vitro* than *in vivo* conditions.

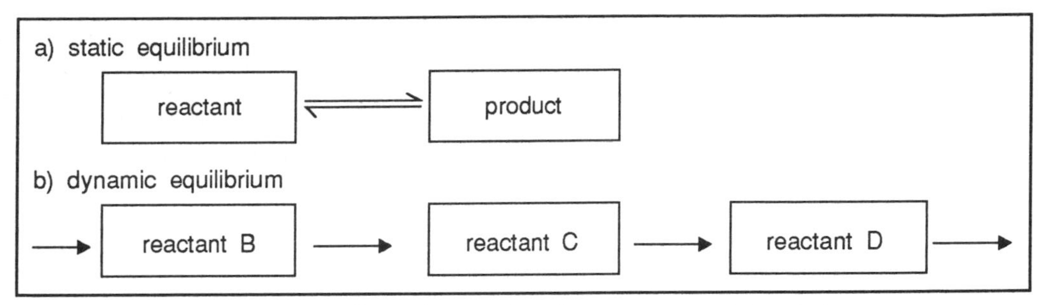

Figure 2.3 Static and dynamic equilibrium. In case a), we have a static equilibrium. This does not mean that no chemical transformation takes place, what this means is that the rate at which reactant(s) form product(s) is matched by the rate at which product(s) form reactant(s). Thus there is no net forward or backward reaction in this kind of equilibrium. In case b), the concentration of reactant C is kept constant by the rate of conversion of B to C matching exactly the rate of conversion of C to D.

For a system in static equilibrium, the rate of the forward and back reactions are equal so that there is no net reaction and under these conditions $\Delta G = 0$.

We know that:

$$\Delta G^{o'} = -RT \ln K_{eq} \qquad \text{(E - 2.12)}$$

We also know that:

$$\Delta G = \Delta G^{o} + RT\ln \frac{[\text{product of initial concentration of products}]}{[\text{product of initial concentration of reactants}]}$$

Now let us turn our attention to the dynamic equilibrium.

Π Is $\Delta G = 0$ for a dynamic equilibrium?

The answer is no. In this case there is a continued flux through the pathway, C is being continually formed and used. For the reaction $C \rightarrow D$ (See Figure 2.3) we can write:

$\Delta G = \Delta G^{o} + RT \ln \frac{[D]}{[C]}$ where [C] and [D] are the steady state concentrations of reactant C and product D.

But $\frac{[D]}{[C]}$ is the mass action ratio (γ) so we can write: $\Delta G = \Delta G^{o} + RT\ln \gamma$.

We may write for a reaction at 298K:

$$\Delta G = -RT \ln K_{eq} + RT \ln \gamma$$

$$= -RT(\ln K_{eq} - \ln \gamma)$$

$$= -RT \ln \left(\frac{K_{eq}}{\gamma}\right)$$

$$= -2.303\ RT \log \left(\frac{K_{eq}}{\gamma}\right)$$

$$= -5.7 \log \frac{K_{eq}}{\gamma}$$

If the flux (or flow of material) through an *in vivo* metabolic sequence leads to a mass action ratio (γ) far removed from the 'static' or 'true' equilibrium constant (K_{eq}) then the term 'non-equilibrium reaction' is used to describe the reaction in question. Such 'non-equilibrium reactions' represent potential control points in metabolic sequences and it is in fact possible to define a regulatory enzyme as one which catalyses a non-equilibrium reaction.

Some examples of mass action ratios (γ) and equilibrium constants (K_{eq}) for reactions in a common metabolic pathway (glycolysis) are given in Table 2.8. You need not worry about the exact chemical changes being brought about by each of the stages. You will meet these later in this text.

enzyme	K_{eq}	[γ]
hexokinase (HK)	4.7×10^3	2.6×10^{-2}
phosphoglucoisomerase (PGI)	0.41	0.23
phosphofructokinase (PFK)	1×10^3	0.63
aldolase (ALD)	8.9×10^{-5}	2.9×10^{-5}
triosephosphate- isomerase (TPI)	4×10^{-2}	1.5
glyceraldehyde-3-phosphate dehydrogenase plus phospho-glycerate kinase (G3P/PGK)	0.8×10^3	17.8
phosphoglycerate (PGM)	0.15	0.43
enolase (ENO)	3.7	1.6
pyruvate kinase (PK)	9×10^3	2.8

Table 2.8 Mass-action ratios (γ) and equilibrium constants (K_{eq}) of reactions in the glycolysis pathway for ascites tumour cells at 25°C (298 K). Data based on Newsholme E.A and Stuart C. (1973). Regualtion in Metabolism published by Wiley

Π We can use the data in Table 2.8 to calculate values for $\frac{K_{eq}}{\gamma}$ and thus ΔG for the reactions of the glycolytic pathway in ascites tumour cells. Attempt this for one or two of the reactions, and compare your results with those in Table 2.9.

It can be seen from Tables 2.8 and 2.9 that many of these reactions are not far removed from equilibrium and hence are likely to be kinetically reversible *in vivo* (ie ΔG is small).

SAQ 2.11

Identify from Table 2.9 which reactions are far removed from equilibrium in the glycolytic pathway for ascites tumour cells and comment on the significance of this.

reaction	$\frac{K_{eq}}{\gamma}$	ΔG J mol^{-1}
HK	$\frac{4.7 \times 10^3}{2.6 \times 10^{-2}} = 1.8 \times 10^5$	-30.0
PGI	$\frac{0.41}{0.23} = 1.8$	-1.5
PFK	$\frac{1 \times 10^3}{0.63} = 1.6 \times 10^3$	- 18.3
ALD	$\frac{8.9 \times 10^{-5}}{2.9 \times 10^{-5}} = 3.1$	- 2.8
TPI	$\frac{4 \times 10^{-2}}{1.5} = 2.7 \times 10^{-2}$	+ 8.9
G3P/PGK	$\frac{0.8 \times 10^3}{17.8} = 44.9$	- 9.4
PGM	$\frac{0.15}{0.43} = 0.35$	+ 2.6
ENO	$\frac{3.7}{1.6} = 2.3$	-2.01
PK	$\frac{9 \times 10^3}{2.8} = 3.2 \times 10^3$	-20.0

Table 2.9 ΔG value for glycolytic enzymes catalysed reactions in ascites tumour cells (ΔG calculated from $\Delta G = -5.7 \log \frac{K_{eq}}{\gamma}$).

2.6.2 Entropy, Gibbs function and structure

third law of thermo-dynamics

The third law of thermodynamics may be stated in the form:

> Taking the entropy of every element, in its most stable form, as zero at a temperature of absolute zero then every substance will have a positive entropy which becomes zero for all perfect crystalline substances (including compounds) at a temperature of absolute zero.

This is clearly consistent with our idea that entropy is a measure of disorder because, for a perfect crystalline solid at absolute zero, when there is no movement, we have complete information about the exact state of the system. Under such conditions there is no disorder.

The value of the third law is that it enables standard entropies (S°_m) and standard reaction entropies ($\Delta_r S^\circ$) to be obtained.

When we compare the S°_m values in Table 2.10 we see that entropy increases, as we would expect, with increasing disorder in simple substances such as carbon, water and carbon dioxide.

substance	S^o_m J $K^{-1}mol^{-1}$	substance	S^o_m J $K^{-1}mol^{-1}$
C (diamond)	2.4	methanol (gas)	239.8
C (graphite)	5.7	ethanol (liquid)	160.7
H_2O (liquid)	69.9	ethanol (gas)	282.7
CO_2 (gas)	213.7	acetic acid (liquid)	159.8
sucrose (solid)	360.2	acetic acid (aqueous)	178.7
methanol (liquid)	126.8	acetate ion (aqueous)	86.6

Table 2.10 Standard entropies (S^o_m) at 298 K.

When we consider solid diamond with its highly ordered structure we see it has a lower entropy than graphite. Sucrose, a more complex solid, has a much higher entropy which reflects the rotations and vibrations within the molecule.

Π Examine Table 2.10 carefully and note the changes in entropy associated with: the vaporisation of methanol and ethanol; the increase in chain length from methanol to ethanol; the changes when ethanol CH_3CH_2OH is converted to acetic acid CH_3COOH; acetic acid being dissolved into aqueous solution; the ionisation to aqueous acetate ion.

entropy changes in denatured proteins

In the molecular biology area, entropy changes are of particular interest with regard to changes in the three dimensional shape of macromolecules. Such changes are often associated with profound changes in the biological activity of the molecules.

A classical 1934 study by Anson and Mirsky involved the reversible thermal denaturation of the enzyme trypsin.

$$Tr_{native} \rightleftarrows Tr_{denatured} \quad K_{eq} = \frac{[Tr_{denatured}]}{[Tr_{native}]}$$

They found that ΔH° for the denaturation was +282 800 J mol^{-1} over the temperature range studied and that the enzyme was 50% inactivated at 317 K.

Since:

$$\Delta G = \Delta H - T\Delta S'$$

or for a defined set of conditions:

$$\Delta G' = \Delta H^\circ - T\Delta S'$$

We know that:

$$\Delta G' = -2.303 \, RT \log_{10} K_{eq}$$

$$K_{eq} = \frac{[\text{inactive enzyme}]}{[\text{active enzyme}]}$$

Thus for 50% inactivation $K_{eq} = 1$:

therefore:

$$\Delta G' = 0 \text{ at 317 K.}$$

Substituting for $\Delta G'$ and ΔH°:

$$0 = +282\,800 - 317\,\Delta S' \text{ at 317 K}$$

therefore:

$$\Delta S' = \frac{282\,800}{317}$$

hence:

$$\Delta S' = +\,892\ JK^{-1}\ mol^{-1}$$

This large increase in entropy can be explained in terms of an unfolding of the protein molecule with a consequent increase in disorder and a fall in information about the exact state of the molecule.

entropy changes in allosteric enzymes

Enzyme activity, particularly for enzymes which regulate metabolic activity is frequently modified by the binding of allosteric effectors. These are molecules which bind at a site removed from that at which the normal substrate binds. They alter enzyme activity by altering the conformation of the protein molecule. If the change in shape is large this should be revealed by an entropy change. We will now look at some data on this in SAQ 2.12.

SAQ 2.12

Adenosine monophosphate (AMP) is found to bind reversibly to a bacterial $NADH_2$ dehydrogenase respiratory enzyme.

$$Enz + AMP \rightleftarrows Enz - AMP$$

$\Delta H'$ and $\Delta G'$ for the nucleotide binding at 30°C were found to be +52.3 and -20 kJ mol^{-1} respectively.

Calculate the value of $\Delta S'$ for this reaction and comment upon the result.

Gibbs function changes give valuable information on protein stability as shown by the data in Tables 2.11 and 2.12.

enzyme	**$\Delta G°$ KJ mol^{-1}**
yeast phosphoglucose isomerase	+22.3
T. thermophilus phosphoglucose isomerase	+49.7
T. thermophilus cytochrome C	+119.4
horse cytochrome C	+53.1
Candida cytochrome C	+58.6

Table 2.11 Gibbs function changes for the denaturation of enzyme at 298 K.

Π Examine the data for the Gibbs function change for the denaturation of enzymes in an extreme thermophilic bacterium such as *T. thermophilus* and in 'normal' organisms and answer the following questions.

1) Which type of organism has the most thermostable enzymes?

2) Cytochrome C is an enzyme of fundamental importance for the existence of respiratory pathways. How does its thermostability compare with that of a less essential enzyme such as phosphoglucose isomerase?

From the data in Table 2.11 we can conclude that: 1) thermophilic organisms have the most thermostable enzymes; 2) the enzymes of fundamental importance appear to be the most thermostable.

SAQ 2.13

Examine the information given in the Table 2.12 and answer the following questions after looking at the structures of the four amino acids.

protein	amino acid 49	$\Delta_{unf}G$
wild type	glutamic acid	+36.8
mutant	glutamine	+26.4
mutant	valine	+50.2
mutant	leucine	+63.0

Table 2.12 Gibbs function change (K J mol^{-1}) for the unfolding of mutants of the α sub-unit of tryptophan synthase at pH7.0 and 298 K. (The mutant sub-units have different amino acids in position 49).

glutamic acid (glu) - $HO_2C . CH_2 . CH_2 . CH . NH_2 . CO_2H$;

glutamine (gln) - $H_2NOC . CH_2 . CH_2 . CH . NH_2 . CO_2H$;

valine (val) - $(H_3C)_2 . CH . CH . NH_2 . CO_2H$

leucine (leu) - $(H_3C)_2 . CH . CH_2 . CH . NH_2 . CO_2H$.

1) What is the effect of replacing the glutamic acid at position 49 (glu 49) with valine or leucine on the unfolding of the sub-unit?

2) What is the effect of substituting glutamine for glu 49?

3) Suggest what type of amino acid should be substituted for glu 49 to give a form of the enzyme which would unfold less readily than the native enzyme.

2.6.3 Exergonic and endergonic processes

In this section we attempt to distinguish between two pairs of terms which are applied to reactions. These are:

- exergonic and endergonic processes;
- endothermic and exothermic processes.

In essence:

- an exergonic reaction is one in which ΔG is negative and which can drive other processes and is therefore capable of doing useful work;
- an endergonic reaction is one in which ΔG is positive and which must be driven;
- an exothermic reaction is one which releases (evolves) *heat* (ie ΔH is negative);
- an endothermic reaction is one which absorbs *heat* (ie ΔH is positive).

The condition for a process to occur spontaneously is that for a system at constant temperature and pressure ΔG must be negative. Such a process is termed exergonic and represents the system moving towards equilibrium.

The value of ΔG is an indication or function of the displacement of the system from equilibrium. When a process moves away from equilibrium and ΔG is positive the process is termed endergonic.

It is clear that if a specific process in a biochemical or biological system is endergonic then the cell must find a way to make the overall process exergonic. This can be done by coupling thermodynamic process to an exergonic process. Thus in nature the occurrence of an endergonic process implies it must be linked to an exergonic process.

It must also be recognised that if a process is exergonic it only means that the process can occur spontaneously, not that it will proceed at a measurable rate. The ΔG for a reaction gives no indication of the speed of the reaction. This is determined by kinetic, as against thermodynamic, factors.

The reason why evolution of heat (ΔH is negative) in a reaction does not determine spontaneity is because ΔG depends upon two factors one of which is the $T\Delta S$ contribution which can in some circumstances balance out the ΔH change, [recall Equation (E - 2.11.)]

$$\Delta G = \Delta H - T\Delta S$$

Equation (E - 2.11) contains terms derived from both the first law (ΔH) and the second law ($T\Delta S$) of thermodynamics.

Both S and H are functions of state as are G and U. Being functions of state all depend solely upon the initial and final states of the system and hence the value of ΔG will be independent of the pathway between the initial and final states. This means that the value of ΔG for a complex series of transformations can be obtained from the algebraic summation of the ΔG changes for each component reaction in the series. This is a Hess's law-type relationship (see Section 2.4.8) and can tempt one into the error of regarding the Gibbs function as an energy term to which the law of conservation of energy applies. The use of terms such as 'free energy' for the Gibbs function, 'energy releasing' for exergonic processes and 'energy consuming' for endergonic processes can reinforce this error and indeed this is the reason why some modern texts tend to speak of Gibbs function rather than Gibbs 'free energy'.

The expression $\Delta G = \Delta H - T\Delta S$ can suggest that the tendency of systems to move to lower G values is due to a tendency to go to states of lower enthalpy and higher entropy. The true driving force for change is the tendency of the universe, the system and its surroundings, to move to a state of higher entropy. The sum of the entropy of the system and its surroundings is maximised. Heat which enters the surroundings as a result of an exothermic reaction ($\Delta H < 0$) increases the entropy of the surroundings. It may puzzle you that spontaneous endothermic reactions do occur with $\Delta H > 0$. In these cases the positive ΔH term is balanced out by a large increase in the entropy of the system.

An important characteristic of ΔG is that changing the relative concentrations of products and reactants can change not only its magnitude but also its sign.

Consider the case where an enzyme catalyses the reaction:

$$A + B \rightleftarrows C + D$$

and $\Delta G^{o'}$ is found to be -5.7 kJ mol^{-} at 298 K at pH 7.

Π Calculate the values of ΔG for the reaction for the following concentrations of substrates and reactants and note the implication of the results in terms of the direction in which the reaction proceeds.

1) $[A] = [B] = [C] = 1 \text{ mmol } l^{-1}; \ [D] = 0.1 \text{ mmol } l^{-1}$

2) $[A] = [C] = [D] = 1 \text{mmol } l^{-1}; \ [B] = 0.1 \text{ mmol } l^{-1}$

3) $[A] = [C] = [D] = 1 \text{mmol } l^{-1}; \ [B] = 0.01 \text{mmol } l^{-1}$

You should have used the following relationship:

$$\Delta G = \Delta G^{o'} + 2.303 \ RT \log_{10} \frac{[C] \times [D]}{[A] \times [B]}$$

thus:

1) $$\Delta G = -5.7 + 5.7 \log_{10} \frac{0.1}{1} = -5.7 + (-5.7) = -11.4 \text{ kJ mol}^{-1}$$

2) $$\Delta G = -5.7 + 5.7 \log_{10} \frac{1}{0.1} = -5.7 + 5.7 = 0 \text{ kJ mol}^{-1}$$

3) $$\Delta G = -5.7 + 5.7 \log_{10} \frac{1}{0.01} = -5.7 + 2\text{x } 5.7 = + 5.7 \text{ kJ mol}^{-1}$$

The sign of ΔG and the direction of spontaneous change have reversed. It follows that if other metabolic reactions can vary the concentrations of A, B, C or D the direction of metabolic flux through this reaction can be reversed.

It is essential to recognise that ΔG not $\Delta G^{o'}$ determines if a reaction is spontaneous under a given set of conditions. It is the concentration of reactants and products which determines if ΔG is smaller than, greater than or equal to $\Delta G^{o'}$.

It can be seen from the above that substrate and reactant concentrations directly affect the value of ΔG. It is not possible to set a value of ΔG for a reaction *in vivo* in the absence of knowledge of the cellular concentrations of the substrates and reactants. ΔS is concentration related in the same way as ΔG.

ΔG is a state function so its value depends not on how work is done but only on the values of G for the defined initial and final states.

Changes in enthalpy (ΔH) are the source of energy to perform work unless ΔG is greater than ΔH when heat or energy is absorbed from the surroundings.

In 'biology', metabolism involves using chemical energy (internal energy) for carrying out the work of the cell and ΔH sets an upper limit on how much work can be done.

Photosynthesis is the basis of sustained life because the radiant energy absorbed does work in the biological system compensating for the unfavourable entropy changes in the reactions taking place in the living cell.

2.6.4 The coupling of reactions

If we consider the following reaction, catalysed by the enzyme pyruvate kinase, which occurs in the glycolytic pathway we see that two molecular transformations have taken place. The overall reaction:

reaction A phosphoenol pyruvate + ADP ⇄ pyruvate + ATP

may be looked at as involving the following two reactions:

reaction B phosphoenol pyruvate + H_2O ⇄ pyruvate + Pi

$\Delta G^{o'} = -55.6$ kJ mol^{-1}

reaction C ADP + Pi ⇄ ATP + H_2O

$\Delta G^{o'} = +32.2$ kJ mol^{-1}

If we add together reactions B and C, we get reaction A.

Likewise by adding together the energy changes the overall change in Gibbs function is:

$\Delta G^{o'}$ is -55.6 + 32.2 = -23.4 kJ mol^{-1}.

The net process is exergonic and could be expected to proceed spontaneously.

It should be recognised that the enzyme pyruvate kinase is a complex protein, a tetramer of 55 kD subunits, which can exist in three forms and that reaction A is the net reaction which takes place when the substrates react with the enzyme. The two reactions B and C do not actually take place as written. Indeed the steps in the conversion of reactants to products in reaction A involve enzyme-bound species so that the equation written for reaction A represents only the net reaction.

However the Gibbs function is a function of state and $\Delta G^{o'}$ for the net reaction will depend only on the initial and final states. It is independent of the actual pathway so that the thermodynamics of the process may be analysed without need for knowledge of the intermediate reaction complexes.

It is clear that the large negative ΔG associated with the loss or transfer of the phosphate group from phosphoenol pyruvate is the exergonic process which drives the endergonic process of transferring a phosphate group to ADP.

We may describe the two processes as being coupled and the concept of 'coupled' reactions enables us to understand the energetics of many biochemical reactions.

Now try to answer the following SAQ which tests your understanding of this point.

SAQ 2.14

Using data from the previous example and given that $\Delta G^{o'}$ is -13.8 kJ mol for the reaction:

glucose-6-phosphate + H_2O → glucose + Pi

calculate $\Delta G^{o'}$ for the reaction:

glucose + ATP ⇄ glucose-6-phosphate + ADP

and comment upon the physiological feasibility of the enzyme catalysed reaction and its mechanism. You will need the data for the change in Gibbs function associated with the synthesis of ATP given in the text above.

2.6.5 'High energy' phosphates

This term is misleading but it is still widely encountered. Its origins lie, in the observation that $\Delta G^{o'}$ for the hydrolysis of ATP and certain other phosphate esters (such as creatine phosphate, phosphoenol pyruvate and 1,3 - diphosphoglycerate) was much higher than, for example, that of glucose-6-, and glucose-1-phosphate or glycerol monophosphate.

The phosphates with high, negative values for $\Delta G^{o'}$ of hydrolysis appeared to be generally associated with metabolic pathways and processes involved in yielding useful energy for biological purposes.

erroneous concept of 'high energy' phosphate

The use of the symbol for the 'high energy' phosphate link as in ADP ~ P (ATP) gave the mistaken impression that unusual amounts of energy are associated with a specific bond in the molecule. The term 'bond energy' is used in chemistry to mean the amount of energy needed to break a chemical bond. Thus a high energy bond should be one which is hard to break. A moment's thought will reveal that the term is used in the opposite sense when talking about a high energy bond in biology. A further misconception is that the energy of a molecule is located in specific bonds as against being associated with the whole molecule.

It should be remembered also that:

$$\Delta G = \Delta H - T\Delta S$$

so that the Gibbs function change depends both on changes in enthalpy and changes in entropy. If the entropy change is relatively large then the Gibbs function change will be sensitive to temperature.

In the case of ATP hydrolysis at 310K, the approximate values of the thermodynamic function changes are:

$$\Delta G' = \text{approximately} - 30 \text{ kJ mol}^{-1}$$

$$\Delta H' = \text{approximately} - 20 \text{ kJ mol}^{-1}$$

$$\Delta S' = \text{approximately} - 34 \text{ J mol}^{-1}$$

The enthalpy change reflects the fact that there are greater electrostatic repulsions between the negative charges in the ATP molecule than in the reaction products, the products of hydrolysis have higher enthalpies of solvation than ATP and that there is resonance stabilisation of the reaction product molecules. The release of the phosphate group on hydrolysis contributes a relatively large entropy change for ATP hydrolysis. but ΔS is negative because of the increased ordering of water molecules around the products of hydrolysis.

The use of terms like 'high energy' phosphate bond, 'energy rich' bond are thus erroneous as the whole molecule rather than a single isolated bond determines the energetics of the system. Nevertheless the use of the term 'high energy phosphate bond' is a convenient shorthand providing the explanations given here are clearly understood.

A further misconception is to view ΔG for the hydrolysis of ATP as exceptionally high by comparison with other phosphate esters in biological systems. The misconception was initially caused by early estimates of $\Delta G^{o'}$ for the hydrolysis of ATP of -50 kJ mol^{-1} whilst much lower values were (correctly) obtained for glucose-6-phosphate and glycerol-1-phosphate.

2.6.6 Substances of high group transfer potential

Table 2.13 below lists the $\Delta G^{o'}$ values for the hydrolysis of a number of substances involved in the transfer of groups in biosynthetic reactions and includes a value for $\Delta G^{o'}$ for ATP which is commonly accepted for biological systems. The reader will recall however from Table 2.6 and Figure 2.2 that $\Delta G^{o'}$ values for ATP hydrolysis are dependant upon temperature, ionic strength and magnesium concentration as well as pH.

donor molecule	group activated for transfer	$\Delta G^{o'}$ kJ mol^{-1}
ATP	phosphate	-32.2
1,3 - diphosphoglycerate	phosphate	-56.9
phosphoenol pyruvate	phosphate	-55.6
creatine phosphate	phosphate	-42.7
acetyl phosphate	acyl and phosphate	-42.3
acetyl-S-coenzyme A	acyl	-33.5
UDP glucose	glucose	-30.5
valyl tRNA	valyl	-35
N^{10} formyl tetrahydrofolate	formyl	-26

Table 2.13 $\Delta G^{o'}$ values for biologically important substances of high groups transfer potential.

Π Below we have listed a variety of phosphates found in biological systems in ascending order of $\Delta G^{o'}$ values for their hydrolysis. What do you notice about the position of ATP in this list and how appropriate do you consider the application of terms such as 'high energy' phosphate to ATP?

substance	$\Delta G^{o'}$(pH7)
glycerol 1-phosphate	-9.6
glucose 6-phosphate	-13.8
glucose 1-phosphate	-20.9
phosphodiesters	-25.1
ATP	-32.2
acetyl phosphate	-42.3
creatine phosphate	-42.7
phosphoenolpyruvate	-55.6

Table 2.14 Gibbs function changes for the hydrolysis of some biologically important phosphates.

ATP has an intermediate group transfer potential

It can be seen from an examination of Table 2.14 that ATP is in fact a compound of intermediate group transfer potential with a $\Delta G^{o'}$ of hydrolysis intermediate between the very high values for substances like phosphoenolpyruvate and the very low values such as for glycerol-1-phosphate.

2.6.7 The measurement of ΔG

We have seen that ΔG values can be obtained from equilibrium data. To illustrate potential problems in obtaining ΔG values let us examine a specific example and consider the likely accuracy of the ΔG value which would be obtained.

The value of $\Delta G^{o'}$ for the reaction $A \rightleftarrows B$ is -57 kJ mol^{-1} at 298 K. It can be shown (Section 2.4.6) that:

$$\log_{10} K_{eq} = -\frac{\Delta G^{o'}}{5.7}$$

thus:

$$K_{eq} = 10^{\left(-\frac{\Delta G^{o'}}{5.7}\right)} \quad \text{with } \Delta G^{0'} \text{ in kJmol}^{-1}$$

so that:

$$K_{eq} = 10^{-\left(-\frac{57}{5.7}\right)} = 10^{10}$$

The equilibrium will be such that the concentrations of reactants at equilibrium will be so small as to be impossible to measure them with any accuracy (if they can be measured at all!)

Other methods for determining ΔG^o include:

1) Substituting known values of ΔH^o and ΔS^o into the equation:

$$\Delta G^o = \Delta H^o - T\Delta S^o$$

2) From the difference between the algebraic sum of the standard Gibbs function of formation of the products and reactants:

$$\Delta G^o = \Sigma\Delta_f G^o_{products} - \Sigma\Delta_f G^o_{reactants}$$

3) Using ΔG^o values for reactions whose net outcome is the reaction for which a ΔG^o value is desired (see Section 2.6.4).

Note that methods 2) and 3) require knowledge of the appropriate ΔG^o and $\Delta_f G^o$ values.

4) In the case of oxidation - reduction reactions ΔG^o may be obtained from electrode potential measurements. These will be discussed in the next chapter.

ΔS^o values may also be obtained if the appropriate data are available using the equation:

$$\Delta S^o = \Sigma\Delta S^o_{products} - \Sigma\Delta S^o_{reactants}$$

ΔH values may be obtained by calorimetry or from heat of combustion data for the reaction components or the application of Hess's law to other appropriate data (see Section 2.3.7).

ΔH° values may also be obtained by using the relationships for a reaction at constant pressure.

$$\Delta G^\circ = - RT \ln K_{eq}$$

and:

$$\Delta G^\circ = \Delta H^\circ - T\Delta S^\circ$$

so that:

$$\Delta H^\circ - T\Delta S^\circ = -RT \ln K_{eq}$$

hence:

$$RT \ln K_{eq} = T\Delta S^\circ - \Delta H^\circ$$

$$\therefore \ln K_{eq} = \frac{\Delta S^\circ}{R} - \frac{\Delta H^\circ}{RT} \text{ or } \log K_{eq} = \frac{\Delta S^\circ}{2.303R} - \frac{\Delta H^\circ}{2.303RT}$$

but R and ΔS° are constants, then

$$\log K_{eq} = \text{constant} - \frac{\Delta H^\circ}{2.303RT}$$

ΔH° may thus be obtained from the slope of a plot of $\ln K_{eq}$ against $\frac{1}{T}$.

SAQ 2.15

Below is a plot of $\ln K_{eq}$ against $\frac{1}{T}$ for a reaction . From this plot determine;

1) the value of ΔH° for the reaction;
2) the value of $\Delta S'$ for the reaction at 333 K;
3) the value of $\Delta G'$ for the reaction at 333 K.

Assume $R = 8.314\ JK^{-1}\ mol^{-1}$.

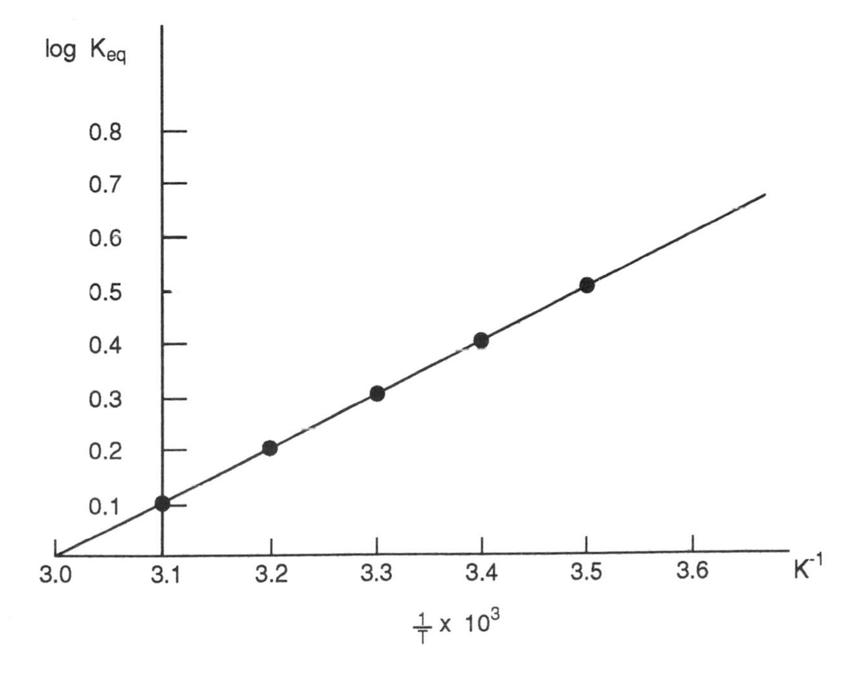

2.6.8 The biochemical role of ATP

It is useful to view ATP and ADP as acting as a donor - acceptor pair for phosphoryl group transfer reactions, accepting phosphate groups from substances of higher group transfer potential and donating them to substances of lower group transfer potential.

The ability to transfer a phosphoryl group to water as indicated by $\Delta G^{o'}$ for hydrolysis of the phosphate compound (Table 2.14) gives a measure of group transfer potential.

Table 2.13 demonstrates how this view of $\Delta G^{o'}$ of hydrolysis group transfer has a wide application.

This view of the role of ATP would be consistent with a rapid interconversion of ATP/ADP within the cell and relatively low ATP concentrations.

The concentrations of ATP found in living cells are in fact of the order of 10^{-3} to 10^{-2} mol l^{-1} of cell water and it has been estimated that the half-time for turnover of the terminal phosphate group is a minute or two in animal cells but only a matter of seconds in bacterial cells.

The hydrolysis of ATP is exergonic ($\Delta G' < 0$) under standard physiological conditions so that about 30 kJ mol^{-1} is available to drive other reactions through coupling. This is a biologically important characteristic of the substance (see Section 2.6.4).

SAQ 2.16

Calculate the maximum concentrations of DL leucylglycine which would be spontaneously produced by direct reaction between 1.0 mmol l^{-1} aqueous solutions of glycine and DL leucine at 37°C and pH 7.0. Discuss, the implications in terms of peptide synthesis. Given $\Delta G' = +14$ kJ mol^{-} for the reaction:

$$\text{glycine} + \text{DL leucine} \rightleftarrows H_2O + \text{DL leucylglycine}$$

Protein biosynthesis is a highly endergonic process not only because of the energy accounted for by the enthalpy change in the formation of a large number of peptide linkages but also because of the large entropy decrease. This decrease is due to the assembly of a large number of amino acids into the protein chain in a precisely ordered sequence. The effects upon entropy of the formation of protein secondary, tertiary and quaternary structure and changes associated with solvation must also be taken into account.

Protein synthesis is driven by linking it to ATP via a complex sequence involving transfer RNA as a carrier of the amino acids (Table 2.13 note $\Delta G^{o'}$ for valyl tRNA for example).

exergonic ATP hydrolysis drives endergonic peptide bond formation

Thus the exergonic hydrolysis of ATP is used to drive the endergonic peptide bond formation which has an average $\Delta G^{o'}$ of about +17 kJ mol^{-1} but because of the indirect nature of the process about 3 ATPs are used per peptide link formed.

Glucose oxidation under aerobic conditions produces 38 ATP molecules per glucose molecule and for a modestly sized protein molecule with about 150 peptide links (about the size of myoglobin); peptide bond formation alone will use about 450 ATP molecules. This means about 12 glucose molecules are needed per 1 protein molecule as a minimum. As will be seen in the next chapter, ATP is involved in cellular transport processes across membranes so that more ATP (and more glucose) may be required to transport the glucose and amino acids into the cell before synthesis can begin.

The cell may produce ATP via a number of mechanisms apart from the oxidation of organic substrates. In particular photosynthesis is a major route whilst oxidation of inorganic substrates is used by some bacteria. Anaerobic processes such as the conversion of glucose to lactate in glycolysis are also linked to ATP production.

$$\text{glucose} + 2\text{Pi}^- + 2\text{ADP} \rightarrow 2\ \text{lactate}^- + 2\text{ATP} + 2H_2O$$

(We will examine the details of this in a later chapter).

At 310 K $\Delta G^{'}$ is about -218 kJ mol^{-1} for the conversion of glucose to two lactate molecules. The standard reaction enthalpy for this reaction is -120 kJ mol^{-1}.

A large increase in entropy due to the breaking of the glucose molecule accounts for the exothermicity being exceeded by the exergonicity. In the production of two ATPs $\Delta G^{'}$ is about +60 kJ mol^{-1} so that the whole process of lactate and ATP production via glucose fermentation has a value of $\Delta G^{'}$ of about -158 kJ mol^{-1}. The process is thus a spontaneous one with the equilibrium strongly in favour of ATP and lactate formation.

Knowledge of $\Delta G^{'}$ for ATP hydrolysis can be exploited in many ways as shown by this final example.

Nitrite is converted to nitrate by micro-organisms which contain an enzyme system able to catalyse the reaction.

$$NO_2^-(aq) \quad + \ {}^1\!/_2O_2 \ \rightarrow \ NO_3^-(aq)$$

Note also that $\Delta_f G^{'}$ for NO_2^- (aq) is -34.5 kJ mol^{-1} and for NO_3^- (aq) is -110.5 kJ mol^{-1} we can answer the question, 'Is it thermodynamically possible for the organism to grow aerobically using nitrite as an energy source?'

$\Delta G^{'}$ for the reaction ADP + Pi $\rightarrow$ ATP + H_2O is about + 32 to 33 kJ mol^{-1}.

The oxidation of nitrite to nitrate is thus sufficiently exergonic to drive the endergonic biosynthesis of at least one ATP per molecule of nitrite oxidised.

All living organisms appear to use ATP so nitrite is certainly a potential substrate, on thermodynamic grounds, for the growth of micro-organisms under aerobic conditions.

Summary and objectives

In this chapter we have examined a wide range of thermodynamic concepts and have learnt how to apply them to a range of biochemical problems.

We have examined the laws of thermodynamics, and learnt of a number of state functions. We have also explained conventions of thermodynamic nomenclature,

Now that you have completed this chapter you should be able to:

- explain why the study of thermodynamics is important;
- define closed, open, and isolated systems;
- list the features of the macroscopic and the microscopic approach to thermodynamics;
- use correctly a wide range of thermodynamic terms, including enthalpy, entropy, Gibbs function, thermodynamic energy, Helmholtz function;
- use correctly the nomenclature used to describe thermodynamic functions;
- calculate a variety of thermodynamics values (eg ΔU, ΔG, ΔS) from supplied data;
- relate energy changes to equilibria and to calculate energy changes from data concerning equilibria and vice versa;
- use thermodynamic data to identify likely regulatory stages in metabolic pathways;
- use thermodynamic data to predict the likely stability of biochemicals, especially proteins;
- use appropriately the expression 'high energy' phosphate.

Thermodynamics and compartmentalisation

Thermodynamics and compartmentalisation

In the previous chapter we looked at the basic concepts of thermodynamics and their application to equilibrium in a non-compartmentalised system. In particular the role of adenosine triphosphate (ATP) and related substances in bioenergetics was examined. Historically the first unifying concept in bioenergetics had been that the generation of ATP was the purpose of metabolism and that it was linked to the performance of biologically useful work through 'coupling' and the hydrolysis of 'high energy' phosphates.

unifying concept

If we move from simple 'test tube' biochemistry to living systems the striking difference is that their cells and organisms are highly compartmentalised. Biochemical reactions in living cells are characterised by the movement of molecules, ions and chemical groups across membranes.

compart-mentalised

These movements are directional and membrane structure is not the same in all directions ie they are anisotropic. This gives metabolism within the cell a vectorial character. Biological membranes are electrically insulating and the movement of ions across them can generate electrochemical potentials of up to about 500 mV.

It has become generally recognised over the last thirty years that the movement of ions, particularly sodium and hydrogen ions, plays a key role in cell energetics. The coupling of metabolism to the performance of biological work is achieved in two ways.

- through chemical 'coupling' agents such as the ATP - ADP system.
- through ion currents across membranes.

In this chapter we will look at the basic electrochemistry which underlies 'ion coupling' in cellular energetics and also at oxidation - reduction reactions which play an important part in the metabolism of living organisms.

3.1 Oxidation and reduction

reduction

Reduction is the term used to describe the addition of one or more electrons to an ion or molecule.

$$A + e^- \rightarrow A^-$$

In this case the oxidised form (oxidant A) is reduced to A^-.

oxidation

Oxidation refers to the removal of one or more electrons from a molecule.

$$B \rightarrow B^+ + e^-$$

In this case the reduced form (reductant B) is oxidised to B^+.

redox reaction

If A and B are mixed we would have an oxidant and a reductant couple and the subsequent reaction would be a redox reaction.

$$A + B \rightleftarrows A^- + B^+$$

SAQ 3.1

State which of the following reactions are reductions:

1) $Cu^{2+} + 2e^- \rightarrow Cu$;

2) $Zn^{2+} + 2e^- \rightarrow Zn$;

3) $Zn \rightarrow Zn^{2+} + 2e^-$.

The oxidation-reduction (redox) reaction consists of a transfer of electrons to an oxidising agent. In consequence there is no reduction without an oxidation taking place.

3.1.1 Biochemical oxidations and reductions

Oxidation-reduction processes are very widespread in biochemical pathways. Typical examples are:

- the reduction of the oxidised form of the iron containing protein cytochrome C which is a component in the respiratory chain.

$$\text{cyt } C_{ox}\ (Fe^{3+}) + e^- \rightleftarrows \text{Cyt } C_{red}\ (Fe^{2+})$$

and:

- the conversion of pyruvate to lactate by the enzyme lactate dehydrogenase using reduced pyridine nucleotide as a cofactor.

$$CH_3CO\ COOH + NADH + H^+ \rightleftarrows CH_3\ CH\ OH\ COOH + NAD^+$$

This last reaction may be broken down into the reduction of pyruvate:

$$\underset{\text{pyruvate}}{CH_3COCO_2^-} + 2H^+ + 2e^- \rightleftarrows \underset{\text{lactate}}{CH_3CHOHCO_2^-}$$

and the oxidation of nicotinamide adenine dinucleotide (NAD):

$$NADH \rightleftarrows NAD^+ + H^+ + 2e^-$$

Oxidation and reduction involve the transfer of electrons. To understand these processes we will need to first appreciate some fundamentals of electrochemistry. To ensure you understand the concepts involved we will first discuss electrodes and electrochemical cells before applying these concepts to biochemical processes.

3.1.2 Electrodes

When a metal rod is immersed in a solution of a salt of that metal there will be a tendency for the metal to dissolve. If we take the metal zinc and add it to a solution of copper sulphate we find that copper precipitates whilst the zinc dissolves. Thus zinc has displaced the copper.

$$Zn + Cu^{2+} \rightleftarrows Zn^{2+} + Cu \downarrow$$

The reaction occurs spontaneously and at 298 K the equilibrium constant for the reaction is about 10^{37}. This indicates that there is a much greater tendency for zinc metal to dissolve to give Zn^{2+} than for copper to dissolve to give Cu^{2+} ions under these conditions.

electrochemical cell

We can set up an electrochemical cell, as shown in Figure 3.1. Here we have a zinc electrode in zinc sulphate solution and a copper electrode in copper sulphate solution. These two half cells have been joined together by a conducting bridge dipping into the two sulphate solutions. We have just seen that zinc has a greater tendency to give up electrons and form Zn^{2+} than copper has to form Cu^{2+}. This means that the copper electrode will be found to be positive with respect to the zinc electrode.

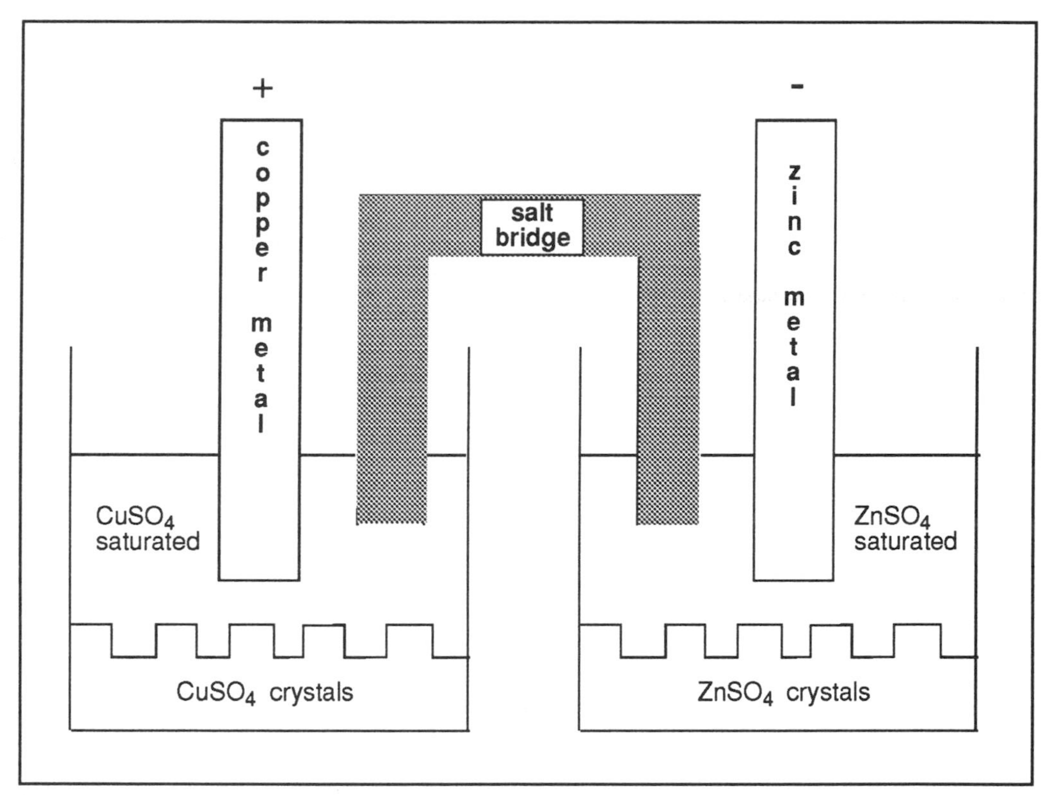

Figure 3.1 A copper zinc (Daniell) cell. The salt bridge is essentially a tube full of saturated KCl solution and agar gel. It serves to more-or-less eliminate liquid function potential effects.

If the zinc and copper electrodes are electrically connected directly to each other a current will flow. Electrons will move from the zinc electrode to the copper as Zn^{2+} ions are formed (oxidation) and Cu^{2+} ions are reduced to copper. Now we can apply an opposing potential to this electrochemical Zn/Cu cell as shown in Figure 3.2. If the applied opposing potential is slowly increased a point will be reached at which it will exactly balance the potential difference between the copper and zinc electrodes. At this point there will be no net current flow between the electrodes.

null point

electromotive force

The point at which no current flows is called the null point. The applied potential difference at the null point, is termed the electromotive force (emf) of the cell.

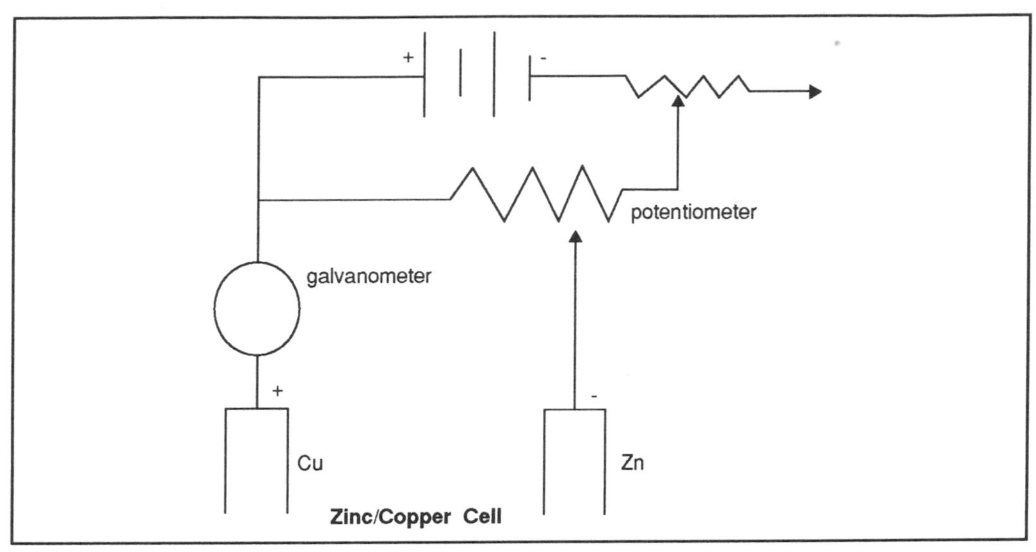

Figure 3.2 A null point circuit. When the electrode potential from the cell is exactly balanced by the opposing potential from the potentiometer no deflection is observed on the galvanometer and no current is drawn from the cell. The cell is now at the null point.

At a slightly lower applied opposing potential the net reaction:

$$Zn + Cu^{2+} \rightarrow Zn^{2+} + Cu$$

occurs whilst at a slightly higher opposing potential difference we get the reverse reaction:

$$Cu + Zn^{2+} \rightarrow Zn + Cu^{2+}.$$

At the null point the system is in chemical equilibrium and we have no net change in composition:

$$Zn + Cu^{2+} \rightleftarrows Zn^{2+} + Cu$$

Under these conditions, with no current flow, the system may be regarded as thermodynamically reversible for all practical purposes (see previous chapter).

3.1.3 Anodes and cathodes

During the process of electrolysis the reaction is driven by an externally generated current and electrons are supplied to the cathode. This electrode is thus negative with respect to the anode. Table 3.1 below summarises the situation.

	electrolytic cell	
	anode	cathode
potential	high (+)	low (-)
reaction	$A^- \rightarrow A + e^-$	$B^+ + e^- \rightarrow B$
process	oxidation	reduction

Table 3.1 Electrolytic cell.

anode
cathode

The cathode is thus the electrode at which reduction takes place and the anode the electrode at which oxidation takes place.

In a discharging galvanic cell, such as the Zn/Cu cell we have been considering, the cathode is still the electrode at which reduction takes place whilst oxidation again takes place at the anode. The relative potential of the anode with respect to the cathode is now negative with the anode acting as a source of electrons as summarised in Table 3.2.

	discharging galvanic cell	
	anode	cathode
potential	low (-)	high (+)
reaction	$Zn \rightarrow Zn^{2+} + 2e^-$	$Cu^{2+} + 2e^- \rightarrow Cu$
process	oxidation	reduction

Table 3.2 Discharging galvanic cell.

In the discharging galvanic cell the copper ion is reduced and withdraws electrons from the copper electrode to leave it with a net positive charge. The zinc electrode in contrast is the site of electrons leaving the cell and going into the external circuit.

3.2 Electrochemical cells

The notation for electrochemical cells is based upon a consideration of the individual electrodes. We have seen that cell reactions involve oxidations and reductions and we can always write the reaction in the standard form:

$$Red_L + Ox_R \rightarrow Red_R + Ox_L \qquad (E-3.1)$$

where L and R signify the left and right hand electrodes of the cell as written. Red and Ox refer to reductant and oxidant respectively. It will be appreciated that each oxidant (electron acceptor) has an associated conjugate reductant (electron donor) form into which it is converted on accepting electrons. The reverse is, of course, true for reductants.

The reaction occurring at the left hand electrode may thus be written as:

$$Red_L \rightleftarrows Ox_L + e^-$$

whilst the right hand electrode reaction is:

$$Ox_R + e^- \rightleftarrows Red_R$$

SAQ 3.2

Identify Red_L, Ox_L, Red_R and Ox_R for the zinc copper electrochemical cell shown in Figure 3.1.

3.2.1 Types of half cells

half cells

There are four common types of half cells.

1) The metal electrode in a solution of its own ions as already considered in the case of a copper electrode dipping into a solution containing cupric and sulphate ions. The half cell reaction is:

$$Cu^{2+} + 2e^- \rightleftarrows Cu$$

2) The ion, insoluble salt, metal electrode such as the chloride, silver chloride, silver electrode which consists of a silver electrode coated with insoluble silver chloride dipping into a solution containing chloride ions. Here the half cell reaction is:

$$AgCl + e^- \rightleftarrows Ag + Cl^-$$

3) The gas, inert metal electrode where gas is bubbled over a suitable inert metal electrode, such as a platinum electrode coated with finely divided platinum (platinum black), dipping into a solution containing ions derived from the gas. The hydrogen electrode, where hydrogen gas is bubbled over a platinum electrode dipping into a solution containing hydrogen ions is an example. The reaction is:

$$H^+ + e^- \rightleftarrows \tfrac{1}{2} H_2.$$

In this electrode the hydrogen gas can be regarded as being absorbed onto the platinum black.

4) The so called redox or oxidation - reduction electrode which consists of an inert metal electrode dipping into a solution containing a species which can exist in solution in two oxidation states such as Fe^{2+} and Fe^{3+}, where the half cell reaction would be:

$$Fe^{3+} + e^- \rightleftarrows Fe^{2+}$$

In cases 3) and 4) the only role of the metal electrode is to conduct electricity; in the other cases it 'transfers' electrons by being itself a component of the redox reaction.

3.2.2 Cell diagrams

cell diagrams

The conventional way to represent cells uses a vertical line | to indicate the physical separation of two phases and a comma to indicate that two species are in the same phase. A double vertical || line is used to denote the junction between two half cells.

A cell made up from two half cells consisting of platinum wires dipping into solutions of substances which can exist in two oxidation states would thus be represented diagrammatically as:

$$Pt \mid Red_L, Ox_L \parallel Red_R, Ox_R \mid Pt.$$

SAQ 3.3

Write down representations of:

1) a zinc copper (Daniell) cell as shown in Figure 3.1;

2) an Fe^{2+}, Fe^{3+} redox half cell (with Pt electrode);

3) a cell combining a hydrogen electrode half cell on the left with a silver, silver chloride half cell on the right.

3.2.3 Electrode potentials

electrode potentials

In Section 3.1.2 we saw that the electromotive force (emf) E of a cell can be measured. This is the potential difference measured in volts between the two half cells under conditions in which no current flows and the system is in thermodynamic equilibrium.

It is the convention to subtract the potential of the left-hand electrode from that of the right as the cell is written.

The emf of the cell is:

$$E = E_R - E_L \qquad (E - 3.2)$$

calculation of emf

The sign of E, the emf, indicates which half cell has the highest electrode potential.

If the right-hand electrode has a positive charge relative to the left-hand electrode, due to the reaction at the right hand electrode tending to withdraw electrons from that electrode, then the emf will be positive.

A value of $E > 0$ therefore indicates that there is a tendency for reduction to occur at the right-hand electrode:

$$Ox_R + e^- \rightarrow Red_R$$

which electrode, by definition, is the cathode. When $E < 0$ there is a relative excess of electrons at the left-hand electrode which indicates that oxidation tends to occur at that electrode:

$$Red_L \rightarrow Ox_L + e^-.$$

A positive emf therefore means that the reaction tends to occur as follows:

$$Red_L + Ox_R \rightarrow Red_R + Ox_L$$

In the same way if $E < 0$, this negative emf indicates that the reaction will tend to be:

$$Red_R + Ox_L \rightarrow Red_L + Ox_R$$

SAQ 3.4

Using the boxes provided fill in the blanks and indicate the direction of the tendency for electron flow.

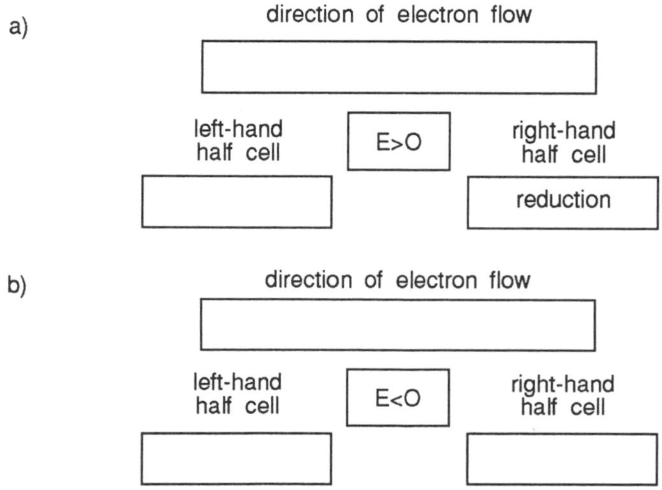

a)

direction of electron flow

[ ]

left-hand half cell	E>O	right-hand half cell
[ ]		reduction

b)

direction of electron flow

[ ]

left-hand half cell	E<O	right-hand half cell
[ ]		[ ]

3.3 Standard electrode potentials

The affinity for electrons of a half cell is only measurable when it is linked to another half cell as in the case of an electrochemical cell such as the copper zinc cell. If a standard electrode potential of one half cell is defined as zero then the relative electrode potential of other half cells can be determined and tabulated. The standard hydrogen electrode (SHE) has been selected as the basis for the measurement of the standard electrode potentials of all other redox systems (or couples). The potential of a standard hydrogen electrode is defined as zero at all temperatures. A standard hydrogen electrode consists of a platinum black coated platinum electrode in contact with hydrogen ions and hydrogen gas in their standard states at the specified temperature.

standard hydrogen electrode

standard electrode potential defined

The standard electrode potential E° of any other redox system is defined as the emf of the cell in which the system forms the right hand half cell and the standard hydrogen electrode the left, with all species being in their standard states.

$$E^\circ_{cell} = E^\circ_R - E^\circ_{SHE} = E^\circ_R \quad \text{(E - 3.3)}$$

3.3.1 Electrochemical series

tendency to undergo reduction

The standard electrode potential or redox potential as defined above is a reduction potential and indicates the tendency for the couple or redox system to undergo reduction. Couples with high standard electrode potentials are reduced by couples with low standard electrode potentials ie low reduces high.

Compounds and elements can be listed in order of their standard electrode potential or redox potentials. Table 3.3 lists some standard electrode potentials and redox potentials of interest. Strong reducing agents such as sodium metal or NADH have negative potentials whilst strong oxidising agents such as oxygen or chlorine have positive potentials. Electrons normally move towards the couple with the more positive electrode or redox potential.

It should be noted that in older and American literature standard oxidation potentials as against reduction potentials were listed. These have the same numerical value but the opposite sign. It is important to be sure which convention has been followed. The standard electrode potential or redox potential should be negative for $Na^+ + e^- \rightarrow Na$ and $NAD^+ + 2H^+ + 2e^- \rightarrow NADH + H^+$ whilst the potentials for $Cu^{2+} + 2e \rightarrow Cu$ and $\frac{1}{2} O_2 + 2H^+ + 2e^- \rightarrow H_2O$ should be +ve. It is always worth checking which convention has been followed. Unfortunately the most authoritative collection of inorganic data (Latimer's 'Oxidation States of the Elements and their Potentials in Aqueous Solution', Prentice Hall, New York, 1938) lists oxidation potentials and follows the 'wrong' convention.

reaction			E^o(V)	E_o' (V)
$Na^+ + e^-$	→	Na	-2.71	
$Zn^{2+} + e^-$	→	Zn	-0.76	
succinate + CO_2 + $2H^+ + e^-$	→	α-ketoglurate		-0.67
$H^+ + e^-$	→	$\frac{1}{2}H_2$ (at pH7)		-0.41
$NAD^+ + 2H^+ + 2e^-$	→	$NADH + H^+$		-0.32
Pyruvate + $2H^+ + 2e^-$	→	lactate		-0.19
$D^+ + e^-$	→	$\frac{1}{2}D_2$	-0.0034	
$H^+ + e^-$	→	$\frac{1}{2}H_2$ (at pH0)	0.00	
cyt C (Fe^{3+}) + e^-	→	cyt C (Fe^{2+})		+0.22
$Cu^{2+} + 2e^-$	→	Cu	+0.34	
$Fe^{3+} + e^-$	→	Fe^{2+}	+0.77	(+0.77)
$\frac{1}{2}O_2 + 2H^+ + 2e^-$	→	H_2O (at pH7)		+0.82
$\frac{1}{2}O_2 + 2H^+ + 2e^-$	→	H_2O (at pH0)	+1.23	
$\frac{1}{2}Cl_2 + e^-$	→	Cl^-	+1.36	
$HClO + H^+ + e^-$	→	$\frac{1}{2}Cl_2 + H_2O$	+1.63	
$O_3 + 2H^+ + 2e^-$	→	$O_2 + H_2O$	+2.07	
$\frac{1}{2}F_2 + e^-$	→	F^-	+2.85	

Table 3.3 Some standard electrode and redox potentials of biological interest. (See text for the meaning of of the symbols E^o and E_o' D = deuterium (^2H), other chemical symbols are standard). (E^o = true standard redox potential at pH0 with all components in their standard states (gas at a partial pressure of 1 atmosphere and unit activity for solutes).

E^o = true standard redox potential at pH 0 or a practical standard redox potential measured at pH 0 but with solute components at 1 mol l^{-1} concentration. E_o is measured under conditions where the oxidised and reduced forms are present in equal concentrations (not necessarily 1 mol l^{-1}) and its value in dilute aqueous solution is assumed equal to E^o. It is sometimes referred to as the midpoint potential and given the symbol E_m.

E_o' refers to E_o values obtained at a pH differing from zero and may be assumed as being at pH 7 unless otherwise stated.

The symbol E_h is used for redox potential values for a couple whose composition does not correspond to E^o, E_o or E_o' but to a stated composition.

Make certain at this stage you are familiar with these symbols, especially E^o, E_o and E_o'

All redox and standard electrode potentials are measured in volts and at 298 K unless otherwise stated.

3.3.2 The significance of redox potential values

The value of E° for Zn^{2+}/Zn is - 0.76 whilst the value for Cu^{2+}/Cu is +0.34 and therefore Cu^{2+} will tend to be reduced when zinc is added to a solution of a copper salt. This happens when zinc is added to an aqueous solution of copper sulphate; copper is precipitated whilst the zinc dissolves.

The fact that there is a tendency for a redox reaction to go in a particular direction does not mean that it does so at a measurable rate in the absence of a suitable catalyst. The $NAD^+/NADH$ couple has a lower (more negative) redox potential than the pyruvate/lactate couple but the reaction will only go at a measurable rate when the enzyme lactate dehydrogenase is added. Since some reactions are very sluggish a reliable redox potential cannot be obtained in all cases.

Redox potentials are frequently pH-dependent as the reactions often involve H^+ ions. It can be seen from Table 3.3 that the standard electrode potential of the $H^+/\frac{1}{2}H_2$ couple changes from zero (by definition) at pH0 (1 mol l^{-1} H^+ concentration) to -0.41 V at pH7 (10^{-7} mol l^{-1} H^+ concentration). The standard electrode potential of the $\frac{1}{2}O_2/H_2O$ couple also falls by 0.41 V on moving from pH0 to pH 7.

Molecular environment

Molecular environment strongly influences the redox potential for an element such as iron. The values at pH 7 of E_o' vary from -0.60 to +0.375 for protein complexes in which the iron is linked to sulphur and from -0.35 to +0.375 for cytochromes where the iron forms part of a protein-linked porphyrin complex. It should be remembered that a change in solution pH will alter the state of ionization of protein molecules and therefore may effect E_o' values of iron-containing protein molecules.

SAQ 3.5

1) Is it possible that the binding of a substrate molecule could effect the redox potential of an iron containing enzyme such as cytochrome P_{450} which is involved in the metabolism of xenobiotics (compounds foreign to most cells)?

2) Would substitution of an atom in a metabolite by an isotope of that atom be likely to produce a significant change in redox potential for a couple involving that metabolite?

(Examine the data in Table 3.3 before answering this question).

3.3.3 Oxidising germicides

germicides

If we examine Table 3.3 it emerges that the normal range of E_o' values for redox couples associated with living cells at pH 7.0 is from about -0.7 to +0.8 V. It is likely that powerful oxidants and reductants outside this reduction potential range will kill cells, including micro-organisms.

SAQ 3.6

Examine the data in Table 3.3 and identify potential oxidising germicides.

3.4 Thermodynamics and reversible cells

In a simple case a reducing agent loses one electron per molecule (atom, ion or radical) during a redox reaction. A mole of any substance contains the same number of particles (the Avogadro number) so that the same number of electrons will be produced. The charge carried by 1 mol of electrons is called the Faraday constant, F, and is 96 487 coulombs mol^{-1}.

Faraday constant

electrical work

The electrical work done when n mol of electrons are transported through a potential difference E is:

$$W_{elec} = n\,F.E.$$

When E is in volts and F is in coulombs per mol, W_{elec} has the units of J mol^{-1}.

If we consider the reaction:

$$Red_R + Ox_L \rightleftarrows Ox_R + Red_L$$

this may be considered as an electrical cell:

$$Pt \mid Red_L, Ox_L \mid\mid Ox_R, Red_R \mid Pt$$

composed of the two half cells:

$$Ox_L + ne^- \rightleftarrows Red_L \qquad E_L$$

and:

$$Ox_R + ne^- \rightleftarrows Red_R \qquad E_R$$

and:

$$E_{cell} = E_R - E_L$$

The oxidation of 1 mol of Red_L would yield nF coulombs of electricity under these conditions from the left hand cell. These electrons on moving to the right hand cell, would be used in the reduction of 1 mol of Ox_R.

The electrical work done in moving these electron will be nFE_{cell} (J mol^{-1}).

3.4.1 Standard potential and ΔG

If we consider an electrochemical cell operating at its null point then from the first law of thermodynamics:

$$\Delta U = \Delta Q + W$$

where W is the work done on the system, ΔQ is the heat absorbed and ΔU is the change in the internal energy of the system.

W will consist of the work done on the system by its surroundings (which for a system at constant pressure will be $-P\Delta V$ where ΔV is the volume change and the electrical work done on the system W_{elec}). We can therefore write:

$$\Delta U = \Delta Q - P\Delta V - W_{elec}$$

The sign of W_{elec} is negative because when the electrons move spontaneously from the left hand half cell to the right the electrochemical cell does the work and in consequence a negative amount of electrical work is done on the cell.

At constant pressure:

$$\Delta H = \Delta U + P\Delta V$$

where ΔH is the enthalpy change.

Substituting in the previous equation gives us:

$$\Delta H = \Delta Q - W_{elec}$$

From the second law of thermodynamics for a thermodynamically reversible change:

$$\Delta Q = T \Delta S$$

We saw in Chapter 2 that:

$$\Delta G = \Delta H - T \Delta S$$

so that for thermodynamically reversible, isothermal, electrochemical processes at constant pressure:

$$\Delta G = - W_{elec}$$

however, we saw in Section 3.4 that:

$$W_{elec} = n F E_{cell}$$

so we can write:

$$\Delta G = - n F E_{cell} \qquad \text{(E - 3.4)}$$

This shows us that we can determine ΔG from redox potential data.

3.4.2 Calculation of ΔG from redox potential data

calculation of ΔG

Let us examine how we can calculate $\Delta G^{o'}$ for the reduction of pyruvate to lactate by NADH using the data given in Table 3.3.

Remember $\Delta G^{o'}$ is the change in Gibbs function under standard conditions at pH7.

The reduction of pyruvate to lactate, which is catalysed by the enzyme lactate dehydrogenase, may be written as:

$$\text{pyruvate} + \text{NADH} + H^+ \rightleftarrows \text{lactate} + NAD^+$$

We can write this as two 'half cell' reactions (although an electrochemical cell as such has not been physically constructed).

Oxidising couple (electron accepting process):

$$\text{pyruvate} + 2H^+ + 2e^- \rightleftarrows \text{lactate} \qquad E_o' = -0.19V$$

Reducing couple (electron donating process):

$$NADH + H^+ \rightleftarrows NAD^+ + 2H^+ + 2e^- \qquad E_o' = -0.32V$$

but:

$$\Delta G^{o'} = -n F E_{cell} \qquad \text{(Equation (E - 3.4) above)}$$

and:

$$E_{cell} = E_R - E_L$$

If we regard the couple with the more positive (less negative) electrode potential as the right hand half cell.

$$E_{cell} = (-0.19 - (-0.32))\ V$$
$$= +0.13V$$

Now n, the number of mols of electrons, is 2 in the reaction under consideration and the Faraday, F, has a value of 96 487 coulomb mol^{-1}. The potential is measured in volts and one volt equals one joule per coulomb.

thus:

$$\Delta G^{o'} = -2 \times 96\,487 \times (+0.13)\ J\ mol^{-1}$$
$$= -25\,087\ J\ mol^{-1} = -25.09\ kJ\ mol^{-1}$$

$\Delta G^{o'}$ is negative and the reaction will consequently proceed spontaneously.

SAQ 3.7

Using data given in Table 3.3 calculate $\Delta G^{o'}$ for the reaction:

$$\tfrac{1}{2} O_2 + NADH + H^+ \rightleftarrows H_2O + NAD^+$$

and decide if the reduction of oxygen to water proceeds spontaneously under these conditions.

3.5 The effect of concentration upon emf

We saw in Chapter 2 that:

$$\Delta G = \Delta G^o + RT\ \ln \frac{[\text{Products}]}{[\text{Reactants}]}$$

and that for most practical purposes in biochemistry we can work in concentrations. We also know that:

$$\Delta G = -n F E$$

and:

$$\Delta G^\circ = -n FE^\circ$$

or since we can regard $E^\circ = E_o$ for practical purposes; $\Delta G^\circ = -nFE_o$; we can therefore write:

$$-n FE = -n FE_o + RT \ln K_{eq}.$$

or more conveniently:

$$-nFE = -nFE_o + RT \ln \frac{[\text{Products}]}{[\text{Reactants}]}$$

dividing each side by -nF gives:

$$E = E_o - \frac{RT}{nF} \ln \frac{[\text{Products}]}{[\text{Reactants}]} \qquad \text{(E - 3.5)}$$

This is called the Nernst equation.

If we consider the redox reaction:

$$Red_R + Ox_L \rightleftarrows Ox_R + Red_L$$

(Remember R and L refer to the 'half cells'/electrodes), this becomes:

$$E = E_o - \frac{RT}{nF} \ln \frac{[Ox_R][Red_L]}{[Red_R][Ox_L]}$$

If we consider a reaction which includes the standard hydrogen electrode (SHE) the reaction equation becomes:

$$\frac{n}{2} H_2 + Ox_R \rightleftarrows Red_R + nH^+$$

and we can write:

$$E = E_o - \frac{RT}{nF} \ln \frac{[Red_R]\,[H^+]^n}{[Ox_R]\,[H_2]^{\frac{n}{2}}}$$

Now for the SHE, $[H^+]$ and $[H_2]$ are both unity so we can write:

$$E = E_o - \frac{RT}{nF} \ln \frac{[Red]}{[Ox]}$$

so that after inverting the ln term, reversing the sign and converting to log to the base ten we get:

$$E = E_o + \frac{2.303RT}{nF} \log \frac{[Ox]}{[Red]} \qquad \text{(E - 3.6)}$$

When the oxidant and reductant of a redox couple are present in equal concentrations the term $\log \frac{[Ox]}{[Red]}$ is zero so that E is equal to E_o. This is why the standard redox potential E^o (or E_o) of a redox couple is taken as being equal to the 'midpoint' redox potential (see Table 3.3). Note that at pH7 we can write $E = E_o' + \frac{2.303}{nF} RT \log \frac{[Ox]}{[Red]}$

3.5.1 Redox potential and percentage reduction

Using the Nernst equation we can calculate the redox potential for a couple in various states of reduction.

Π Examine Table 3.3 and, using the Nernst equation, calculate the redox potential at 298 K and pH 7 corresponding to cytochrome C being a) 80% in its reduced form and b) 20% in its reduced form (Note $R = 8.314\ JK^{-1}\ mol^{-1}$ and $F = 96\ 487$ coulombs mol^{-1})

From Table 3.3:

$$\text{cyt C } (Fe^{3+}) + e^- \rightarrow \text{cyt C } (Fe^{2+}) \quad E_o' = +0.22V$$

We can employ the equation:

$$E = E_o' + \frac{2.303RT}{nF} \log \frac{[Ox]}{[Red]} \qquad \text{(E - 3.6)}$$

n = 1 for the reaction we are considering and at 298 K; $\frac{2.303RT}{nF}$ is $\frac{0.059}{n}$

- At 80% reduction $\frac{[\text{cyt C }(Fe^{3+})]}{[\text{cyt C }(Fe^{2+})]}$ is equal to $\frac{20}{80}$, that is 0.25, so that:

$$E = +0.22 + 0.059 \log 0.25$$
$$= +0.22 + 0.059\ (-0.602)$$
$$= +0.185\ V$$

- At 20% reduction $\frac{[\text{cyt C }(Fe^{3+})]}{[\text{cyt C }(Fe^{2+})]}$ is equal to $\frac{80}{20}$, that is 4, so that:

$$E = +0.22 + 0.059 \log 4$$
$$= +0.22 + 0.059\ (0.602)$$
$$= +0.256\ V$$

The Nernst equation resembles in form the Henderson-Hasselbalch equation.

$$pH = pKa + \log \frac{[base]}{[acid]}$$

Just as we can use a base/acid mixture to buffer the pH of a solution so we can use a redox couple to buffer the redox potential to a value close to the E_o' value of the buffering couple.

From the Nernst equation a tenfold change in the oxidised to reduced ratio causes a change of about 0.059/n in the redox potential.

redox buffer

A pH buffer has a significant buffering capacity within a range of 1 pH unit on either side of the pKa and the buffering range of a redox buffer is about (0.059/n) V around the E_o'. Thioglycollate is often used in this way to maintain growth media for anaerobic bacteria at a low redox potential around -0.34 V which is required for the growth of these bacteria.

3.5.2 Temperature and redox potential

The potential of the standard hydrogen electrode is defined as zero at all temperatures.

temperature and redox potential

The Nernst equation enables us to calculate the effect of temperature upon E. Note that the $\frac{2.303\,RT}{F}$ factor varies by about 0.0002 per degree over the physiological temperature range.

3.5.3 The effect of pH upon redox potential

pH and redox potential

If a redox couple includes species which are capable of ionisation then it will be effected by pH changes. In many biologically important couples the reductant tends to dissociate as a weak acid and proteins such as metal containing enzymes will also tend to dissociate. This can lead to quite complex relationships between pH and E. It is generally wise to use a pH-buffered medium for work involving redox processes.

In cases where hydrogen ions are involved as part of a redox couple the Nernst equation may be applied.

If we consider the reaction:

$$B + aH^+ + ne^{-1} \rightleftarrows B.H_a$$

we can write:

$$E = E_o + \frac{RT}{nF} \ln \frac{[B]\,[H^+]^a}{[B.H_a]}$$

and the H^+ term can be taken out thus:

$$= E_o + \frac{RT}{nF} \ln \frac{[B]}{[B.H_a]} + \frac{RT}{nF} \ln [H^+]^a \qquad \text{(E - 3.7)}$$

SAQ 3.8

At 298 K $E_o = +1.23$ for the redox couple $\frac{1}{2}\,O_2 + 2H^+ + 2e^- \rightarrow H_2O$

Calculate E_o' for the couple at pH 7.0

3.5.4 The measurement of pH

pH electrode

It is, in principle, simple to measure pH by using a hydrogen gas electrode dipping into the solution of unknown hydrogen ion concentration as one half cell and measuring the potential against a suitable reference electrode. The use of hydrogen gas is not convenient and so the familiar glass electrode is normally used. This electrode involves a glass membrane enclosing a solution of constant hydrogen ion activity into which an internal silver/silver chloride electrode dips. When this assembly is dipped into a solution of unknown hydrogen ion concentration a potential develops across the glass membrane.

In the pH meter this glass electrode forms one half cell of an electrochemical cell with a standard reference cell as the other half cell. The electrode measures hydrogen ion (H^+) activities by means of the relationship.

$$pH = \frac{E'_G - E_{const}}{0.059} \qquad \text{(E - 3.8)}$$

where E'_G is the measured electrode potential and E_{const} is a characteristic value for the electrode. The meter is set for E_{const} by using a buffer solution of known pH and will read pH directly for any solution, of between about pH 1 and pH 10, in which the electrode is immersed.

3.5.5 Deviations from ideal behaviour

In electrochemistry and thermodynamics generally we should strictly be dealing with activities and not concentrations. Errors will result if concentrations are used but as we can see below they are not so large as to make the use of concentrations unacceptable for most biochemical purposes.

In a study on the cell reaction:

$$\frac{1}{2} H_2 + \frac{1}{2} Hg_2Cl_2 \rightleftarrows Hg + H^+ + Cl^-$$

the Nernst equation was used to calculate apparent E_o values from the measured emf values for various HCl concentrations. The true E° value was obtained by extrapolation to infinite dilution. Table 3.4 below lists some of the values obtained.

concentrations	observed		
HCl(mol^{-1})	E(v)	apparent E_o	apparent ΔG°
10^{-1}	0.4046	0.2866	-27.65
10^{-2}	0.5099	0.2739	-26.43
10^{-4}	0.7406	0.2686	-25.92
		E°	ΔG°
infinite dilution		0.2680	-25.86

Table 3.4 Influence of concentration on apparent E_o determined from the Nernst equation. Note that the differences observed in the apparent E° values are because concentrations and not activities were used in the Nernst equation (see text for further discussion).

error due to concentration

It can be seen from this data that the error in E_o and ΔG values introduced by using concentration is about 0.5% for concentrations of around 10^{-4} mol l^{-1}. The concentrations within the compartments of a living cell are usually difficult, if not impossible, to measure accurately but most are probably 10^{-4} mol l^{-1} or less with only a few about 10^{-2} mol l^{-1}. Even in this case the error introduced by using concentrations rather than activities will be in the order of 2% which is probably less than the uncertainty in many concentration measurements for metabolic intermediates in living systems.

3.6 Electron acceptors in biology

Figure 3.3 gives a compilation of E_o' values at pH 7.0 and 298 K for a number of biochemical substrates and the 'natural' electron carriers which form the electron transport pathway in living cells. This electron transport pathway carries electrons from substrates to the terminal oxidants or electron acceptors such as oxygen (or nitrate and sulphate for some bacteria).

natural electron carriers

Providing a suitable catalyst, if necessary, is present, a redox reaction will go almost to completion between two redox systems which differ by 0.25V or more in their redox potentials. We, therefore, can arrange the 'natural' electron carriers in a way which helps us to predict the electron transport pathway reaction sequence. We may also predict which substrates are likely to pass electrons to which carriers.

SAQ 3.9

Examine Figure 3.3 and answer the following:

1) Are flavoproteins or pyridine nucleotides, such as NADH or NADPH, likely to be natural cofactors for the reaction:

$$\text{succinate} \rightarrow \text{fumarate} + 2H^+ + 2e^-$$

2) Place the list of redox couples below in the likely order in which the flow of electrons runs in the transport of electrons from reduced substrates to oxygen the terminal electron acceptor during aerobic metabolism:

$\frac{1}{2}O_2/H_2O$

cytochrome C Fe^{3+}/Fe^{2+};

flavoproteins ox/red;

cytochrome A_3 Fe^{3+}/Fe^{2+};

$NAD^+/NADH + H^+$.

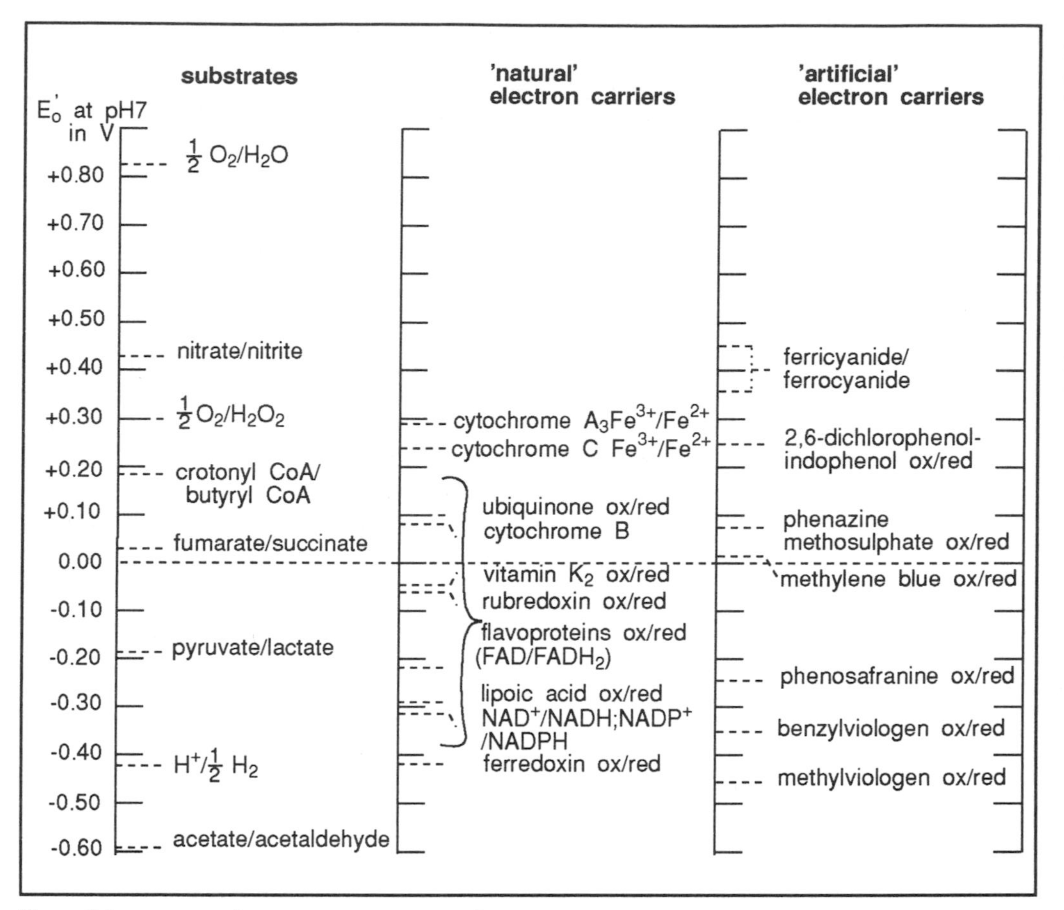

Figure 3.3 Values of E_o' at pH7 and 298 K (ie 25°C) for several redox couples of especial interest to biologists. Redrawn from J G Morris 'A Biologist's Physical Chemistry', 2nd Edition, Edward Arnold, London. Note the wide range of E_o' values for flavoproteins. The physical/chemical environment of the flavin moiety greatly influences the E_o' value in these molecules.

3.6.1 Artificial electron acceptors

artificial electron carriers

Many of the natural electron acceptors are proteins and in consequence, not very stable in a chemical sense. They also often do not show suitable colour changes in the visible region and are very expensive to purchase or time consuming to prepare. A number of artificial electron acceptors exist and are extremely useful tools in biochemistry. A number of these are listed in Figure 3.3.

methylene blue

Methylene blue (MB) is a typical substance of this type and has an intense blue colour when in its oxidised state but is colourless when reduced.

$$\underset{\text{(blue)}}{MB} + 2H^+ + 2e^- \rightleftarrows \underset{\text{(colourless)}}{MB.H_2} \qquad E_o' = 0.011V$$

Methylene blue is an autoxidisable dye which means that the reduced form will react with molecular oxygen and be converted back to the oxidised form thereby regenerating the blue colour:

$$MB.H_2 + \frac{1}{2} O_2 \rightarrow MB + H_2O$$

The dye is widely used in metabolic studies as it can replace the normal physiological electron transport system from the flavoproteins to oxygen.

DCPIP

The dyestuff dichlorophenol-indophenol (DCPIP) is another dye which is coloured in its oxidised form and colourless in its reduced form.

$$DCPIP + 2H^+ + 2e^- \rightleftarrows DCPIP.H \qquad E_o' = +0.217V$$

(blue above pH) colourless

In this case the dye is not autoxidisable and does not react with molecular oxygen in the absence of a suitable catalyst even though such a reaction is thermodynamically favoured and would be 'spontaneous'. The fact that a process is thermodynamically spontaneous does not mean that it will necessarily occur at a measurable rate in the absence of a catalyst.

The phenazine methosulphate (PMS) molecule is also a useful artificial electron acceptor and is used in the laboratory, for example, in the assay of the flavoprotein enzyme glycolate oxidase.

$$\text{glyoxalate} + 2H+ + 2e- \rightleftarrows \text{glycolate} \qquad E_o' = -0.085V$$

PMS will accept electrons from the enzyme and pass them on to cytochrome C thus coupling the oxidation of glycolate to the reduction of added cytochrome C. The

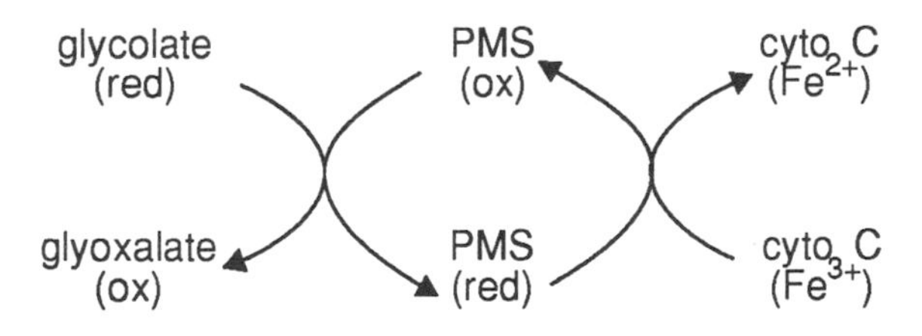

SAQ 3.10

The dyestuff DCPIP will react with ascorbic acid (vitamin C) and oxidise it to dehydro-ascorbic acid. At pH 7 and 298 K:

$$\text{dehydroascorbate} + 2H^+ + 2e^- \rightleftarrows \text{ascorbate} \qquad E_o' = +0.08V$$

Under the same conditions $E_o' = +0.217V$ for the reduction of the oxidised form of DCPIP.

Calculate the percentage of ascorbic acid which will be present in the reduced form when enough oxidised DCPIP has been added to a solution of reduced ascorbic acid to give $E = +0.217V$. Comment upon the practicability of measuring reduced ascorbic acid colourimetrically by the decolourisation of oxidised DCPIP. Note: $R = 8.314\ JK^{-1}\ mol^{-1}$; $F = 9.648 \times 10^4$ coulombs mol^{-1}

3.6.2 Sluggish redox couples

Not all redox reactions are chemically reversible under normal reaction conditions and a thermodynamically-reversible electrode reaction can not always be established.

tetrazolium salts

The tetrazolium salts are examples of substances of this type and are widely used in biology. They are converted on reduction to highly coloured insoluble formazans according to the reaction:

$$[\text{tetrazolium}]\ 2H^+ + 2e^- \rightarrow [\text{formazans}] + H^+$$

They act as useful artificial electron acceptors for enzyme catalysed reactions and are used in histochemistry to demonstrate enzyme activity. Flavoprotein intermediates can transfer electrons to the tetrazoliums which have been reported as having a redox potential of - 0.08 V.

The redox potentials quoted should be treated as suspect and are very dubious because of the uncertainties arising from the formation of an insoluble product.

In other cases redox couples which act in a perfectly reversible way chemically may not give stable electrode potentials because they are slow in reaching electronic equilibrium with a metal electrode. The redox potential of these 'sluggish' couples may be determined by combining them with a redox couple which does react rapidly with a metal electrode and with which they readily react.

3.7 Chemical potential and membrane phenomena

When considering multi-component systems and systems with internal barriers, such as biological membranes, it is useful to divide the total ΔG into the contributions made by the individual components. For systems in which temperature and pressure are kept constant, if the amount of an individual component i changes by dNi whilst the other components are unchanged then the resultant change in Gibbs function is $\mu_i\ dN_i$. μ_i is called the 'chemical potential' of component i.

chemical potential

In ideal solutions, the chemical potential may be defined in terms of the concentration of the species X_i and a constant 'standard state' chemical potential.

$$\mu_i = \mu_i^\circ + RT \ln Xi \qquad \text{(E - 3.9)}$$

Chemical potential is essentially equal to the molar Gibbs function for that species.

electro-chemical potential

When we consider an ion, the chemical potential μ_i alone is not enough to describe its thermodynamic behaviour because the ion, will interact with the electrostatic field Ψ within which it finds itself. Other charged species in the vicinity will generate this field.

The electrochemical potential $\tilde{\mu}_i$ may thus be defined as:

$$\tilde{\mu}_i = Z F\Psi + \mu_i$$

$$= Z\ F\Psi + \mu_i^\circ + RT \ln X_i \qquad \text{(E - 3.10)}$$

where $ZF\Psi$ is an additional electrostatic term and Z is the charge on the ion.

3.7.1 The chemiosmotic hypothesis

Mitchell in the 1960s produced a revolutionary change in our thinking on the coupling of electron transport to ATP synthesis. He proposed, in the chemiosmotic hypothesis, that the flow of electrons to carriers of higher redox potential is coupled to the movement of protons across a membrane barrier. This hypothesis has proved very illuminating in the study not only of oxidative metabolism but also of photosynthesis and other aspects of cell physiology.

If we consider for example, oxidative phosphorylation in the mitochondria, as this process involves the oxidation of reduced substrates being linked to the synthesis of ATP in the mitochondria. The mitochondria are sub-cellular organelles surrounded by a double membrane which separates the matrix inside from the outside. This membrane is of limited and selective permeability to small molecules and ions.

In the chemosmotic hypothesis hydrogen ions (protons) are moved from the mitochondrial matrix to the outside of the mitochondria during electron transport to create a pH gradient across the membrane. This causes the outside of the mitochondrion to become positively charged and acidic whilst the inside becomes negatively charged and alkaline. The membrane is normally impermeable to protons and energy is stored in this electrochemical gradient.

The electrochemical gradient is generated by the hydrogen ion potential difference so:

$$\tilde{\mu}_{i(in)} - \tilde{\mu}_{i(out)} = Z F\Psi_{(in)} - Z F\Psi_{(out)} + RT \ln \frac{[H^+]_{(in)}}{[H^+]_{(out)}}$$

or

$$\Delta \tilde{\mu}_i = Z F \Delta \Psi - 2.3 RT \Delta pH \quad (E - 3.11)$$

membrane potential

where $\Delta\Psi$ is the potential across the membrane (by convention, inside minus outside) and ΔpH is the pH difference across the membrane (pH inside - pH outside). For the proton Z = 1 and converting to electrical units by dividing throughout by F we get:

$$\Delta p = \Delta \Psi - \frac{2.303RT}{F} \Delta pH \quad (E - 3.12)$$

proton motive force

where $\Delta\Psi$ is the membrane potential and Δp is termed the 'protonmotive force' and both are measured in volts.

3.7.2 Protonmotive force and ATP synthesis

It has been proposed that ATP synthesis occurs when definite numbers of protons re-enter the mitochondrion down their electrochemical gradient via a mitochondrial ATPase which synthesizes ATP (Figure 3.4).

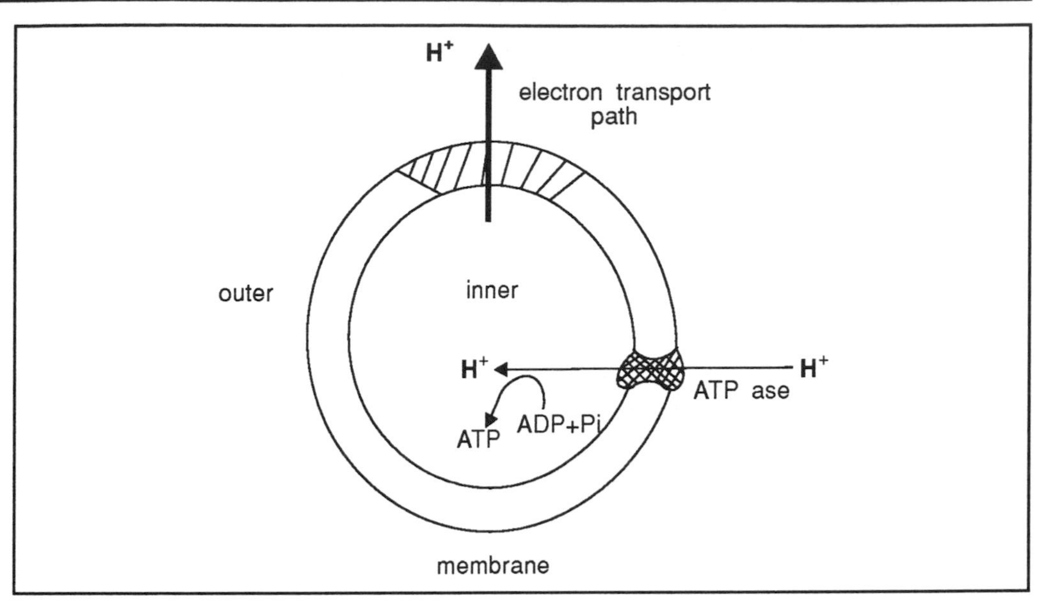

Figure 3.4 Mitochondrial oxidative phosphorylation in terms of the chemiosmotic hypothesis.

If mitochondria are synthesizing ATP and the concentration has reached a steady level during respiration, the concentrations of ATP, ADP and Pi can be measured. This allows ΔG to be obtained for the reaction ADP + Pi $\rightarrow$ ATP provided the value of $\Delta G^{o'}$ for the reaction is known.

$$\Delta G = \Delta G^{o'} + RT \ln \frac{[ATP]}{[ADP]\,[Pi]} \qquad \text{(E - 3.13)}$$

ΔG may then be used to calculate the protonmotive force using the relationship ΔG = -nFE to give:

$$\Delta G = -n\,F\,\Delta\,p \qquad \text{(E - 3.14)}$$

where n is the number of hydrogen ions passing through the ATPase per ATP molecule synthesized.

The protonmotive force Δp can be obtained from a knowledge of the values of $\Delta\Psi$ and ΔpH. The value of ΔpH can be measured using the distribution of a non-metabolised weak acid or base between the inside and outside of the mitochondrion. The membrane is not generally permeable to charged particles and it is assumed that only the uncharged form of the acid or base crosses the membrane. An acid probe is used when the internal pH is expected to be higher than that of the external medium.

Membrane potential $\Delta\Psi$ is measured using the distribution of a suitable ion across the membrane. The membrane becomes permeable to certain ions such as Rb^+ in the presence of substances called ionophores of which valinomycin is an example.

The expression:

$$\Delta\,\tilde{\mu}_i = Z\,F\Delta\,\Psi + RT \ln \frac{[X_i]_{in}}{[X_i]_{out}} \qquad \text{(E - 3.15)}$$

gives the electrochemical potential across a membrane. At electrochemical equilibrium $\Delta \tilde{\mu}_i = 0$ as there is no tendency for X_i to move from one side to the other. Under these conditions:

$$\Delta \Psi = -\frac{RT}{ZF} \ln \frac{[X_i]_{in}}{[X_i]_{out}} \qquad \text{(E - 3.16)}$$

From the chemiosmotic theory we would expect the outside of the mitochondrion to be positive relative to the inside in consequence of the hydrogen ion translocation. The positively charged Rb^+ ion will enter the mitochondrion until it is in electrochemical equilibrium across the mitochondrial membrane.

Under those conditions $\Delta\Psi$ can be calculated using Equation (E - 3.16) by measuring $[Rb^+]$ inside and $[Rb^+]$ outside the mitochondrion.

It is interesting to note that because biological membranes have a low electrical capacitance the net transfer of a few thousand protons across the plasma membrane of a bacterial cell will generate a $\Delta\Psi$ of - 0.06 V. Because of the pH buffering capacity of the cytoplasm the number of protons which must be moved to generate a ΔpH of one unit is about 100 times greater. In consequence of this, membrane potentials can develop quickly but store little energy whilst pH gradients take longer to develop but store much more energy. It is important to remember that ionisation to release protons and also recombination takes place very rapidly so protons are always potentially available even at pH 7 when the average bacterial cell, with a volume of only about one cubic micrometre, contains only a few hydrogen ions. It is instructive to calculate how many hydrogen ions will be present in one cubic micrometre of solution of pH8!

SAQ 3.11

A suspension of mitochondria, actively synthesising ATP, was incubated with $[^3H]$ acetate and $^{86}Rb^+$ in the presence of valinomycin (an 'ion carrier' or ionophore which allows Rb^+ to cross the membrane). The following results were obtained at 298 K. (Note using the radioactively labelled Rb^+ to measure Rb^+ concentration).

relative concentrations

Inside mitochondria	outside mitochondria	ion
108	1	$^{86}Rb^+$
1	65	H^+

1) Calculate ΔpH, $\Delta\Psi$ and Δp

2) Analysis of the mitochondrion suspension showed that the following concentrations were reached.

ATP = 1.1 m mol l^{-1}

ADP = 0.021 m mol l^{-1}

Pi = 1.01 m mol l^{-1}

It has been variously proposed that 2, 3 or 4 protons must cross the mitochondrial membrane per ATP synthesized.

Using the above information and a value of +30.5 kJ mol^{-1} for $\Delta G^{o'}$ for the reaction ADP + Pi $\rightarrow$ ATP comment upon which of the above proposals this experiment supports. Hint; you will need to use equation 3.16, 3.13 and 3.14 to solve this.

Summary and objectives

This chapter has focussed on developing your knowledge of oxidation reduction potentials and membrane potentials and has introduced you to the application of these to biochemical thermodynamics. We began by explaining what is meant by oxidation and reduction and, by considering electrochemical cells. This introduced you to the concepts and conventions used in considering electrochemical events. This enabled us to develop your knowledge of standard electrode and redox potentials and to enable you to relate redox potentials to ΔG. This, together with associated studies on artificial electron acceptors, membrane potential and protonmotive force, has extensive application in the study of cell metabolism and energetics.

Now that you have completed this chapter you should be able to:

- identify oxidation and reduction reactions;
- use appropriate nomenclature to describe oxidation and reductions;
- predict the direction of oxidation reduction reactions from supplied data;
- describe some of the factors which effect the redox potential of biological molecules;
- calculate ΔG values from supplied redox and concentration data;
- sequence oxidation reduction reactions from supplied data;
- use supplied data to calculate protonmotive force, membrane potential and pH gradients across a membrane.

Enzymes

Enzymes

Enzymes as mediators of metabolism.

4.1 Introduction

Enzymes are specialised protein molecules which are the essential catalysts of biochemical reactions. The word enzyme (meaning 'in yeast') was first used in 1877, many years after Berzelius had published a theory of chemical catalysis (in 1835) which included, for the first time, an example of what we now know was an enzyme reaction. Buchner, in 1897, first extracted functional enzymes from cells but it was not until 1927 that Sumner crystallised and purified the first enzyme, urease. Sumner postulated that all enzymes were proteins, a view not accepted until the 1930s when several other enzymes were purified and crystallised. All enzymes are now considered, by definition, to be protein molecules even though they may require small non-protein molecules to assist or promote activity.

Currently over 2000 different enzymes are known. Only a relatively small number of these, especially those involved in central metabolism reactions, have been thoroughly investigated. Many more have received much less attention and there is still much more to be learnt about the ways in which enzymes work and how their activities are regulated.

A detailed discussion of the structure and properties of enzymes is given in the BIOTOL text, 'The Molecular Fabric of Cells', and a more advanced treatment is given in the BIOTOL text, 'Principles of Enzymology for Technological Applications'. Here we have chosen to provide a brief reminder of the important features of enzymes, necessary to understand their roles in metabolism. We begin by describing the general properties of enzymes, with particular emphasis on the general basis of catalysis. The major part of the chapter is however devoted to a consideration of the kinetics of enzyme catalysed reactions and the factors which may influence these kinetics.

4.2 General properties of enzymes

All enzymes involved in metabolism are proteins. As such, they consist of one or more polypeptide chains and display properties typical of proteins. Thus we should anticipate that their structure and, therefore, their activities will be influenced by physical parameters, such as pH and temperature, which influence protein structure.

The essential feature of enzymes is that they catalyse reactions. They are essential to the metabolic activities of cells and have certain properties which make them unique and vastly superior to any non-biological catalysts.

4.2.1 General basis of catalysis

activation energy

Enzymes accelerate reactions by lowering the activation energy of the reaction. The following treatment is not a strict thermodynamic description but should enable the significance of the change in activation energy to be grasped.

The explanation of catalysis is most easily appreciated if the reaction is depicted in a graph in which the vertical axis represents the energy content and the horizontal axis represents the progress of the reaction (Figure 4.1).

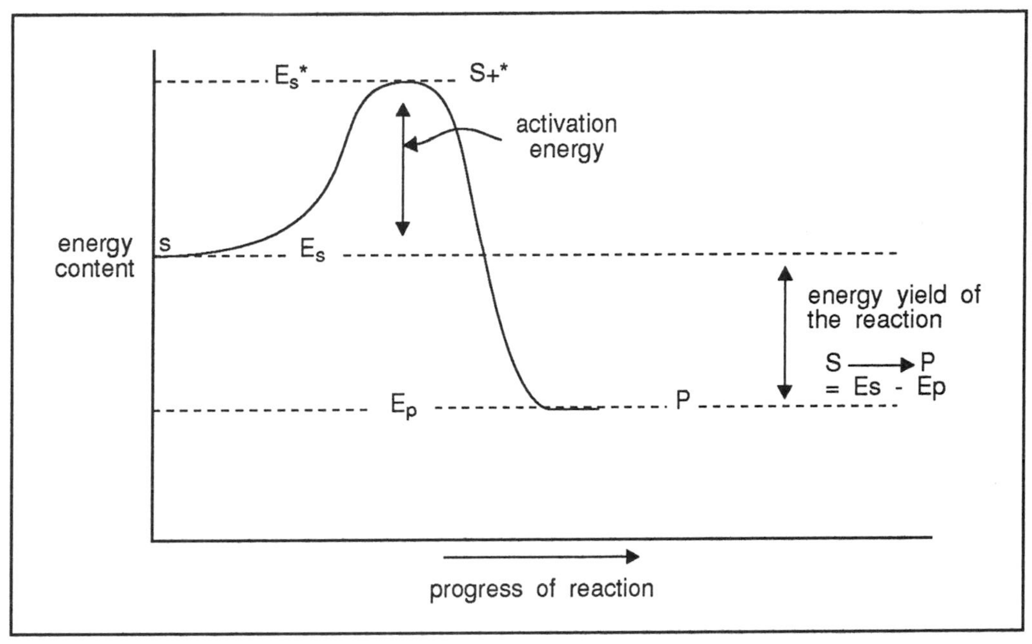

Figure 4.1 Energy relationship of substrate and product. For a reaction to actually occur, an activation energy barrier must be overcome. Es = energy of substrate, Ep = energy of product S = substrate P = product (see text for details).

We can see that the products are at a lower energy level than the reactants: when reactants (substrates) are converted to products, energy will be released: the forward reaction is thus spontaneous and the reaction can proceed. Whether it does proceed depends on a 'barrier' between S and P which is known as the activation energy. For any molecule to actually react, it must posses energy corresponding to the top of this barrier (ie S + *).

In the absence of an enzyme, only a small proportion of the substrate molecules contain sufficient energy to overcome the activation energy barrier.

How do enzymes speed up such reactions? They do not provide the additional energy but instead provide an alternative route, for which less energy is needed (Figure 4.2 a)). The overall effect of this is to lower the activation energy barrier.

The effect of this lowering of the required activation energy can be seen most easily by examining the energy contents of molecules (Figure 4.2 b)). In the absence of enzyme, very few substrate molecules had sufficient energy to react (only those with energy greater than Es + *). With the enzyme, the required energy is lowered so that substantially more molecules have sufficient energy and hence can react. This is how enzymes achieve their catalytic effect. Note that the enzyme only affects the activation energy; the energy given out in the reaction is unaltered (this is the difference in energy levels of S and P). When glucose is converted to carbon dioxide and water by organisms, using enzymes, precisely the same amount of energy is released ($2880 kJ\ mol^{-1}$) as in the bomb calorimeter.

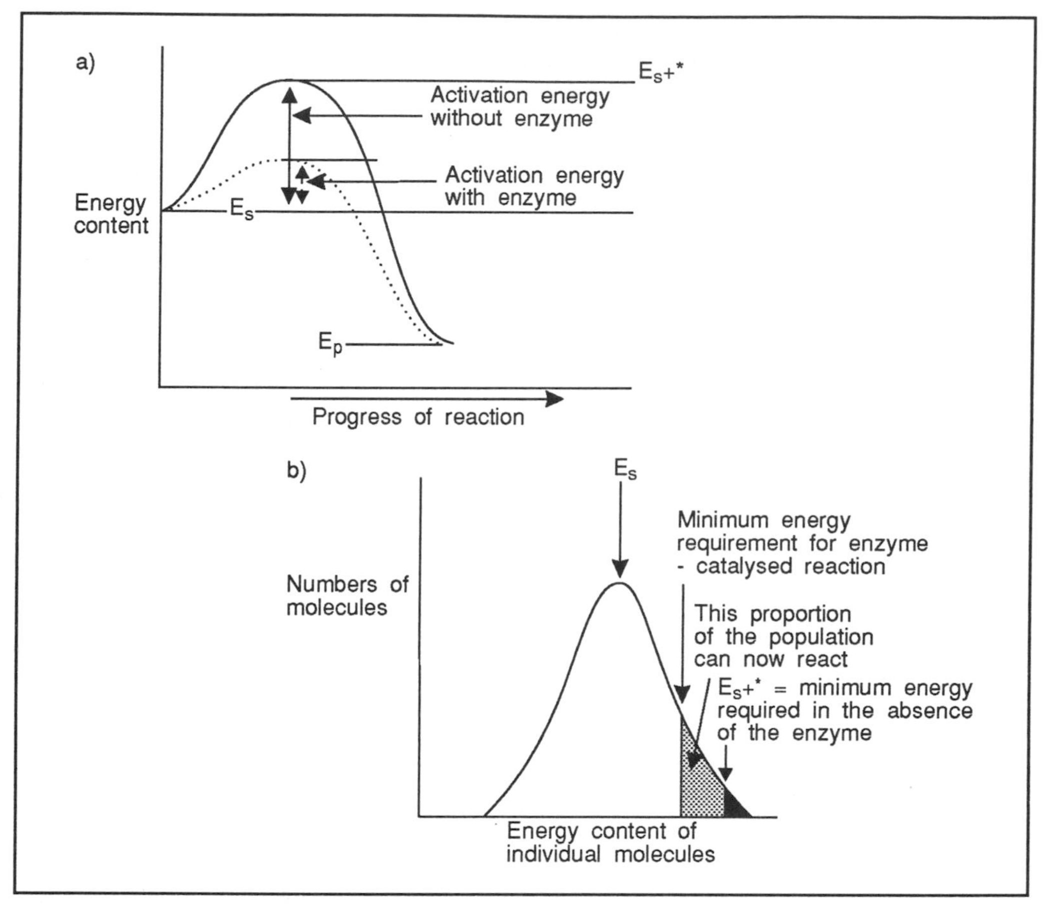

Figure 4.2 a) Enzymes cause catalysis by providing an alternative reaction path, for which the activation energy requirement is lower. b) Energy profile of the substrate molecules; the enzymic reaction route means that a greater proportion of the molecules now posses the necessary energy (shown hatched).

catalase

The catalytic power of enzymes can be enormous. One of the most remarkable enzymes, in terms of rate enhancement, is catalase. This enzyme is very widely distributed in nature (grass, microbes, most animal tissues) and catalyses the breakdown of hydrogen peroxide

$$2\ H_2O_2 \rightarrow 2\ H_2O + O_2$$

One molecule of catalase can react with about 5 million molecules of hydrogen peroxide per minute: this means that catalase is accelerating the reaction rate by at least 10^{14} times compared to the uncatalysed reaction.

mechanisms for lowering activation barrier

Whilst we have seen, in overall terms, how enzymes achieve catalysis, the precise mechanisms by which the activation energy is lowered should be mentioned. The main mechanisms are described below. Note that different enzymes are believed to use combinations of different ways to achieve rate increases.

Proximity: most reactions involve more than one substrate. In order to react, these substrates must collide. If an enzyme binds both substrates simultaneously, their proximity is increased - and they are much more likely to react. An alternative way of thinking about this is to consider the concentration of the substrates. When they are bound to the enzyme, their localised concentrations are much higher than they would be in free solution. As we shall see later, the rates of chemical reactions increase when substrate concentrations are increased.

Orientation: since the active site of an enzyme (the part of the enzyme to which the substrates bind) usually closely matches the shape of its substrates (this is how specificity is obtained) and therefore, substrates are bound in a particular position. This means that enzymes can bring substrates together so that their orientation is optimal. In free solution, many collisions (in which the substrate molecules become close enough together to react) would nonetheless be unproductive because the orientation is inappropriate - the 'wrong' parts of the molecules meet. Even a small departure from the optimal alignment may mean that the energy required to enter the transition state (an intermediate form of higher energy content between substrate and product) is greatly increased.

transition state

Strain: one way by which the activation energy requirement is lowered is by the enzyme only binding a distorted ('strained') form of the substrate. This is accomplished by the 'complementarity' between active site and substrate being correct except in the region of the bond to be reacted upon. We can postulate that, for binding to occur, the substrate's shape must be altered. This distortion has the effect of raising the energy level of the reactant, such that the difference between this level and that of the transition state is lessened. This is certainly one of the mechanisms used by the enzyme lysozyme: in this case it is considered to contribute a rate enhancement of approximately 3000.

Acid-base catalysis: this concept helps to demonstrate the importance of the correct positioning of catalytic functional groups in the active site. These groups may be able to donate or accept protons (or electrons) to and then from the substrate. This may contribute to the reaction. This acid or base catalysis is an important component of rate enhancement. Thus, with lysozyme at pH 5 (at which pH lysozyme is most active), the sidechain of an active site glutamic acid residue (Glu-35) is still protonated (sidechain is -COOH). It is also perfectly positioned to donate this proton to the substrate, thereby promoting bond cleavage. This is an example of acid catalysis.

There is also here an element of the effect of concentration discussed earlier. Proton concentrations are not high in cells (at pH 7.0, the $[H^+]$ is only 10^{-7} mol l^{-1}). The ability of an active site group to provide a proton in this way may mean that non-availability of protons is eliminated.

covalent bond formation

Whilst these are the main mechanisms which enzymes use, others do occur. Thus, covalent bonds may be formed, permitting an unstable (ie rapidly reacting) reaction intermediate to be formed. In some cases, enzymes may stabilise a transition state intermediate: this will have the effect of promoting its formation.

rate constant varies exponentially with activation energy

Although we have now identified the ways by which enzymes lower the activation energy requirement (in general terms, by providing an alternative reaction route), does this reduction in activation energy really account for the observed rate enhancements? In Figure 4.2 a), the enzyme-catalysed route is shown with an activation energy which is about a third of the size of the uncatalysed activation energy. Does this relatively modest change in activation energy really lead to the massive increases in reaction rates which enzymes achieve? It can be shown that the relationship between the rate constant k for breakdown for the enzyme substrate (ES) complex to products (whereby rate of reaction is given by k[ES], see Section 4.4) and activation energy is exponential, such that a small reduction in activation energy produces a large increase in rate. This is shown in Table 4.1. The rate of reaction increases by a factor of about 2.4×10^4 for every drop of 25 kJ mol^{-1} in activation energy. An enzyme able to reduce the activation energy from 100 to 50 kJ mol^{-1} would increase the reaction rate by a factor of about 570 million! Further, the sorts of rate enhancements typically achieved by enzymes can be realistically explained by the catalytic mechanism which we have identified.

activation energy , J mol^{-1}	relative k (s^{-1})
100,000	1.84×10^{-5}
75,000	4.4×10^{-1}
50,000	1.06×10^{4}
25,000	2.5×10^{8}

Table 4.1 Effect of change in activation energy on enzymatic rate enhancement.

In summary, enzymes accelerate only those reactions which are proceeding spontaneously, and they do not change equilibrium, only the speed of achieving equilibrium. They achieve this catalytic effect by lowering the activation energy.

SAQ 4.1

Inspect the diagram below which shows the energy profile for a hypothetical reaction involving possible conversion of A to B. The solid line shows the uncatalysed reaction; the dashed line depicts that in the presence of an enzyme. Identify which of the following statements are True and which are False, giving brief reasons.

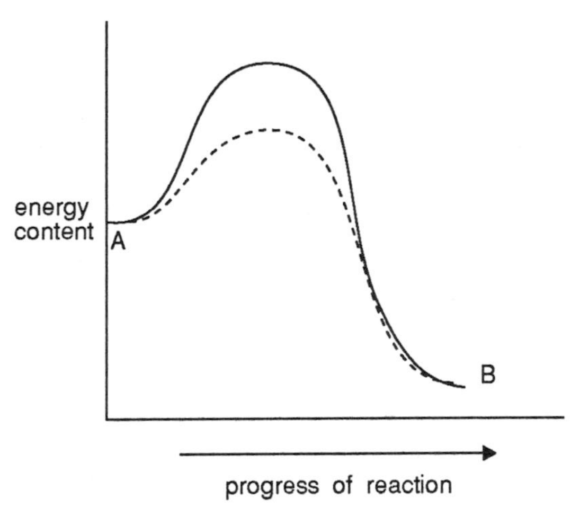

1 Conversion of A to B is rapid in the absence of any enzyme.

2 Addition of enzyme accelerates the reaction by lowering the activation energy; it also lowers the energy yield of the reaction.

3 The activation energy barrier precludes conversion of A to B under any circumstances.

4 The reaction of A to B is spontaneous but, in the absence of enzyme, occurs to a negligible extent, because of the activation energy barrier.

5 The enzyme accelerates the reaction by lowering the activation energy, with no effect on the overall energy yield.

4.2.2 Specificity

Whilst enzymes can achieve remarkable rate increases, rate enhancement is only one of the crucial properties and functions of enzymes. A second vital property is that of specificity , in terms of what compound(s) an enzyme acts on, and what product(s) it produces. This is of enormous importance in cells because it allows the fate of compounds (ie what molecules they are converted to) to be controlled and directed. Material is accurately and precisely passed along the pathway via a series of enzymes: routinely, no side-products are formed (unless there is a fork in the pathway, which would require an additional enzyme) and all the material will be converted to the product of the pathway. This high specificity is also vital for many of the commercial, medical and analytical areas in which enzymes are used.

specificity of substrate

Enzyme specificity thus has two separate but important aspects. The first concerns which molecules any particular enzyme acts on: enzymes have ability to discriminate between different molecules. Different enzymes show different degrees of discrimination. Urease, which hydrolyses urea according to the equation

$$CO(NH_2)_2 + H_2O \rightarrow CO_2 + 2\ NH_3$$

is highly specific, failing to act even on closely related compounds. In other cases, specificity is less absolute, permitting one enzyme to act on closely related substances. Thus the proteolytic enzyme trypsin, which hydrolyses peptide bonds, acts on the peptide bond to the C- terminal side of either arginine or lysine. Trypsin will tolerate minor differences in the precise shape of its substrate, providing the amino acid side-chain possesses a positive charge. Restricted specificity, as displayed by trypsin, occurs frequently. For some proteolytic enzymes, however, specificity is much tighter. Thrombin, which is involved in blood-clotting, only cleaves the peptide bond between arginine and glycine, and only then when this dipeptide occurs in particular sequences.

There are also enzymes which display a relative lack of specificity. These are often hydrolytic enzymes, whose role is to break down proteins (or esters), perhaps to enable their constituent amino acids (or carboxylic acids/alcohols) to be used as a foodstuff. In the same way that there are clear circumstances in which absolute specificity is desirable, so too there are circumstances (eg release of amino acids for nutrition) in which looser specificity has advantages.

specificity of reaction

The second important aspect of enzyme specificity concerns what reaction is performed. Enzymes tend to be highly specific in terms of which groups and bonds are acted upon and what products are formed. This is clearly vital for metabolism. The enzyme hexokinase catalyses the addition of phosphate groups onto glucose. There are 5 hydroxyl groups which could, in principle, be phosphorylated, yet only one, (at C-6) actually is: thus only glucose-6-phosphate is formed. This shows the absolute positional specificity as far as this reaction is concerned.

selection of stereoisomers

Enzymes are highly discriminative of stereoisomers. Thus proteolytic enzymes such as trypsin only act on peptides made of L-amino acids; peptide bonds in synthetic peptides made of D-amino acids are not hydrolysed. Similarly, hexokinase is inactive on L-glucose; this absolute specificity for either D- or L-isomers is normally observed with enzymes which act on optically active compounds. Figure 4.3 a) depicts the D- and L-forms of the amino acid alanine.

What is the origin of enzyme specificity? Enzymatic catalysis involves the substrates binding transiently to the surface of the enzyme thereby 'selecting' only one stereoisomer. In a purely chemical reaction taking place in free solution no comparable 'selecting' procedure is available and therefore no discrimination between stereoisomers can occur. Whereas in free solution both stereoisomers could react (eg both D- and L- glucose; D- and L- amino acids), the requirement to bind to an enzyme enables the enzyme to discriminate between them. This is easily seen with simple models, which you could make (with plasticine and matchsticks if atomic models are not available).

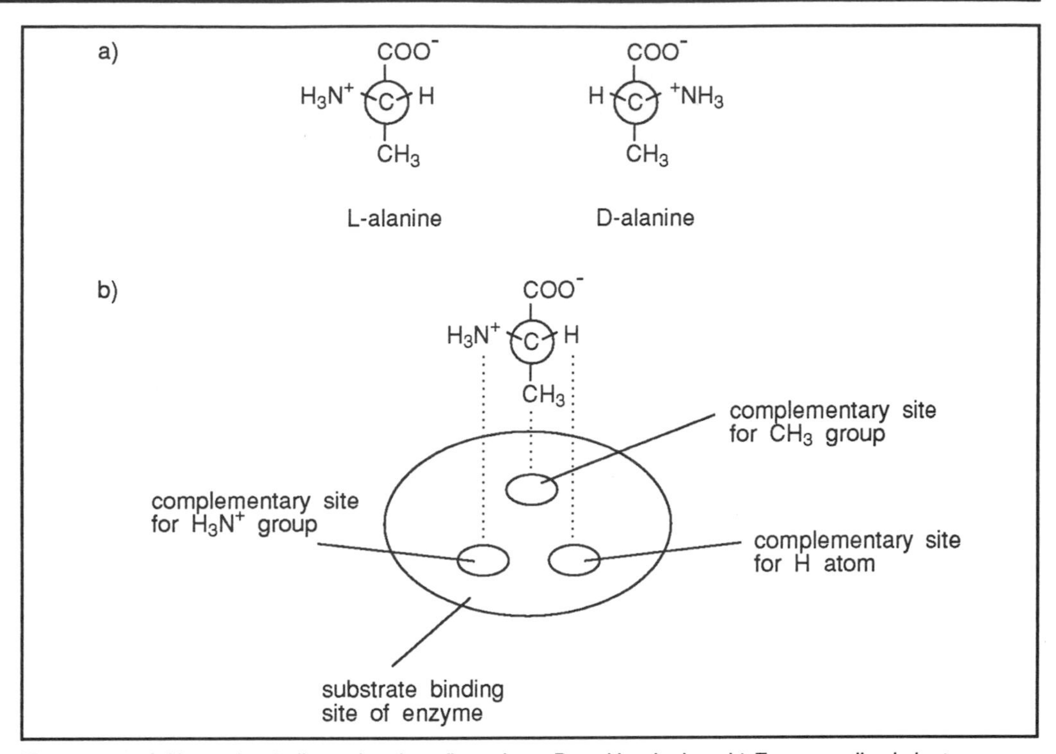

Figure 4.3 a) Alternative 3-dimensional configurations, D-and L- alanine. b) Enzymes discriminate between stereoisomers by providing complementary binding sites to several groups of the substrate. If 3 such sites are used, only one of the stereoisomers, L-alanine, can bind.

only 'correct' isomers will bind

In free solution, isomers will react in a similar manner. However, if they have to bind to the surface of an enzyme, and there are complementary sites for 3 of the substituent groups, (eg for $-CH_3$, $-COO^-$, and $-NH_3^+$ groups) then the position of these complementary sites dictates which isomer will bind - and hence may react. In the example shown, (Figure 4.3 b)), only L-alanine contains correctly positioned groups for binding to the enzyme; D- alanine will show no tendency to bind. Thus the prerequisite for a molecule to bind to an enzyme before catalysis can occur also underpins the absolute specificity for the 'correct' stereoisomer.

substrates are usually much smaller than enzymes

Only a small part of the enzyme directly interacts with the substrate(s): this part is known as the active site. If you think about the relative size of substrates and enzymes, this is not very surprising: molecules of intermediary metabolism are typically small, with relative molecular masses of up to several hundreds, whilst enzymes are usually much larger (with a molecular mass in excess of 20 000 kD). Thus catalytic activity is likely to be confined to only part of the enzyme molecule. Note that the remaining parts of the enzymes nonetheless have a vital role, including maintaining the shape of the active site and providing binding sites for regulatory molecules. It may also ensure that the enzyme sticks to a membrane or associates with other proteins.

active site has precise shape and charge distribution

The active site has to accomplish two objectives. Firstly, it must enable the correct substrate(s) to bind, whilst rejecting other molecules, and ensure correct positioning of the substrate. This is required for the second task - catalysis - to take place. Binding of substrate, and its correct positioning for catalysis, are achieved by the precise shape and charge distribution of the active site.

lock and key hypothesis

The classical depiction of enzyme-substrate interaction involves a lock and key analogy (Figure 4.4), in which only the correct substrate binds.

This is useful as a simple model, since we can readily see how some molecules are rejected and cannot bind (Figure 4.4 b)) whereas others possess the required shape (Figure 4.4 a)). Further, we can see how minor alterations in substrate structure may be acceptable (Figure 4.4 c)), although there may be differences in the strength of binding of different substrates. The tendency to bind results from the complementarity of shape and charge on the substrate and the active site. Particularly good binding results when there are a lot of close contacts but no positions where atoms are forced too close together. Substrates bind to enzymes through numerous relatively weak bonds, including ionic and hydrogen bonds, and hydrophobic and van der Waals interactions. Only occasionally are covalent bonds formed and this is usually as part of the catalysis.

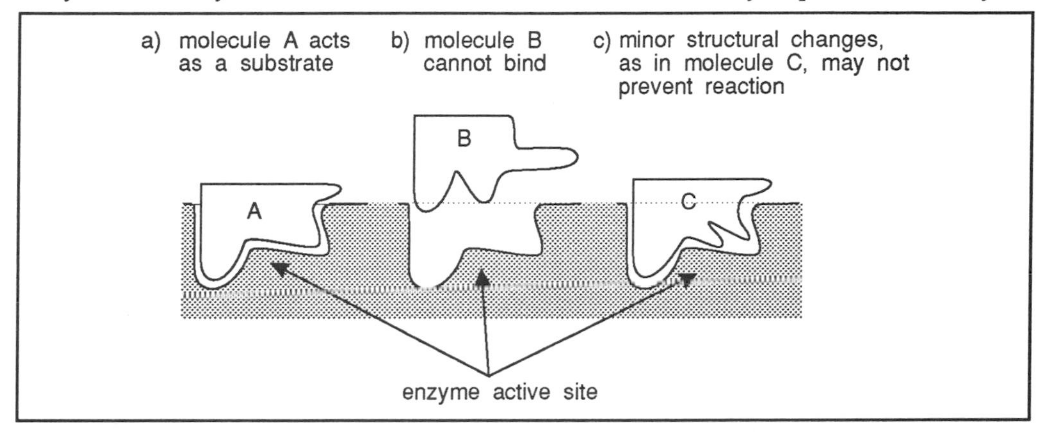

Figure 4.4 Models of the active site. According to the lock and key model, some molecules fit and will bind and react (Figure 4.4 a)), whereas other molecules (Figure 4.4 b)) do not possess the required shape. Molecules closely resembling the optimal shape may also be acted on (Figure 4.4 c)).

induced fit hypothesis

The lock and key hypothesis of the active site implies that the enzyme is rigid and that substrate binds to it as a key fits a lock. There is, however, extensive evidence that enzymes are flexible. An alternative model for enzyme action is based on induced fit (Figure 4.5 a)).

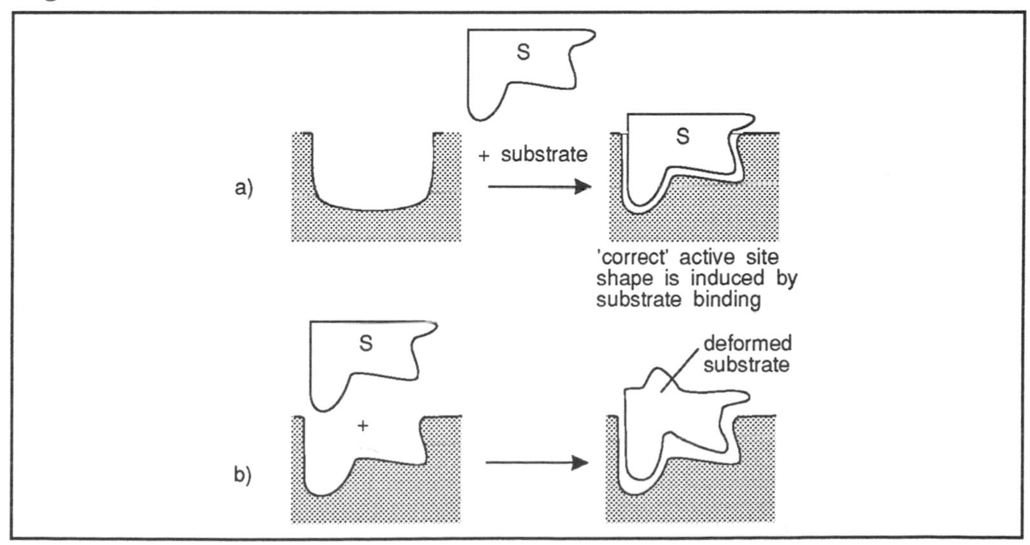

Figure 4.5 a) The induced fit model for the active site. The active site changes shape when substrate binds: the active conformation, which is responsible for catalysis, is induced by the substrate and only occurs when substrate is bound. b) In addition, the enzyme may deform the substrate.

In this hypothesis, the enzyme itself undergoes conformational changes as a result of substrate binding. The active site thus changes shape as the substrate binds and the active form of the enzyme only exists when substrate is bound. Thus the substrate can be said to induce the correct fit. This idea may be usefully extended in terms of the enzyme deforming the substrate (Figure 4.5 b)); some enzymes, notably lysozyme, can only bind their substrate if it is strained or forced into an abnormal configuration. This provides one of the mechanisms by which rate enhancement is achieved.

Both the lock and key and induced fit models have relevance and usefulness in thinking about enzymes. The complementarity of enzyme and substrate are easily seen through the lock and key model, with obvious consequences for substrate specificity. However, enzymes are not rigid and the induced fit model is a useful reminder of the importance of flexibility.

SAQ 4.2

Comment on the truth, or otherwise, of the following statements. In each case write no more than 2 sentences to support your conclusion.

1) Enzymes alter the equilibrum position of the reaction they speed up.

2) An enzyme-substrate complex must be formed before enzymatic catalysis can occur.

3) When an enzyme binds substrate, binding sites for 2 groups on the substrate confer stereospecificity on the enzyme (ie the ability to distinguish stereoisomers).

4) Enzymes are specific to only one compound and hence only one reaction.

5) Substrate specificity is a consequence of the particular shape and charge distribution of the active site.

4.3 Enzyme assays

The measurement of enzyme activity is of enormous practical importance. Amongst the many reasons why we may wish to assay enzymes, we can cite the following:

- any detailed analysis of metabolism requires knowledge of the amounts of each component enzyme. Rate-limiting steps, which serve as control points in metabolic pathways, are often identifiable as the step with lowest enzyme capacity;
- development changes in an organism are reflected in the enzymes they contain and their relative levels;
- many disease states are characterised by abnormal levels of particular enzymes, thus appropriate assays may be relevant to both understanding of the condition, its diagnosis, and perhaps its cure;
- finally, if we wish to purify an enzyme (whether as part of a scientific investigation to improve our understanding of the biochemistry of a system or, in a technological context, to use the enzyme in analysis or industry), we must be able to determine how much is present in any given sample.

This important aspect of practical biochemistry is covered in several BIOTOL texts ('The Molecular Fabric of Cells'; 'Techniques used in Bioproduct Analysis Acids' and 'Principles of Enzymology for Technological Application') so we will not repeat the discussion here. If you are faced by the need to measure the activities of enzymes, we would recommend you examine one of the above texts to ensure you develop a sound strategy and avoid the many pitfalls inherent in such determinations.

We will now examine the effects of substrate concentration and the physical environment on the activities of enzymes.

4.4 Effects of substrate concentration on enzyme activity.

4.4.1 Michaelis Menten kinetics

Enzyme-catalysed reactions show a complex relationship between the velocity of the reaction and the concentration of substrate. The most usual relationship is shown in Figure 4.6.

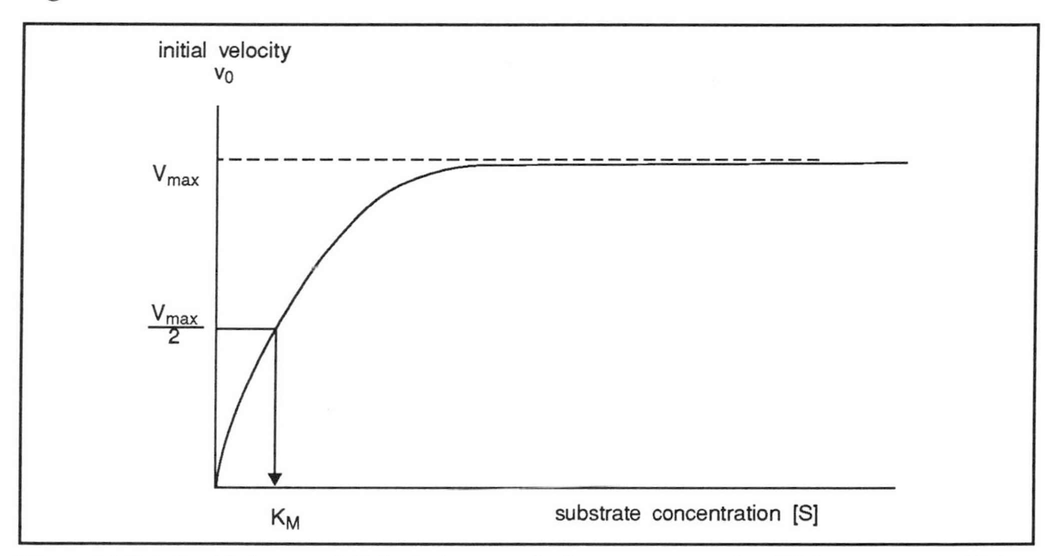

Figure 4.6 Relationship between initial velocity, v_o, and substrate concentration. [S], for an enzyme which obeys the Michaelis-Menten equation. The Michaelis constant, K_M, is the [S] at which half maximum velocity (½ Vmax) is displayed. (See text for an explanation).

zero order reaction

At low substrate concentrations, rate increases in proportion to substrate concentration (that is, first order kinetics); at higher concentrations, the increase in rate for each unit increase in substrate concentration decreases, eventually levelling off to a plateau. At high substrate concentration the rate is said to be zero order, as rate is no longer affected by a change in substrate concentration. Between these two regions we find mixed first and zero order kinetics.

enzyme-substrate complex

The mathematical relationship between initial velocity and substrate concentration was analysed in the early part of the twentieth century by two scientists, Michaelis and Menten. They analysed the enzyme distribution on the basis of the following reaction scheme:

$$E + S \underset{k_2}{\overset{k_1}{\rightleftharpoons}} ES \underset{k_4}{\overset{k_3}{\rightleftharpoons}} E + P$$

where k_1, k_2 etc are rate constants. Catalysis was assumed to require the combination of enzyme (E) with substrate (S) to form an enzyme-substrate complex (ES). This reaction was reversible, but ES could also break down to form free enzyme (E) and product (P). In deriving an equation to describe the observed relationship between initial velocity and substrate concentration, they made several assumptions.

- the concentration of enzyme, (E), was low relative to that of substrate. This meant that formation of ES did not significantly alter the concentration of free substrate, and hence this assumption could be used to simplify the analysis. Typically, this assumption is acceptable;

- product was absent; as a consequence, the possible recombination of E and P to form ES could be ignored (thus any term involving rate constant k_4 was eliminated). This is time only for the initial phase of the reaction;

- the concentration of ES was constant and was in equilibrium with E and S; this is referred to as the equilibrium assumption. Subsequent analysis showed this to be a restricted case. A wider generalisation nonetheless remains that (ES) is constant: a steady state exists between its formation ($E + S \rightarrow ES$) and its breakdown ($ES \rightarrow E + S$ and $ES \rightarrow E + P$) and this steady state is established very soon after enzyme and substrate are mixed. The rate of the overall reaction is given by k_3 [ES] and thus velocity is proportional to [ES]. The existence of ES was initially deduced by Michaelis, on the basis of the hyperbolic relationship between rate and substrate concentration; it has since been observed by electron microscopy and confirmed by spectroscopic techniques.

Michaelis Menten equation

Using these assumptions Michaelis and Menten were able to establish the relationship between the initial velocity of the reaction and substrate concentration, usually referred to as the Michaelis Menten equation. This is usually given as:

$$v_0 = \frac{V_{max} \times [S]}{K_M + [S]}$$

where v_0 = initial velocity of the reaction: [S] = substrate concentration in mol l^{-1};

V_{max} = maximum velocity, seen at (infinitely) high substrate concentration; V_{max} is a constant for a given enzyme:substrate combination;

K_M = Michaelis constant, in mol l^{-1}.

We can now analyse the observed relationship between initial velocity and substrate concentration (Figure 4.6) in terms of the Michaelis Menten equation.

- at low substrate concentration, where $[S] < K_M$, the denominator approximates to K_M (because the contribution of [S] to $K_M + [S]$ is negligible). Under these conditions, the equation becomes:

$$v_0 = \frac{V_{max} \times [S]}{K_M}$$

ie $v_0 \propto [S]$, which is consistent with our earlier comments at low substrate concentrations, where first order kinetics are observed.

- At high substrate concentration, the contribution of K_M to the denominator can be ignored, since $K_M + [S]$ will approximate to [S].

The equation then becomes:

$$v_0 = \frac{V_{max} \times [S]}{[S]}$$

ie velocity is maximal and zero order kinetics (with respect to substrate) will be displayed. Thus both features which we described qualitatively are consistent with the mathematical model.

meaning of K_M

What is the meaning and significance of K_M, the Michaelis constant? We can answer this by analysing a third situation, when [S] is made equal to the K_M. The Michaelis Menten equation then becomes:

$$v_0 = \frac{V_{max}}{2} \quad \left(\text{via } v_0 = \frac{V_{max} \times [S]}{[S] + [S]} \right)$$

and the Michaelis constant, K_M, can be seen to be the substrate concentration which gives half-maximal velocity. This will be when half the enzyme active sites have substrate bound and are in the ES form (remember that we said that rate was proportional to [ES]). This is a useful operational definition for K_M.

The significance of the Michaelis constant is that knowledge of it enables us to predict over what substrate concentration range the activity of an enzyme will vary. The next ITA will amplify this point. Knowledge of the K_M of an enzyme also facilitates optimisation of assay conditions.

Π What is the initial velocity of an enzyme-catalysed reaction at the following substrate concentrations, if $K_M = 1\ \mu mol\ l^{-1}$ and $V_{max} = 100\ \mu mol\ min^{-1}$? Assume that the enzyme obeys Michaelis Menten kinetics.

	[S], $\mu mol\ l^{-1}$
(A)	0.1
(B)	0.2
(C)	1.0
(D)	2.0
(E)	10.0
(F)	100.0

When you have calculated the initial velocities for each [S], it would be useful to plot a graph of v_0 against [S].

The initial velocities at the various concentrations are obtained by substitution into the Michaelis Menten equation. For case (A), this gives

$$v_0 = \frac{V_{max} \times [S]}{K_M + [S]} = \frac{100 \times 0.1}{1 + 0.1}\ \mu mol.min^{-1}$$

$$v_0 = \frac{10}{1.1} = 9.1\ \mu mol.min^{1}$$

Corresponding calculations give:

	[S], $\mu mol\ l^{-1}$	v_0, $\mu mol\ min^{-1}$
(A)	0.1	9.1
(B)	0.2	16.7
(C)	1.0	50.0
(D)	2.0	66.7
(E)	10.0	90.9
(F)	100.0	99.0

If plotted as a graph, this will give the characteristic hyperbolic plot (Figure 4.6), with near proportionally between rate and [S] at low [S], and diminishing proportionally at higher [S], particularly above the K_M. Notice that the rate shows only a very small response to a 10-fold increase in [S] when it is substantially (10x) above the K_M value. Note also that at very low [S], whilst the rate may change nearly proportionally with any change in [S] (look at (A) and (B)), it is nonetheless a small percentage of the maximum activity of the enzyme.

A low K_M means that an enzyme will bind substrate at low substrate concentrations. It is said to have a high affinity for its substrate. For example, if the K_M is 10^{-5} mol l^{-1} the rate will be 91% of V_{max} at substrate concentration of 10^{-4} mol l^{-1}.

high affinity and low affinity

A high K_M, in contrast, indicates that a relatively high substrate concentration will be required to achieve (near) saturation. This is described as low affinity. As an example, a K_M of 10^{-2} mol l^{-1} would require a substrate concentration of 10^{-1} mol l^{-1} to display 91% of its maximum rate. At a substrate concentration of 10^{-4} mol l^{-1}, it will only be working at 1% of its maximum rate!

Knowledge of K_M values is thus useful in understanding and predicting the behaviour of metabolic systems. If a substrate can be acted upon by more than one enzyme, that with the lowest K_M value will 'win' the substrate. These two enzymes may, however, be in different tissues, in which case a different result occurs. As an example of the latter, consider the fate of a glucose molecule in the bloodstream. All cells obtaining glucose from the blood posses the enzyme hexokinase, which phosphorylates glucose according to the following scheme:

$$\text{glucose} + \text{ATP} \rightarrow \text{glucose-6-phosphate} + \text{ADP}$$

The phosphorylated glucose is now trapped within the cell (because of the negatively charged phosphate group it is no longer able to pass through the plasma membrane), this reaction is also the first step in glycolysis, leading to the generation of ATP. Some cells, notably those of liver, possess a second enzyme which carries out this reaction; it is known as glucokinase. Glucose phosphorylated by this enzyme is predominantly stored (as glycogen) in the liver. Thus there are two alternative enzymes for the same reaction, one leading directly to energy generation, the other to energy storage. What is the consequence or importance of this?

effect of differing K_M values

If we look at the relationship between glucose concentration and rate of reaction (Figure 4.7), we find marked differences between the two enzymes.

This is reflected in the K_M values for glucose (typically 0.75 x 10^{-5} mol l^{-1} for hexokinase and 1.8 x 10^{-2} mol l^{-1} for glucokinase). Hexokinases are used for energy needed continuously and are routinely saturated by the normal blood glucose concentration; if the blood glucose concentration; rises (for example, after a meal), their activity is not affected. The substrate concentration had already made the reaction zero order with respect to glucose ([S] > K_M). Activity of glucokinase, in contrast, does change as blood glucose concentration rises, leading to the excess glucose being taken up by the liver and stored (until needed) as glycogen.

K_M values vary widely, typically from 10^{-6} to 10^{-2} mol l^{-1}. Note also that they depend markedly on a variety of factors such as pH and temperature, as well as the particular substrate. V_{max} is the maximum rate that a given amount of enzyme can operate at, achieved when substrate is saturating. A useful concept here is the turnover number of an enzyme. This is the number of substrate molecules transformed per second by a single enzyme molecule, when substrate is saturating; it is equal to the rate constant k_3.

Turnover numbers vary widely, from, for example, 2 molecules transformed per second for tryptophan synthetase to a remarkable 6000 000 molecules per second for carbonic anhydrase. We know little about how this is controlled but it would be potentially very useful if we understood it.

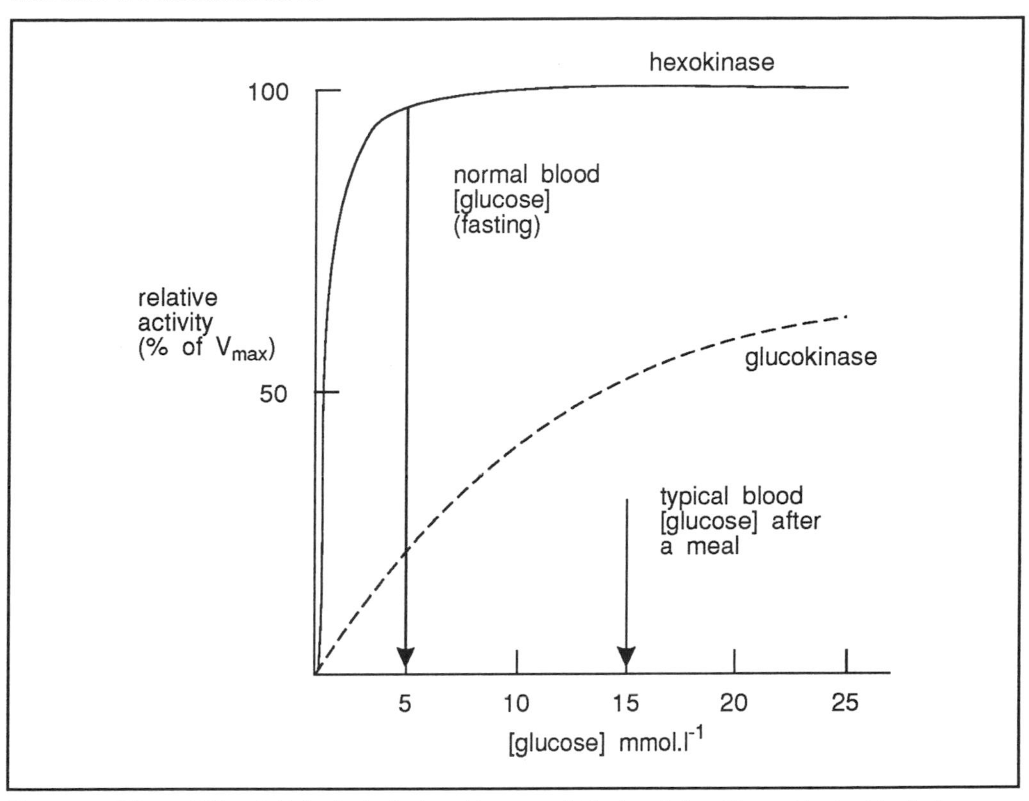

Figure 4.7 Effect of [S] on initial velocity for hexokinase (solid line) and glucokinase (dashed line). Arrows show typical blood glucose concentrations during fasting (ie not immediately after a meal) and after a meal.

4.4.2 Determination of K_M and V_{max}

direct plot of v and (S)

When enzymes are being characterised and described, K_M and V_{max} are routinely determined. How may this best be done? Experimentally, one measures initial velocity at a variety of substrate concentrations (having fixed other parameters such as pH and temperature). The data could then be plotted as in Figure 4.6. From this direct plot, Vmax can, in principle, be obtained; K_M is then the substrate concentration which gives half maximal velocity. This strategy is not a good one! Whilst plotting the data in this way is useful as a check that Michaelis Menten kinetics are occurring (we will meet some exceptions later), satisfactory values for K_M and V_{max} are unlikely to be obtained. Look back at the previous ITA. If the substrate concentration is 10x higher than the K_M, the rate is only 10/11th of V_{max}; even when the substrate concentration is 100x that of K_M, the rate is still only 99% of V_{max}! It may thus be extremely difficult to experimentally obtain V_{max}, and any error in estimating V_{max} will result in error in estimating K_M.

linear plots of v and (S)

These problems are avoided by various mathematical transformations of the data which result in linear plots (providing Michaelis Menten kinetics are obeyed). These are obtained by taking reciprocals of the Michaelis Menten equation, giving:

$$\frac{1}{v_0} = \frac{K_M + [S]}{V_{max} \times [S]} = \frac{K_M}{V_{max}} \times \frac{1}{[S]} + \frac{1}{V_{max}}$$

Lineweaver-Burk plot

This is the equation of a straight line ($y = mx + c$). Although various ways of plotting data have been devised, the most widely used plot is the Lineweaver-Burk plot (Figure 4.8), in which $1/v_0$ is plotted against $1/[S]$. For an enzyme which conforms to Michaelis Menten kinetics, this will give a straight line, with a slope of $K_M V_{max}$.

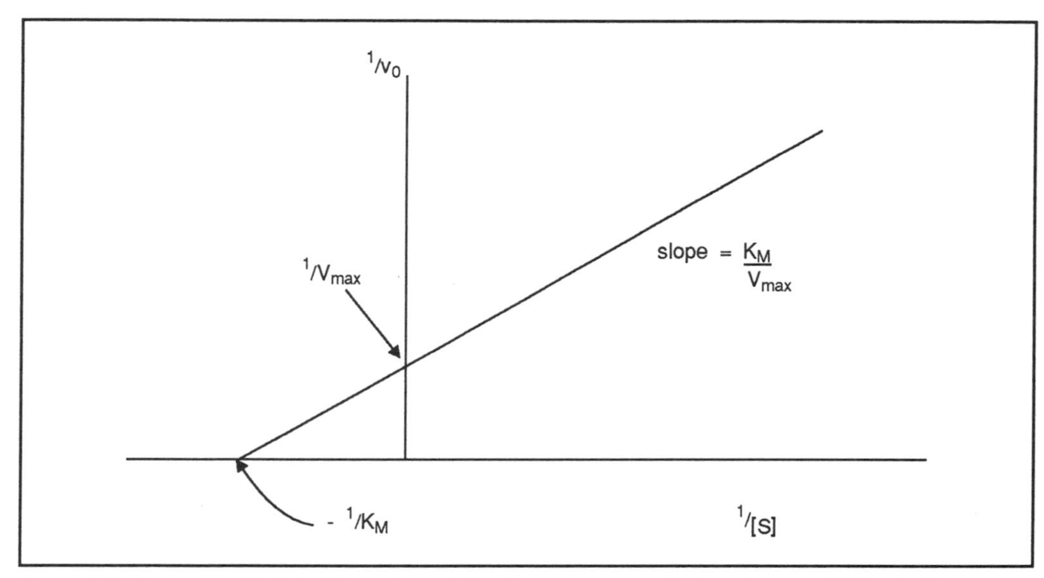

Figure 4.8 Lineweaver-Burk double reciprocal plot. For an enzyme which conforms to Michaelis-Menten kinetics, a straight line will be obtained. The intercept on the $1/v_0$ axis is $1/V_{max}$, whilst that on the negative side of the $1/[S]$ axis equals $-1/K_M$.

The intercept on the $1/v_0$ axis equals $1/V_{max}$, that on the $1/[S]$ axis equals $-1/K_M$. Since a linear relationship has been established, it is much easier and more accurate to extrapolate and the line of best fit can be obtained by regression analysis; this avoids the need to experimentally obtain Vmax. Indeed, data for a Lineweaver-Burk plot are generally most satisfactorily gained over the substrate concentration range $0.2\text{-}5.0 \times K_M$.

SAQ 4.3

1) Calculate K_M and V_{max} for an enzyme, for which the following data is provided:

	[S], mmol l^{-1}	Initial velocity, μmol min^{-1}
(A)	0.6	1.14
(B)	1.0	1.70
(C)	1.5	2.36
(D)	3.0	3.64
(E)	6.0	5.0
(F)	15.0	6.46

2) Assume that you have to accurately measure the K_M of an enzyme, for which preliminary estimates indicate that the K_M is about 4×10^{-3} mol l^{-1}. Recommend what concentrations of substrate should be used for the careful analysis.

4.4.3 Multisubstrate reactions

The kinetics we have described above apply to enzymes which involve a single substrate. Single-substrate reactions are, however, relatively rare. Most enzymes involve more than a single-substrate. For example, enzymes such as trypsin which hydrolyse proteins have two substrates (protein and water). In such cases, however, the second substrate (water) remains at a constant saturating concentration in the reaction and for practical kinetic purposes the reactions catalysed by these enzymes can be considered as being single-substrate reactions.

If additional substrates are used, then the mathematical relationships between substrate concentrations and reaction velocity are quite complex and are dependent upon the mechanisms by which the substrates and enzymes interact. Here we will not deal in depth with the mathematics of these systems (the reader is referred to the BIOTOL text, 'Principles of Enzymology for Technological Application'). We will, however, focus on the general mechanisms of multisubstrate enzyme systems.

To help us to describe these mechanisms we have used the following symbols:

S_1, S_2 and S_3 represent different substrates.

P_1, P_2 and P_3 designate different products

E_1, E_2 and E_3 represent different forms of the enzyme.

It is generally accepted that three general mechanisms exists. These are:

- random order;
- compulsory order;
- ping pong.

Let us examine each in turn.

Random order mechanism, with ternary complex

With some enzymes the two substrates S_1 and S_2 may be added to the enzyme in any order, the products P_1 and P_2 are also released in any order.

Thus we could represent this mechanism for the reaction:

$$S_1 + S_2 \rightleftharpoons P_1 + P_2$$

as

$$E_1 + S_1 \rightleftharpoons E_1S_1 \quad \xrightarrow{S_2} \quad E_1S_1S_2 \rightleftharpoons E_1P_1P_2 \quad \xrightarrow{P_1} \quad E_1P_2 \rightleftharpoons E_1 + P_2$$

$$E_1 + S_2 \rightleftharpoons E_1S_2 \quad \xrightarrow{S_1} \quad E_1S_1S_2 \rightleftharpoons E_1P_1P_2 \quad \xrightarrow{P_2} \quad E_1P_1 \rightleftharpoons E_1 + P_1$$

We might call this system a random order Bi, Bi mechanism (Bi represents two substrates or two products; note also Uni indicates a single substrate as product and Ter represents three substrates or products). It is described as involving a ternary complex, since one(s) involving three components (E_1, S_1, S_2 and $E_1P_1P_2$) are involved.

An example of an enzyme displaying a random order Bi, Bi mechanism is glycogen phosphorylase.

Π Consider the reaction $S_1 + S_2 \rightleftarrows P_1$ in which S_1 and S_2 can be added to the enzyme in any order. Such a system can be regarded as having a random order Bi, Uni mechanism. Draw a scheme to represent this system.

The sort of figure we anticipate you would draw is:

$$E_1 + S_1 \rightleftharpoons E_1S_1 \xrightleftharpoons{S_2} E_1S_1S_2 \rightleftharpoons E_1P_1 \rightleftharpoons E_1 + P_1$$

$$E_1 + S_2 \rightleftharpoons E_1S_2 \xrightleftharpoons{S_1} E_1S_1S_2$$

Note two substrates (= Bi) and one product (= Uni)

Compulsory order mechanisms with ternary complex

In this case, there is a defined order in which the substrates associate with the active sites of the enzyme and all of the products are released in a precise order. The products are released only after all of the substrates have reacted.

We can represent these reactions in the following way:

$$E_1 + S_1 \rightleftharpoons E_1S_1 \xrightleftharpoons{S_2} E_1S_1S_2 \rightleftharpoons E_1P_1P_2 \xrightleftharpoons{P_2} E_1P_1 \rightleftharpoons E_1 + P_1$$

The above reaction could be described as having an ordered Bi, Bi mechanism.

Π Draw a scheme representing a compulsory order Ter, Bi mechanism.

The scheme we anticipate you would draw is:

$$E_1 + S_1 \rightleftharpoons E_1S_1 \xrightleftharpoons{S_2} E_1S_1S_2 \xrightleftharpoons{S_3} E_1S_1S_2S_3 \rightleftharpoons E_1P_1P_2 \xrightleftharpoons{P_2} E_1P_1 \rightleftharpoons E_1 + P_1$$

S_1, S_2 and S_3 represent three substrates (hence Ter) being added in a defined order, forming two products (hence Bi) also being released in a defined order.

Ping pong mechanism

We will first draw out a scheme and then explain what it represents.

$$E_1 + S_1 \rightleftharpoons E_1S_1 \rightleftharpoons E_2P_1 \xrightleftharpoons{\nearrow P_1} E_2 \xrightleftharpoons{S_2 \searrow} E_2S_2 \rightleftharpoons E_1P_2 \rightleftharpoons E_1 + P_2$$

At the first stage, the enzyme (E_1) combines with the first substrate (S_1), to form an enzyme substrate complies (E_1S_1). The substrate is modified to form product (P_1) at the same time the enzyme is modified (E_2). It might be, for example, that a phosphate group is removed from substrate S_1 and attached to the enzyme. We can represent this as:

$$\text{Enzyme} \xrightleftharpoons{A - \textcircled{P} \searrow} \text{Enzyme A} - \textcircled{P} \rightleftharpoons \text{Enzyme} - \textcircled{P}\ A \xrightleftharpoons{\nearrow A} \text{Enzyme} - \textcircled{P}$$

The modified enzyme (E_2) then forms a complex with the second substrate (E_2S_2) and modifies this substrate to form a new product (P_2) and regenerate the original form of the enzyme (E_1).

In the case of our phosphorylated enzyme, we can represent this by:

$$\text{Enzyme} - \textcircled{P} \xrightleftharpoons{B \searrow} \text{Enzyme} - \textcircled{P}\ B \rightleftharpoons \text{Enzyme B} - \textcircled{P} \rightleftharpoons \text{Enzyme} + B - \textcircled{P}$$

The overall reaction ($S_1 + S_2 \overset{E_1}{\rightleftarrows} P_1 + P_2$) described above can be described as a Bi, Bi ping pong mechanism. Notice that no ternary complex, involving both substrates and the enzyme, exists.

Enzymes can, however, be quite complex. Consider the reaction:

$$\text{acetyl CoA} + \text{ATP} + HCO_3^- \rightleftarrows \text{malonyl CoA} + \text{ADP Pi}$$

We can represent the mechanism of this reaction in the following way:

$$E_1 \xrightleftharpoons{\text{ATP} \searrow} (E_1\text{ATP}) \xrightleftharpoons{HCO_3^- \searrow} \left(\begin{matrix} E_1\text{ATP} \\ | \\ COO^- \end{matrix}\right) \xrightleftharpoons{\nearrow \text{ADP}} \left(\begin{matrix} E_2 - \text{Pi} \\ | \\ COO^- \end{matrix}\right) \xrightleftharpoons{\nearrow \text{Pi}} \begin{matrix} E_2 \\ | \\ COO^- \end{matrix} \xrightleftharpoons{\text{acetyl CoA} \searrow} \begin{matrix} E_2\ \text{acetyl CoA} \\ | \\ COO^- \end{matrix} \rightleftharpoons \begin{matrix} E_1 \\ | \\ \text{malonyl CoA} \end{matrix} \xrightleftharpoons{\nearrow \text{malonyl CoA}} E_1$$

(Note we have simplified this by ignoring the evolution or consumption of water in the various stages).

Π Describe the mechanism of this reaction sequence as drawn.

The answer we hope you would give is that the mechanism is a Bi, Bi, Uni, Uni, ping pong mechanism since first two substrates are added, then two products are released, then an additional substrate is added and a final product is released. During this process the enzyme becomes transiently modified.

From the above description it is not surprising that the kinetics of these types of reactions may be complex. Elucidation of these mechanisms depends upon conducting detailed kinetic experiments., in which all of the variables (substrates, enzyme and product concentrations) except one are fixed and then analysing the kinetics and equilibrium constants for the binding of the substrates and products in the presence and absence of the other participating molecules. We will illustrate this approach to the analysis of multisubstrate kinetics in the next section.

4.4.4 Kinetics of compulsory order and ping pong two substrate reactions

The initial rate equation for the compulsory order mechanism is given by

$$v_0 = \frac{V_{max}[S_1][S_2]}{K_{S2}[S_1] + K_{S1}[S_2] + K_{ES}K_{S2} + [S_1][S_2]}$$

in which V_{max} is the maximum rate when both S_1 and S_2 are saturating, K_{S1} and K_{S2} are the Michaelis constants for the substrates S_1 and S_2 (thus K_{S1} is the substrate concentration of S_1 giving half V_{max} when S_2 is saturating), and K_{ES} is the dissociation constant of the E_{S1} complex. If $[S_1]$ is varied, whilst $[S_2]$ is held at a constant value, data may be obtained for a double reciprocal plot. The pattern obtained for 4 fixed concentrations of S_2, with varied $[S_1]$, is shown in Figure 4.9 a).

The initial rate equation for the ping pong mechanism is:

$$v_0 = \frac{V_{max}[S_1][S_2]}{K_{S2}[S_1] + K_{S1}[S_2] + [S_1][S_2]}$$

If $\frac{1}{v_0}$ is plotted against $\frac{1}{[S_1]}$ for four different values of $[S_2]$, the pattern presented in Figure 4.9 b) is obtained. These patterns are highly informative regarding the mechanism used by enzymes involving multiple substrates.

We have included this brief description of the kinetics of two substrate reactions as an example of how the mechanism of a reaction can be established from kinetic data. We do not expect you to remember these equations at this stage. If, however, you become directly involved in determining the mechanisms of enzyme catalysed reactions or in the kinetics of such processes we would recommend the BIOTOL text, 'Principles of Enzymology for Technological Applications'

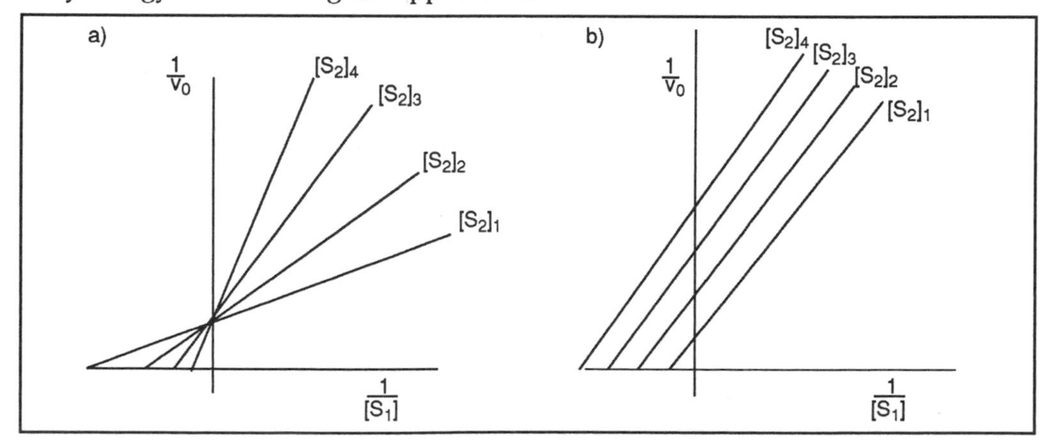

Figure 4.9 Plots of $1/v_0$ against $\frac{1}{[S_1]}$ for a) compulsory order two substrate reactions b) ping pong two substrate reactions.

4.5 Effect of inhibitors on enzyme activity

reversible and irreversible inhibitors

Enzyme inhibitors are compounds which diminish or eliminate the activity of an enzyme. For convenience, they are categorised as either reversible or irreversible inhibitors. As the name implies, reversible inhibition may be reversed, such that full enzyme activity is restored, when the inhibitor is removed. In contrast, in irreversible inhibition, enzyme activity cannot be restored.

4.5.1 Reversible inhibition

dialysis removes reversible inhibitors

Any compound which can bind to an enzyme in a reversible manner, and thereby reduce the activity of the enzyme, is classified as a reversible inhibitor. Such compounds may interact with different parts of the enzyme; they are designated, and distinguished, in the following ways. Reversibility is confirmed by recovery of enzyme activity on removal of the inhibitor, by dilution or by dialysis. Dialysis involves placing the enzyme solution in a sac made of a semi-permeable membrane (like cellophane) which allows low molecular weight compounds to cross, whilst retaining high molecular weight compounds like enzymes (the membrane acts as a molecular sieve). If this sac is placed in a large volume of buffer the inhibitor will dissociate from the enzyme and become dispersed throughout the buffer in the vessel.

Competitive inhibition

competitive inhibition

If substrate and inhibitor compete for the active site of an enzyme, competitive inhibition results. This competition may be visualised as depicted in Figure 4.10, whereby if one compound is bound, then the other cannot bind.

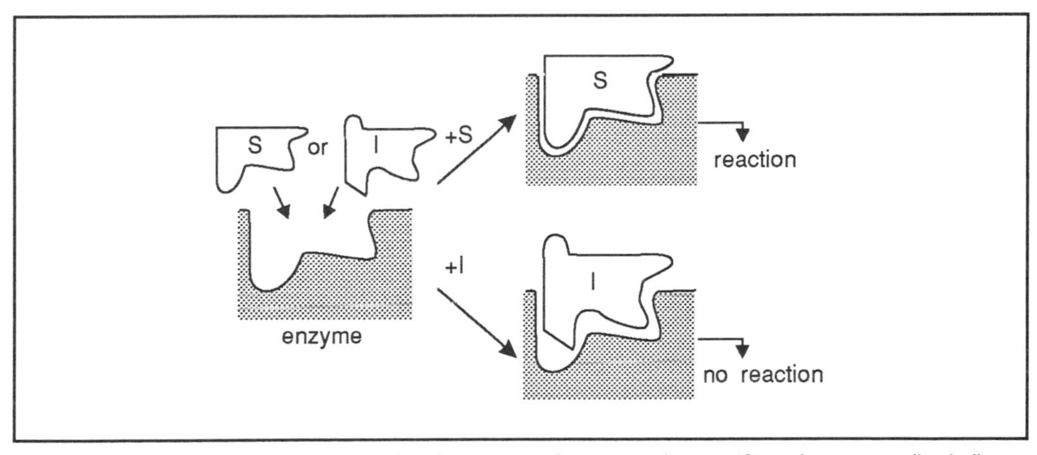

Figure 4.10 Competition for the active site of an enzyme between substrate (S) and a structurally similar compound (I).

It is helpful to think of the distribution of enzyme molecules in this way. In the absence of inhibitor, only the following occurs.

$$E + S \rightleftharpoons ES \longrightarrow E + P$$

With a competitive inhibitor present, an enzyme-inhibitor complex (which does not form product) is formed.

$$\begin{array}{ccccc} E + S & \rightleftharpoons & ES & \longrightarrow & E + P \\ {\scriptstyle +I}\updownarrow & & & & \\ EI & & & & \end{array}$$

In the presence of I, some E is present as EI, thus [E] and hence [ES], is lowered, therefore rate is decreased.

Thus some of the enzyme is 'siphoned off' into a non-productive form which cannot form the ES complex whilst inhibitor is bound to it. Less enzyme is available for substrate binding, and [ES] is lowered. Since the rate is proportional to [ES], the reaction is inhibited.

What factors determine the extent of inhibition? Looking at the enzyme distribution, increasing [I] will result in increased [EI], whereas increasing [S] will, conversely, result in increased [ES]. Thus both [S] and [I] will influence the enzyme distribution and hence the rate. In addition, the affinity of the enzyme for the inhibitor and substrate is important. The K_M serves, in general terms, as an indication of the enzyme-substrate affinity (ie how tightly do E and S bind to each other). That for the inhibitor is equivalent to the dissociation constant of the enzyme-inhibitor complex. This is given by the expression: $K_I = \frac{[E]\,[I]}{[EI]}$.

High substrate concentration will 'pull' the enzyme distribution over towards the ES form. Hence, even though inhibitor is present in the solution, it may still be possible for all enzyme molecules to bind substrate (ie form ES), thus V_{max} is not affected by a competitive inhibitor, although higher [S] will be needed to obtain it than in the absence of the inhibitor. The need for higher [S] will be reflected in an increase in the K_M value. This is shown in Figure 4.11.

Experimental determination of the type of inhibition requires a similar approach to that used to obtain the Michaelis constant, K_M. The reaction rate is measured at various substrate concentrations (perhaps 6, ranged around the K_M value) in the presence and absence of a constant inhibitor concentration (or at 2 inhibitor concentrations). Direct analysis of velocity against [S] (Figure 4.11 a) should not be used.

effect of competitive inhibitors on double reciprocal plots

Instead, a Lineweaver-Burk double reciprocal plot is made (Figure 4.11 b)). The intercepts formed are characteristic of the type of inhibition; with competitive inhibition, the intercept on the 1/v axis is not altered (V_{max} remains the same), whereas that on the 1/[S] is changed (K_M is increased).

competitive inhibitors often mimic the natural substrate

Competitive inhibitors are usually compounds which have structures similar to the substrate. They can thus bind to the active site but do not undergo reaction. Krebs made use of such an inhibitor in his classical studies on the TCA cycle; the step involved is that catalysed by succinate dehydrogenase:

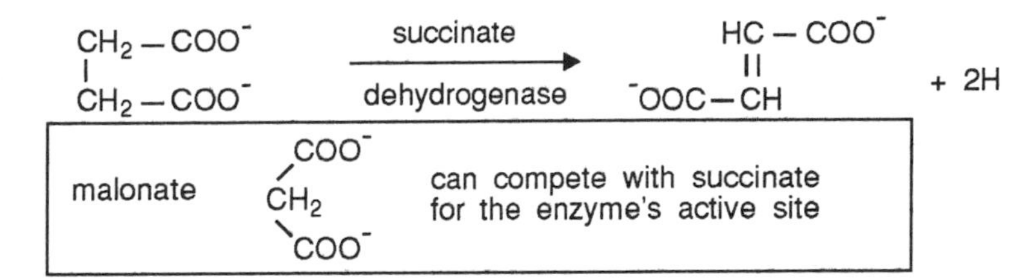

Malonate has only one methylene ($-CH_2^-$) group but nonetheless binds to the active site of succinate dehydrogenase. There is no CH_2-CH_2 group from which a pair of hydrogen atoms can be removed (ie oxidation), therefore, no reaction can occur. It is, however, a potent inhibitor because it prevents succinate binding and thus no ES complex is formed.

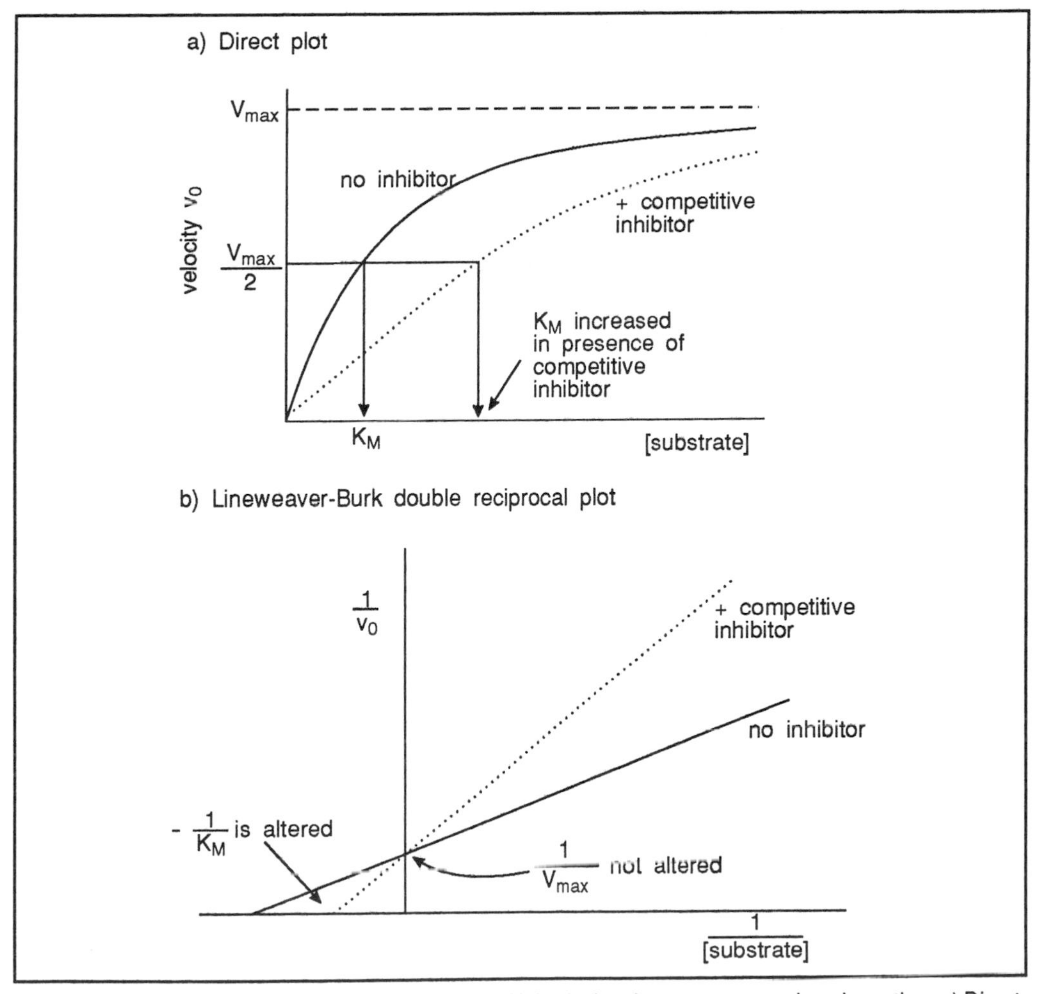

Figure 4.11 Effect of a competitive inhibitor on the initial velocity of an enzyme-catalysed reaction. a) Direct plot of v_o against [S]. b) Lineweaver-Burk double reciprocal plot.

Non-competitive inhibition

non-competitive inhibition

An alternative type of reversible inhibition occurs when a compound binds to the enzyme, at a site away from the active site, but still prevents reaction. A non-competitive inhibitor does not prevent substrate binding. Indeed, they can both be bound at the same time. As a consequence, substrate cannot displace the inhibitor and high [S] does not alleviate the inhibition. Inhibition is presumably caused by interfering with the catalytic steps, rather than with substrate binding. The enzyme distribution can be summarised as follows for non-competitive inhibition:

$$\begin{array}{ccccc} E + S & \rightleftharpoons & ES & \longrightarrow & E + P \\ +I \updownarrow & & +I \updownarrow & & \\ EI + S & \rightleftharpoons & EIS & & \end{array}$$

Since substrate cannot displace the inhibitor, [S] no longer has any influence on the extent of inhibition. This is determined by the concentration of the inhibitor, [I], and its affinity for the enzyme (shown by the dissociation constant, K_I).

effect of non-competitive inhibitor on linear plots

A plot of velocity against [S] for an enzyme-catalysed reaction in the presence of a non-competitive inhibitor is shown in Figure 4.12. When inhibitor is present, some enzyme molecules are non-productive. It does not matter whether they are in the form EI or EIS. Thus less active enzyme is present and V_{max} is lowered.

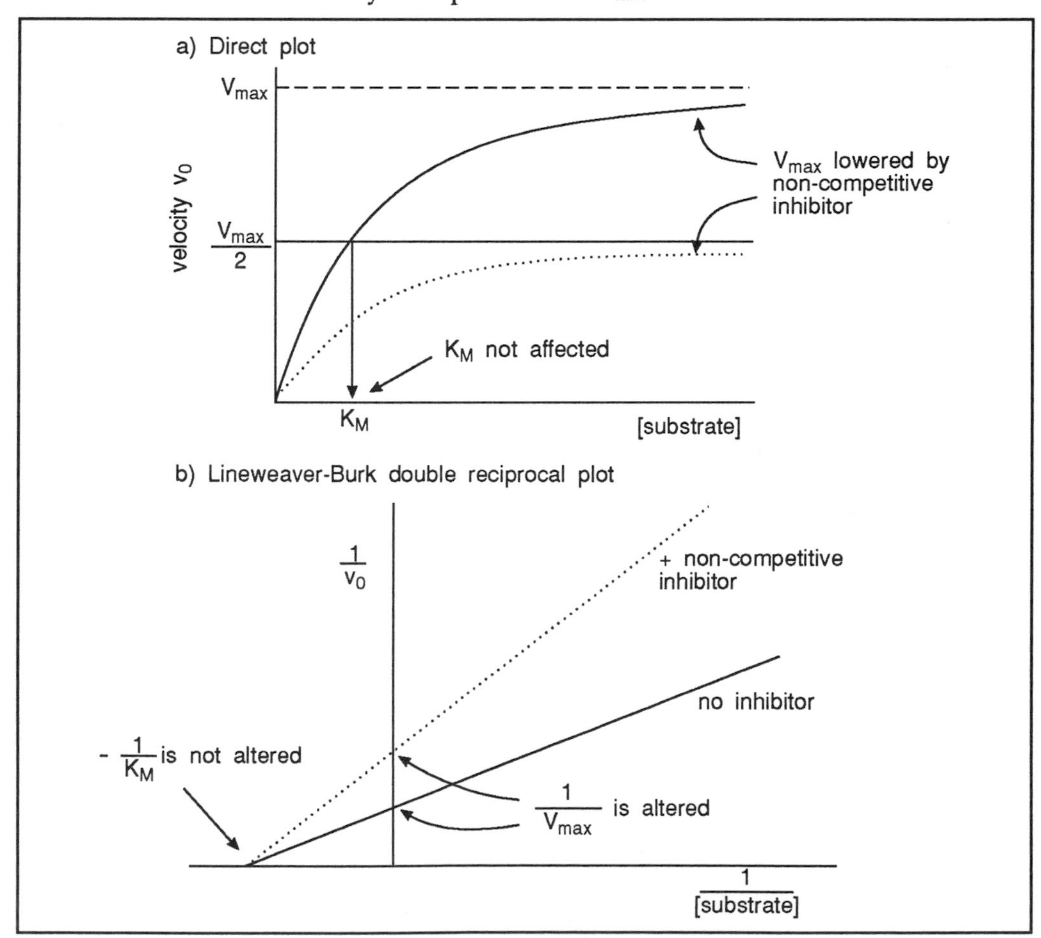

Figure 4.12 Effect of a non-competitive inhibitor on the initial velocity of an enzyme-catalysed reaction. a) Direct plot of v_o against [S]. b) Lineweaver-Burk double reciprocal plot.

Substrate binding is not affected (K_M is not altered).

As with competitive inhibition, identification of non-competitive inhibition is based on the pattern given by the Lineweaver-Burk double reciprocal plot (Figure 4.12 b)). In this case the 1/v intercept is altered (V_{max} is lowered), reflecting the fact that some enzyme molecules are out of action or are working less efficiently. The intercept on the 1/[S] axis, from which K_M is obtained, is unchanged: this is consistent with the inhibitor having no effect on substrate binding.

Non-competitive inhibitors are less easily visualised than competitive ones. One way is to think of the inhibitor causing a conformational change (Figure 4.13 a)) such that the enzyme is less active or in-active. Heavy metal ions, especially Hg^{2+}, are often inhibitory because they bind tightly to sulphydryl groups (ie cysteine sidechains). This can occur away from the active site yet still disrupt catalysis, giving non-competitive inhibition.

mixed type inhibitors

Whilst the two types of reversible inhibitors described are widespread, mixed inhibition, involving a combination of the two types, can occur. For example, imagine an enzyme which has an important sulphydryl group (Figure 4.13 b)); if an Hg^{2+} ion binds to it, the enzymatic activity is diminished. If the bound Hg^{2+} also interferes with substrate binding, then the apparent K_M will also be raised. Thus both V_{max} and K_M will be affected, and the observed Lineweaver-Burk pattern will not be consistent with either 'pure' competitive or 'pure' non-competitive inhibition. This is usually referred to as mixed inhibition.

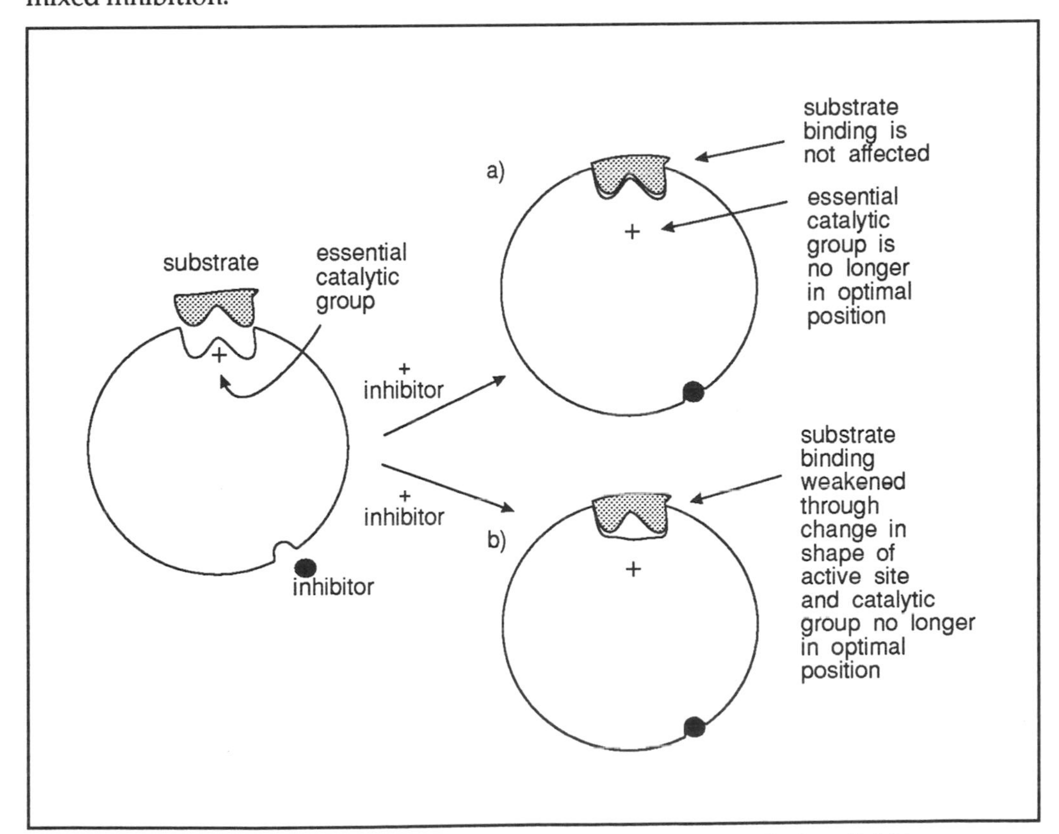

Figure 4.13 Possible mechanisms of non-competitive and mixed inhibition. a) Binding of inhibitor causes a conformational change which does not affect substrate binding, but adversely affects catalysis. b) Binding of inhibitor influences both substrate binding and catalysis, although substrate binding is not prevented.

SAQ 4.4

Which of the following statements is correct; jot down reasons for rejecting the other statements as incorrect. A non-competitive inhibitor:

1) lowers V_{max} and raises K_M;

2) lowers both K_M and V_{max};

3) raises K_M and leaves V_{max} unaltered;

4) lowers V_{max}, whilst K_M is unchanged;

5) results in complete abolition of enzyme activity.

4.5.2 Irreversible inhibition

covalent binding results in irreversible inhibition

Irreversible inhibition involves interaction between enzyme and inhibitor such that dissociation of the inhibitor does not occur. This is accomplished by agents which react covalently with an important functional group of the enzyme. Since the reaction results in covalent bonds, neither dilution nor dialysis will remove the inhibitor, nor will addition of excess substrate affect the situation. If active site residues are involved, causing complete inactivation upon reaction, then the proportion of active enzyme molecules is reduced. Non-reacted enzyme molecules show normal kinetics: thus K_M is unaffected whist V_{max} is lowered.

dialysis does not remove irreversible inhibitors

Irreversible inhibition is distinguished from reversible non-competitive inhibition by failure to restore full activity upon dialysis. In addition, reversible inhibition is usually instantaneous; irreversible, covalent inhibition is often progressive and may take minutes or hours to reach its maximum effect.

iodoacetate

A classical example of irreversible inhibition is that caused by the compound iodoacetate, which alkylates sulphydryl groups: if a cysteine sidechain which is essential for activity is modified, enzyme activity will be irreversibly lost. It has been known for many years that the glycolytic enzyme glyceraldehyde-3-phosphate dehydrogenase is subject to inhibition by alkylation, due to an essential, active site, sulphydryl group. This was used experimentally to inhibit glycolysis and thereby facilitated studies on intermediary metabolism.

SAQ 4.5

A compound, is known to inhibit an enzyme. However, the type of inhibition is not known. Enzyme assays were conducted in the presence and absence of a fixed concentration of the inhibitor at various substrate concentrations. In all other respects, the assays were identical. The data shows the [S] used and the initial velocity in the absence (column A) and presence (column B) of inhibitor.

[S] (mmol l^{-1})	initial velocity A (- Inhibitor)	(μmol. min^{-1}) B (+ Inhibitor)
0.02	31.3	16.1
0.03	41.7	22.2
0.04	47.6	27.8
0.06	57.1	35.7
0.12	71.4	52.6

1) Using whatever procedure you consider to be most appropriate, determine what type of inhibition is occurring.

2) Are there any additional experiments which you would recommend to confirm your conclusions. Briefly outline these.

4.5.3 Significance of enzyme inhibitors

product inhibition

The activity of numerous enzymes will be inhibited by the presence of the product of the reaction. This type of inhibition is competitive, reversible and is known as product inhibition. It means that if the product is not being utilised (and as a consequence its concentration builds up), then further synthesis is inhibited. A special type of inhibition, called feedback inhibition, involves inhibition of the first enzyme of a pathway by the end-product of the pathway: this is an example of allosteric regulation, which is considered further in Section 4.7.

many drugs inhibit enzymes

Many antibiotics, drugs and toxins, whether natural or synthetic, have their effects through inhibition of enzymes. A selection of these are grouped in Table 4.2, with comments on their mode of action and significance.

compound	enzyme inhibited	comments
glyphosate	5-enolpyruvylshikimate -5-phosphate synthase (in pathway of aromatic amino acid synthesis)	active ingredient of Tumbleweed and Roundup weed-killers, non-toxic to humans
penicillin	glycopeptide transpeptidase (synthesis of peptidoglycan in cell wall of bacteria)	penicillins and derivatives form the basis of many bactericidal treatments
iproniazid	monoamine oxidase (MAO)	anti-depressant, through inhibition of monoamine oxidase
6-mercaptopurine	various enzymes of purine biosynthesis	anti-cancer agent
malathion, and other organo-phosphorus compounds	anticholinesterase	potent insecticides and nerve poisons, through covalent modification of acetylcholinesterase

Table 4.2 Examples of commercially useful enzyme inhibitors. NB Tumbleweed and Roundup are trademarks used by the manufacturers of these weedkillers.

Pharmaceutical companies spend considerable sums of money on discovering, designing and developing enzyme inhibitors. Enzyme inhibitors which are themselves proteins also occur widely in nature; examples are shown in Table 4.3. Whilst their precise function is often unknown, they are presumed to either be part of a metabolic regulation system (eg invertase inhibitors preventing breakdown of sucrose by invertase) or have a defence role against pests. For example, protease (trypsin) inhibitors in bean seeds may stop pests from digesting the food they have eaten and they will then die of starvation. In humans, elastase activity in the lungs is normally inhibited by a protein inhibitor called (confusingly!) α-1-anti-trypsin. If the level of α-1-anti-trypsin is low (either through inheritance or by inactivation of the inhibitor by smoking), elastase is no longer controlled and causes destruction of the elastic tissue of the alveoli of the lungs: this is the basis of emphysema.

enzyme inhibited	source of inhibitor
trypsin, chymotrypsin and other proleases	legumes and cereal seeds
α-amylse	cereal seeds
invertase	sugar beet
elastase	α1-antitrypsin, serum
thrombin	anti-thrombin III, plasma

Table 4.3 Examples of protein-based enzyme inhibitors.

Π Glyphosate is one of the most commercially successful weedkillers and is the active ingredient of Tumbleweed and Roundup. We noted in Table 4.2 that a side-attraction of glyphosate was that it was of very low toxicity to man and other mammals. Write down in the margin why this might be so. This is a very open ended question, which should allow your developing biochemical knowledge to show itself! If you don't know where to begin, just speculate - try to come up with at least two possible reasons.

There are several possible explanations, of which the following are the more obvious:

- the glyphosate is metabolised (ie detoxified) before it can do any harm;
- it is not absorbed by humans;
- it is bound by a different protein, without adverse effects;
- the corresponding enzyme in humans is insensitive to the inhibitor;
- the corresponding enzyme in humans is absent.

In fact, the last of the possibilities above is the correct explanation: we lack this enzyme (and the rest of the pathway): as a consequence we require tryptophan and phenylalanine in our diet (they are essential amino acids), but at least we are safe from glyphosate.

This exercise is not entirely frivolous. Two points are significant: (A) comparative biochemistry and physiology can help us to design 'safe' drugs/toxic agents. (B) our list above contained, as a possibility, a form of the enzyme which was insensitive to glyphosate; possession of such an enzyme by a plant would confer resistance against glyphosate. Precisely such an approach has been used to produce (by genetic engineering) glyphosate-resistant crop plants.

4.6 Effect of pH and temperature on enzyme activity

Both pH and temperature have profound influences on enzyme activity. The effect of pH is primarily related to the state of ionisation of particular groups in the enzyme and/or the substrate. We have already indicated that substrate specificity and substrate binding involve 'matching' or complementarity of charged groups. If the pH is such that the complementarity of charge no longer occurs, then some effect on enzyme activity should be expected. Whilst every enzyme has its own relationship between pH and activity, a typical profile is the bell-shaped curve (Figure 4.14), in which there is a pH optimum at which the ionic states of enzyme and/or substrate are most appropriate. As the pH is changed, a change in the ionic state of a key group occurs, such that it is no longer in the required form: for example, the sidechain of histidine will change from being 90% protonated (and hence positively charged) at pH 5.0 to being 50% protonated at pH6.0 (since pKa = 6.0) and 90% deprotonated (and thus uncharged) at pH7.0. Thus a decline in activity from the maximum at pH5 to virtually zero at pH 7-7.5 indicates the involvement of a group with pKa around 6 and could mean that a protonated histidine is involved in catalysis.

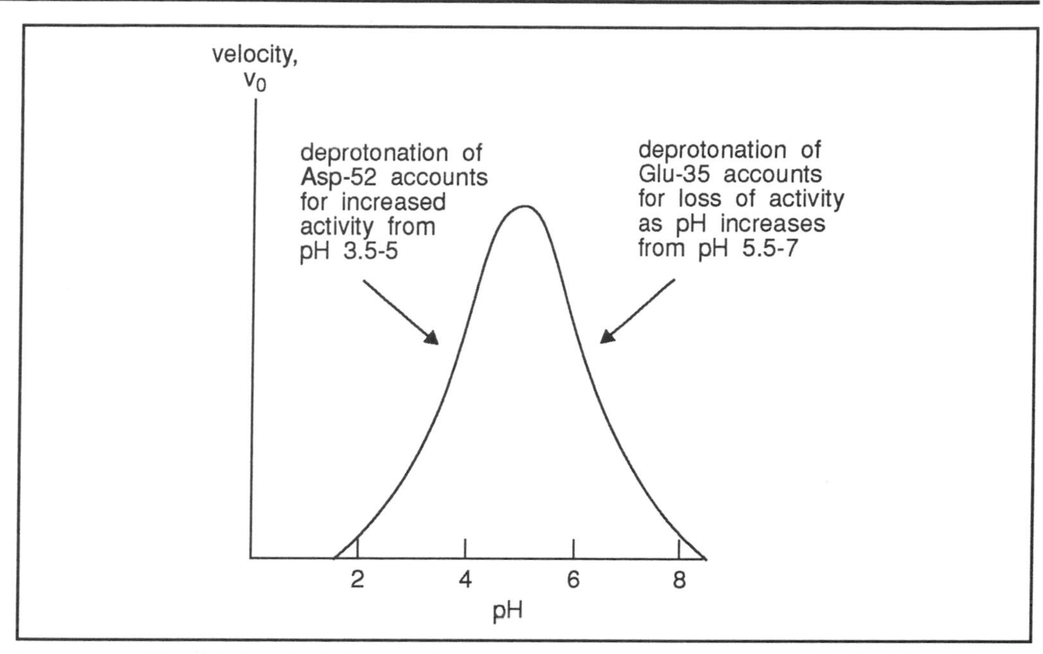

Figure 4.14 Effect of pH on the initial velocity of an enzyme-catalysed reaction. Frequently a distinct optimum pH is seen, although different enzymes have different pH optima. The data shown are for the activity of lysozyme. The increase in activity from pH 3.5 to 5.0 is attributable to deprotonation of the sidechain carboxyl group of Asp-52. The loss of activity from pH 5.5 to 7.0 is explained by deprotonation of the sidechain carboxyl group of Glu-35.

The bell-shaped activity curve we have shown is thus, in its simplest form, the consequence of changes in the ionic status of two crucial groups (see legend to Figure 4.14).

Not all enzymes display bell-shaped pH activity curves with distinct pH optima; some, such as papain (a plant protease) are active over a wide range of pH. This simply indicates that no crucial changes in ionic state occur over this pH range (ie changes may occur but they do not affect residues involved in substrate binding and/or catalysis).

change in conformation may lead to denaturation

The effects of pH referred to so far concern changes in ionic state having little effect on the overall conformation of the enzyme. At extreme pH values, protein denaturation will occur, with loss of enzyme activity. This is frequently irreversible and represents a major change in the three dimensional shape (conformation) of the enzyme. Substantial changes in charge occur, leading to unfolding, and ultimately exposure of previously hidden (internal) parts of the protein, aggregation and precipitation. For enzymes which act at roughly neutral pH, very low or very high pH values may be expected to cause denaturation and should be avoided. Because of the dependence of enzyme activity on pH and the need to prevent denaturation, enzyme-containing solutions are routinely buffered. Care must be taken to ensure that the buffer does not interfere with the enzyme under study. It is equally clear why it is important that intracellular pH should be strictly controlled - as indeed it is!

temperature and activation energy

Temperature affects the rate of most chemical reactions, so it is not surprising that the rate of enzyme-catalysed reactions increases with temperature. However, whilst increase in temperature accelerates the reaction (more substrate molecules possess the necessary activation energy (Section 4.2.1)), it can also speed up denaturation of the enzyme, such that with time fewer active enzyme molecules are present. One must be careful not to confuse these two effects. Examine the data shown in Figure 4.15, this shows how the amount of product formed increases with time at various temperatures.

At 30°C, the rate is low, at 40°C it is about double that seen at 30°C (this is sometimes expressed as the temperature quotient, Q_{10}, which is the increase in rate over a 10°C rise in temperature, in this case $Q_{10} = 2$). However, denaturation is beginning to occur.

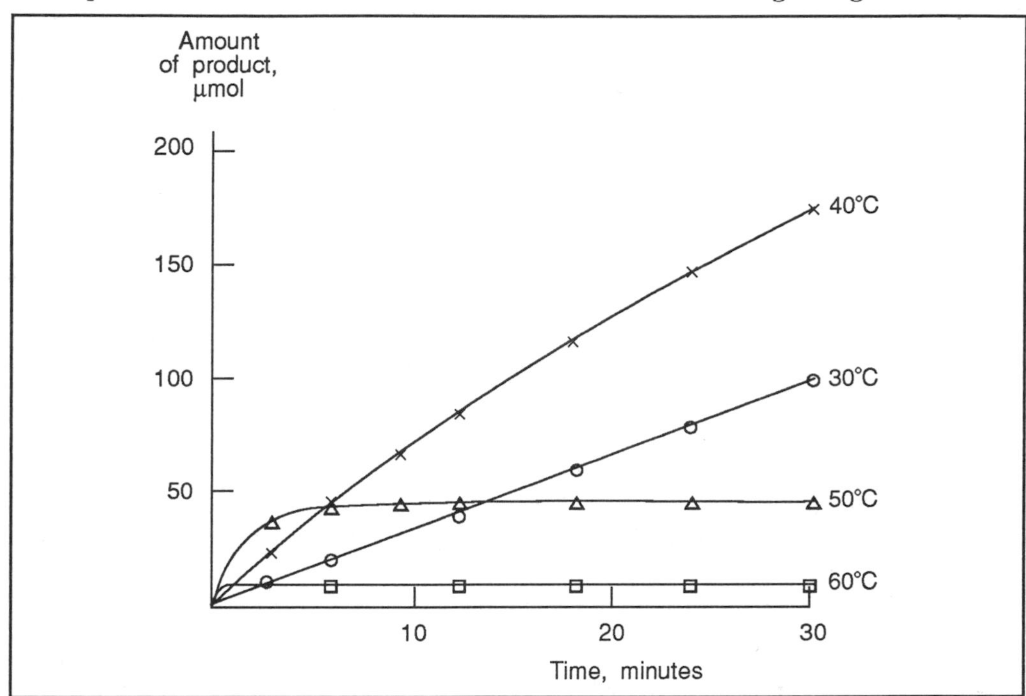

Figure 4.15 Effect of temperature on the amount of product formed in an enzyme-catalysed reaction.

Π What is our reason for saying that denaturation is beginning to occur at 40°C? Look carefully at Figure 4.15, and write the reason in the margin.

The rate of reaction is no longer linear with time at 40°C. The increase in amount of product, per minute, is less towards the end of the experiment than at the beginning, at 40°C. At higher temperatures, denaturation is even faster. There is apparently no denaturation at 30°C over the period of assay, as shown by the linear increase in product with time.

At 50°C the reaction is faster still, but complete denaturation has occurred within 10 minutes; at 60°C the reaction was extremely rapid -and extremely short-lived.

The purpose of this lengthy description is to emphasise the importance of measuring initial velocity when assaying enzymes and to establish the different effects of temperature on enzyme activity. At elevated temperatures, unless great care is taken, an enzyme assay may not measure the true initial rate but may be affected (dominated) by enzyme denaturation.

many agents can denature proteins

We have seen in this section that extremes of pH and temperature can cause denaturation of enzymes. Other agents which can cause denaturation of enzymes include organic solvents such as ethanol and acetone, trichloroacetic acid, detergents and high concentrations of urea. Whilst major changes in conformation will eventually occur, loss of enzyme activity is frequently one of the first indications that denaturation is occurring. In working with enzymes, one tries to avoid conditions which will cause denaturation.

SAQ 4.6

Carefully consider the 5 statements which follow, decide whether they are true or false and jot down any comments or qualifying remarks which you think are relevant.

1) Enzymes are only active at a particular temperature.

2) Change in enzyme activity with pH is caused by change in the charge of active site groups.

3) The only effect that increasing temperature has on enzymes is to increase their activity.

4) In assaying an enzyme, you do not need to worry about denaturation.

5) Raising the temperature means that more substrate molecules possess the necessary activation energy and can therefore react.

4.7 Allosteric enzymes and regulation of enzyme activity

The metabolic pathways and interconversions described in other chapters of this book rely on enzymes; the existence and operation of pathways is routinely totally dependant on the relevant enzymes. Thus the relative abundance and activity of each enzyme can have important consequences for metabolism. Regulation of enzyme activity is often described at two levels.

Coarse control

This operates at the level of gene expression and protein synthesis, and determines how many enzyme molecules are synthesised in a cell. This, together with the rate of degradation of each enzyme, clearly affects the enzymic capability of a cell, but is thought of as a fairly 'crude' and long-term control mechanism. Changes in gene expression and hence synthesis of proteins is often switched on (and off) rapidly in prokaryotes, for example, in response to changes in nutrients in the growth medium, where the presence of a nutrient may induce the synthesis of enzymes which enable the nutrient to be utilised. The great advantage of this is that the organism only synthesises particular enzymes when they are of benefit to it and avoids wasteful use of valuable resources and energy. However, eukaryotes do not routinely respond so rapidly. The alteration of the amount of each enzyme is inappropriate for second-by-second regulation of metabolic pathways. Coarse control, reflected in the presence (or absence) of particular enzymes is none-the-less important in developmental changes in organisms. It also, of course, produces the enzymatic complement upon which sophisticated and sensitive fine-tuning is conducted.

prokaryotes switch protein synthesis on and off rapidly

Fine control

Cells need to be able to adjust rapidly the relative activities of different pathways. This is done by a variety of mechanisms which regulate the activity of crucial enzymes, often in a progressive manner, by which is meant that enzyme activity can be finely adjusted, rather then abruptly turned on or off. This section addresses how this fine control is achieved.

fine control acts through enzyme activity not synthesis

4.7.1 Allosteric enzymes

sigmoidal plots of v_0 and [S]

Whilst many enzymes display a hyperbolic relationship when initial velocity is plotted against substrate concentration, and thus obey Michaelis Menten kinetics, others do not. Instead, a plot of initial velocity against [S] gives a sigmoidal curve (Figure 4.16).

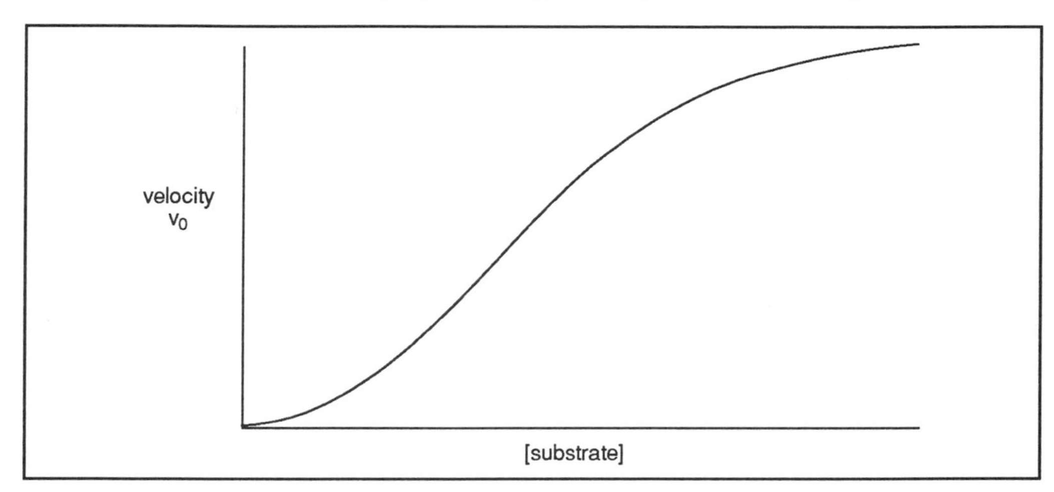

Figure 4.16 Initial velocity against substrate concentration for an allosteric enzyme.

co-operativity

Such enzymes usually consist of several subunits, although possession of multiple subunits is not a prerequisite for the sigmoidal relationship. What must occur, however, is communication between binding sites. This is termed a co-operative effect or co-operativity, and it is this which leads to the sigmoid curve. What co-operativity means is that the binding of one substrate molecule facilitates or enhances the binding of subsequent substrate molecules. Enzymes which display sigmoidal velocity against substrate concentration curves are called allosteric enzymes.

relaxed and tight conformations

How is co-operativity achieved? Let us assume that several subunits (often 4) occur, each of which has a binding site for substrate. Each subunit is believed to be able to adopt two alternative conformations, originally called R (for relaxed) and T (for tight, Figure 4.17 a)). These alternative conformations differ in their kinetic properties, particularly in terms of their affinity for substrate(s); the T form has a low affinity for the substrate, whereas the R form has high affinity. The sigmoid curve results if the enzyme normally exists in the T form, but can switch to the R form.

concerted symmetry of MWC model

Two models have been proposed to explain allosteric behaviour. The 'concerted symmetry' model proposed by Monod, Wyman and Changeux (MWC model) assumes that hybrid states (ie a mixture of R and T forms in one tetramer) do not occur (Figure 4.17 b)). Although the low-affinity T form predominates in the absence of substrate, binding of a substrate molecule to one subunit causes this subunit to switch to the high affinity R form. Alternatively, although the T form predominates in the absence of substrate, it is in equilibrium with the high-affinity R form, which therefore occurs, albeit transiently. When substrate binds, the enzyme is locked into the R form and the equilibrium will then shift towards the R form. The presence of one subunit locked into the R form by the bound substrate forces adjacent subunits to also adopt the high affinity R form. The R form has a higher affinity for substrate than the T form, it is now easier for other substrate molecules to bind. This results in a sigmoid velocity against substrate concentration plot. The degree of sigmoidicity depends on the relative affinities of the two forms of the enzyme for substrate and the equilibrium between T and R in the absence of substrate.

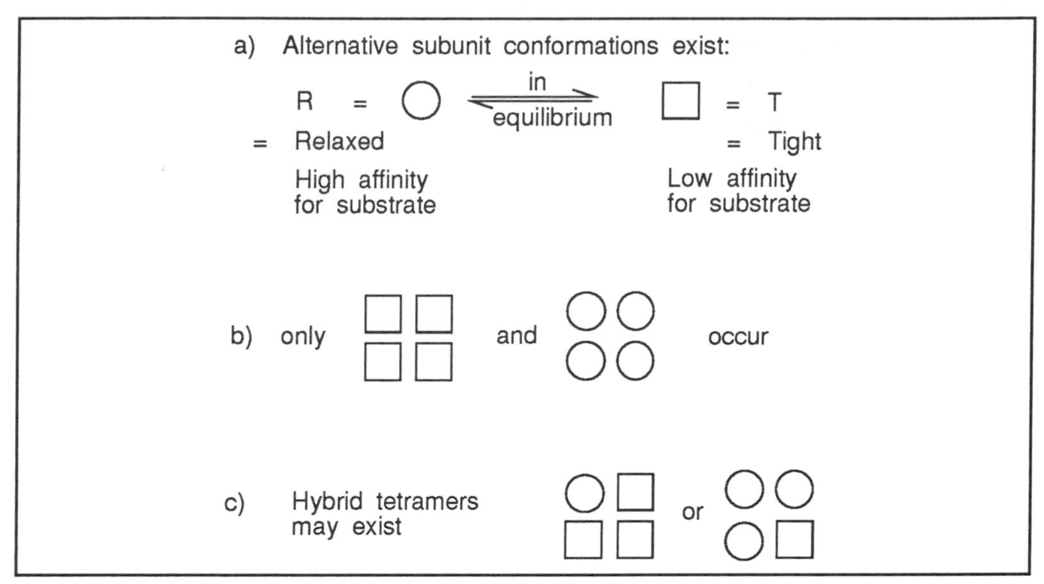

Figure 4.17 Models for allosteric enzymes a) Two subunit conformations can exist, R and T. b) In the MWC model, no hybrid states are permitted. c) The KNF model permits hybrid states. (See text for details)

KNF model

The second model, proposed by Koshland, Nemethy and Filmer (the KNF model), allows both conformations (R and T) to exist within a tetramer (Figure 4.17 c)). This makes for a more complicated, but also more general model, which can accommodate observations which the MWC model cannot (for example, in the MWC model, substrate binding leads to the high affinity form being adopted, it cannot explain cases (which exist) where initial substrate binding impedes additional binding).

significance of sigmoidal binding

What is the significance of sigmoidal binding? Let us examine the degree of saturation of an enzyme at various substrate concentrations (Figure 4.18). With Michaelis Menten kinetics, to increase the rate from 20% to 80% of V_{max} requires a concentration increase of 16 fold - you can verify this for yourself, using the Michaelis Menten equation! Inspection of Figure 4.18, which includes a 'typical' sigmoid curve, shows that the same activity increase only requires a 2-fold increase in substrate concentration. This means that modest increases in [S] cause more marked rate increases with allosteric enzymes; allosteric enzymes are thus more responsive to change in [S].

Π Jot down two ways by which an allosteric enzyme might be distinguished from an enzyme which conforms to Michaelis Menten kinetics.

1) For an allosteric enzyme, a plot of initial velocity against [S] is sigmoid whereas an enzyme which conforms to Michaelis Menten kinetics gives a hyperbolic plot.

2) As a consequence, allosteric enzymes do not give linear double reciprocal plots.

3) Note that the possession of several subunits is not a distinguishing factor for allosteric enzymes. Many enzymes possess more than one subunit, yet they still display Michaelis Menten kinetics.

sigmoidal binding of O_2 by haemoglobin

Sigmoidal binding is very useful in oxygen transport. Haemoglobin behaves like the sigmoid curve in Figure 4.18 (with the rate axis being % saturation with oxygen and the [S] axis being oxygen partial pressure). The oxygen partial pressure in the alveoli would correspond to the right hand side of Figure 4.18, so haemoglobin will be essentially fully oxygenated. After transport to muscle, the oxygen tension will be lower (around 4-5 on the [S] axis of Figure 4.18) and oxygen will therefore be released by haemoglobin - thereby being made available for muscle. Myoglobin, which stores oxygen in muscle and does not have 4 subunits, displays a hyperbolic binding curve (much like that shown in Figure 4.18 for the Michaelis Menten-type enzyme), and is thus able to bind the oxygen released by haemoglobin. Note that if single subunits of haemoglobin were analyzed, their oxygen binding would be hyperbolic and much less oxygen would be released.

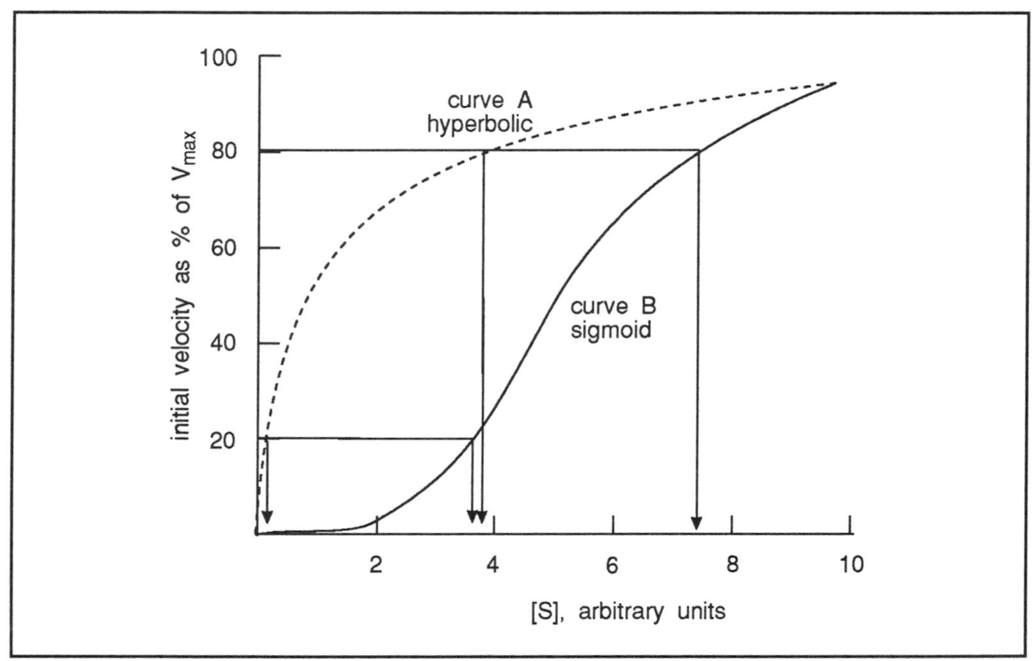

Figure 4.18 Influence of sigmoid and hyperbolic v against [S] relationship on the responsiveness of the enzyme to change in [S]. To increase from 20% to 80% of V_{max} requires a 16-fold increase in [S] with Michaelis-Menten kinetics (curve A), with the sigmoid plot depicted (curve B) for an allosteric enzyme, only a 2-fold increase in [S] is needed.

Π The three-dimensional structure of deoxy-haemoglobin was deduced by X-ray diffraction analysis of crystals. If small crystals of deoxy-haemoglobin were placed in buffer on a slide under a coverslip and were then placed in air, the crystals, particularly towards the edges of the coverslip, cracked and broke down. Suggest an explanation for this phenomenon.

As we have seen, haemoglobin becomes oxygenated in the lungs, and is transported to the peripheral tissues, where it releases oxygen. It thus exists in deoxy- and oxy- forms. Oxygenation leads to a change in protein conformation. When crystals of deoxy-haemoglobin are exposed to air, they bind oxygen and undergo a conformational change. This altered conformation is incompatible with the crystal structure of deoxy-haemoglobin, so the crystals break up.

A second aspect of allosteric enzymes is also important, especially for regulation. The term allostery literally means 'other site' and refers to the occurrence of binding sites other than substrate binding sites. With many enzymes, these bind compounds which alter or regulate the activity of the enzyme.

both allosteric activators and inhibitors occur

Allosteric regulation can include activation and inhibition, compounds which have such effects are then referred to as allosteric activators or allosteric inhibitors. Their effect is to shift the velocity against [S] plot as shown in Figure 4.19. Since the v against [S] plot can be changed by inhibitors and/or activators, and the curve is not hyperbolic, Michaelis Menten kinetics do not apply and K_M is not very meaningful. The sigmoid relationship means that a non-linear Lineweaver-Burk double reciprocal plot will be obtained. Rather then attempting to obtain a meaningless K_M value, an indication of the relationship between velocity and [S] under particular conditions is given by $[S_{0.5}]$, which is the [S] giving half maximal velocity.

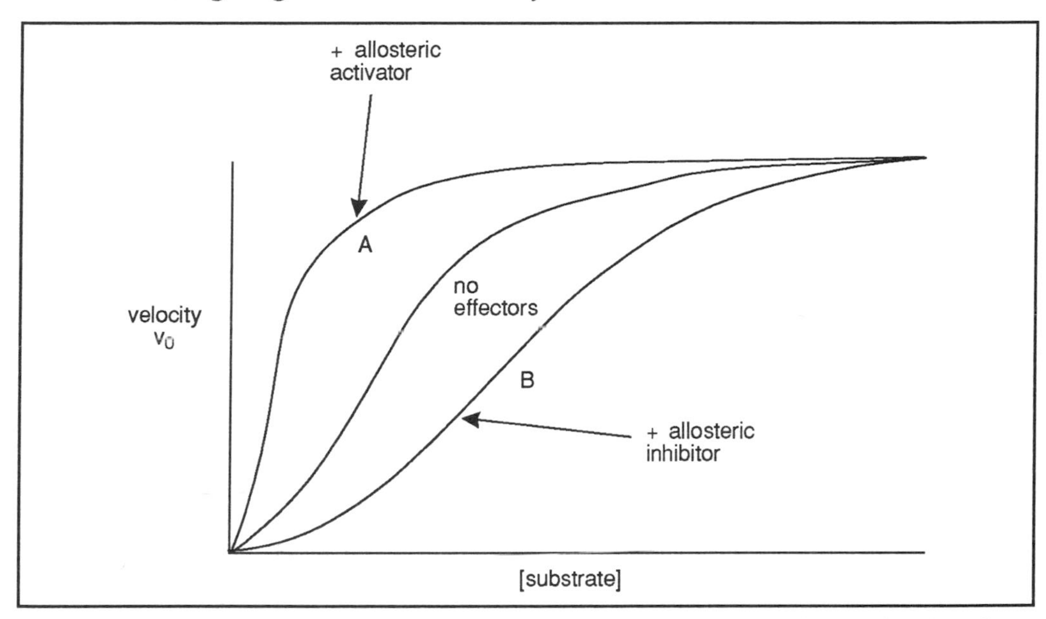

Figure 4.19 Effect of activators (A) and inhibitors (B) on the activity of an allosteric enzyme. An allosteric activator gives a higher rate at a given [S], an allosteric inhibitor lowers the activity at a given [S].

4.7.2 Feedback inhibition

the first reaction in a sequence is called the committing step

The benefit of allosteric regulation is that compounds other than substrate or product can regulate the activity of a crucial enzyme. Typical allosteric inhibitors are the products of a metabolic pathway acting to inhibit the first step (often called the committing step) of the pathway. When adequate levels of the end-product of the pathway (which is why the pathway exists!) are present, it switches off or lessens further synthesis (Figure 4.20). This is obviously beneficial to an organism, preventing an unnecessary build-up of compounds.

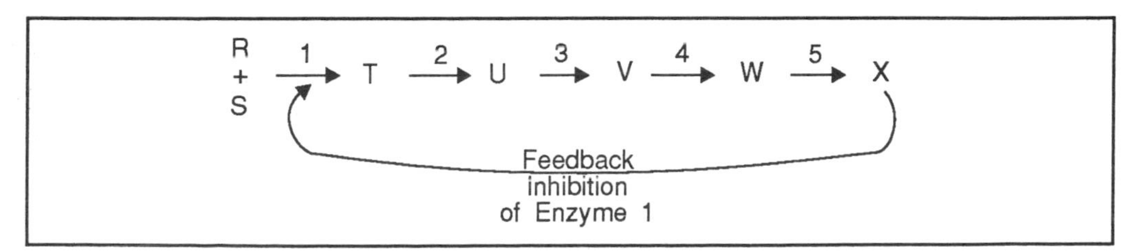

Figure 4.20 Feedback inhibition of the first enzyme of a pathway by the product of the pathway.

allosteric inhibitors lock an enzyme into the low affinity form

This type of regulation is widely used and is referred to as feedback inhibition. The end product binds to a site other than the substrate binding site (it would thus not compete with substrate for the enzyme); this is perhaps not surprising since, if the pathway is long, the end product (X in Figure 4.20) may bear little resemblance to the substrate of the first enzyme of the pathway (S in Figure 4.20). In terms of the models to explain allosteric behaviour (Figure 4.17), allosteric inhibitors are seen as 'locking' the enzyme into the low-affinity T form, thereby making substrate binding less likely.

allosteric activators lock an enzyme into the high affinity form

There may be cases where the end product of one pathway is needed to react with the product of a second pathway (Figure 4.21). What if there is an imbalance in the levels of these two compounds? One resolution of this is if the more abundant compound activates the synthesis of the compound which is deficient. This can be accomplished by an allosteric activator, acting on the first (and rate-limiting) enzyme of the less active pathway (Figure 4.21). The effect of this is to raise the rate at a given [S], as shown in Figure 4.19. This is achieved by binding to the enzyme and 'locking' it into the high-affinity R form. The type of feedback inhibition shown in Figure 4.21 occurs with aspartate transcarboxylase, the first enzyme in the biosynthetic pathway of pyrimidines.

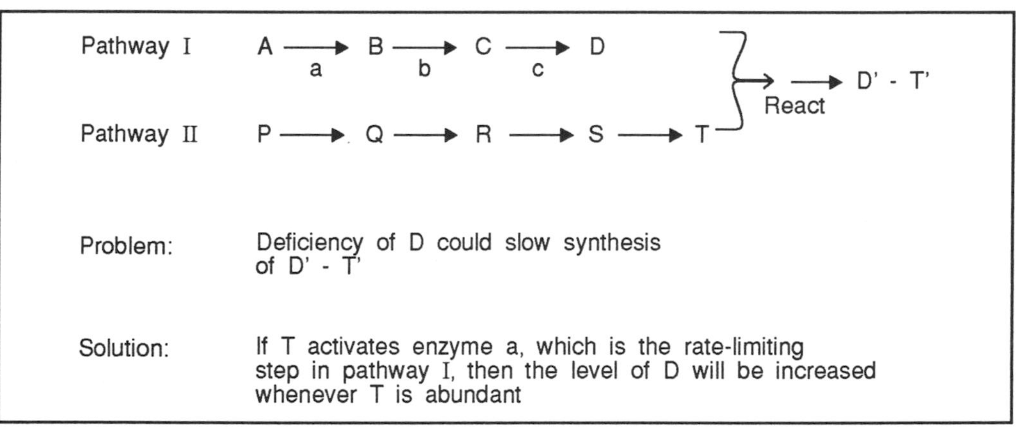

Figure 4.21 The activity of two pathways may need to be co-ordinated. Deficiency of the end-product of one pathway can be rectified by activation of the first enzyme of this pathway by the product of the second pathway.

SAQ 4.7

Consider the following statements. Write down, with brief reasons, whether they are true or false.

1) Allosteric enzymes are likely to be more rigid than enzymes which display Michaelis Menten kinetics.

2) Allosteric enzymes give linear double reciprocal (Lineweaver-Burk) plots, from which the Michaelis constant can be deduced.

3) Allosteric enzymes are ideally suited to a regulatory role.

4) A common feature of allosteric enzymes is possession of more than one subunit.

SAQ 4.8

Why do allosteric enzymes display a sigmoid rate against substrate concentration plot, whereas other enzymes display a hyperbolic relationship?

Choose appropriate responses from the following to explain this phenomenon.

1) Because they have sites for inhibitors to bind to, distinct from the substrate binding site.

2) Because they have more than one substrate binding site per active enzyme molecule.

3) Because they are tetramers of dissimilar subunits.

4) Because they have several subunits, which can exist in two forms of differing affinity; substrate binding influences the equilibrium between these two forms.

Feedback inhibition, involving allosteric regulation at a site other than the substrate binding site, is a widely used strategy in metabolism. It commonly acts at the first enzyme of a pathway. This allows a pathway to be regulated without having excessive concentrations of intermediates. This, of course, only works if the regulated enzyme catalyses the rate-limiting step. One simple way in which this can be visualised is via a pipe analogy, which we will draw in the next section.

4.7.3 Regulation of metabolic pathways

If you think of fluid flowing through a pipe, which has regions of varying diameters, it is the region of narrowest diameter which determines the flow rate (Figure 4.22). If the water pressure remains the same, then varying the diameter anywhere other than at point X will have minimal effect. If we look at a metabolic pathway, thinking of the enzyme capacity at each stage as the diameter of a pipe, then step X in Figure 4.22 is clearly the 'rate-limiting step'. Activation of this enzyme (widening of the pipe) will allow faster throughput, whereas activating any other step will affect the overall flux through the pathway to a more limited extent. We can thus summarise this area by stating that:

- in a metabolic pathway, one catalysed step will be the rate-limiting one. Activity will reflect the level of gene expression (ie how many enzyme molecules are present) and the efficiency of the enzyme as a catalyst;

- if one wishes to regulate the flux of material through the pathway, it is this step which would be expected to be regulated. We should, however, note that this model is a simplification. Actual pathways are often found to have more complicated and subtle control mechanisms than this;

- allosteric regulation (whether inhibition or activation) is a widely used method for achieving 'second by second' adjustments in enzyme activity. This is known as 'fine control' or 'fine tuning' and it is distinct from altering the rate of synthesis of the enzyme, which, as we have seen, is called 'coarse control'.

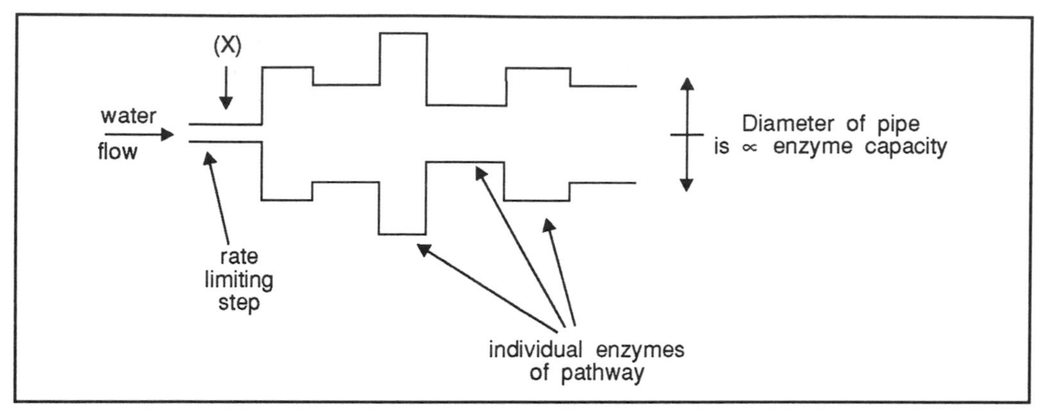

Figure 4.22 The pipe analogy. Pipe diameter represents enzyme capacity, the step with the smallest capacity will be rate-limiting. Affecting this step (either activation or inhibition) will have most effect on the overall throughput of the pathway. For an explanation of X, see text.

Π Imagine an unbranched pathway, involving 7 enzymes, which results in the formation of an amino acid as its end product, as shown:

$$A \xrightarrow[1]{enz} B \xrightarrow[2]{enz} C \xrightarrow[3]{enz} D \xrightarrow[4]{enz} E \xrightarrow[5]{enz} F \xrightarrow[6]{enz} G \xrightarrow[7]{enz} H$$

1) Where is this pathway likely to be regulated?

2) What mechanism could be used for regulation of the pathway that is particularly useful when compound H is abundant?

3) If a single enzyme is the main regulatory site, what properties is it likely to possess?

The answer to 1) is 'early on', typically at the first reaction, the step which commits material to the pathway, in this case, enzyme 1, converting A to B. The answer to 2) is that the compound H is likely to be markedly different in structure to compound A, yet it is preferable to regulate the pathway at its beginning (this prevents the build-up of high concentrations of intermediates B-G). Inhibition of enzyme 1 by H is thus attractive yet it is unlikely that H would bind to the active site of enzyme 1 because of its lack of resemblance to substrate A. Allosteric regulation, in which H binds elsewhere on enzyme 1 yet none-the-less causes inhibition, is likely to be the most effective mechanism. This is known as feedback inhibition. The answer to 3) is that as already noted, it is likely to be an allosteric enzyme, such that an abundance of H causes inhibition. Of the enzymes of the pathway, it is also likely to be present at lowest activity, so that it is the rate-limiting step - there is no point in enhancing the activity of reaction 1, if reaction 5 is the real bottleneck! Delays caused by a 2-lane section of a motorway will not be solved by making an existing 3-lane section into 4 lanes!

covalent modification of enzymes

Other methods of regulating enzymes exist. One approach is by covalent modification of the enzyme, frequently by phosphorylation. This is used in a reversible manner to convert an enzyme from a high-activity form to one of lower activity. One example of this is phosphorylase, which catalyses the release of glucose-1-phosphate from glycogen. This enzyme exists in two forms (Figure 4.23), one of low activity (known as phosphorylase b) and a second of higher activity (phosphorylase a). A specific enzyme (protein kinase) converts phosphorylase b to phosphorylase a by phosphorylating the hydroxyl side-chains of particular serine residues. The reverse process (conversion of phosphorylase a to b) is catalysed by a phosphatase which removes the phosphate groups. This is part of an elaborate cascade control mechanism by which adrenaline and insulin regulate glycogen metabolism.

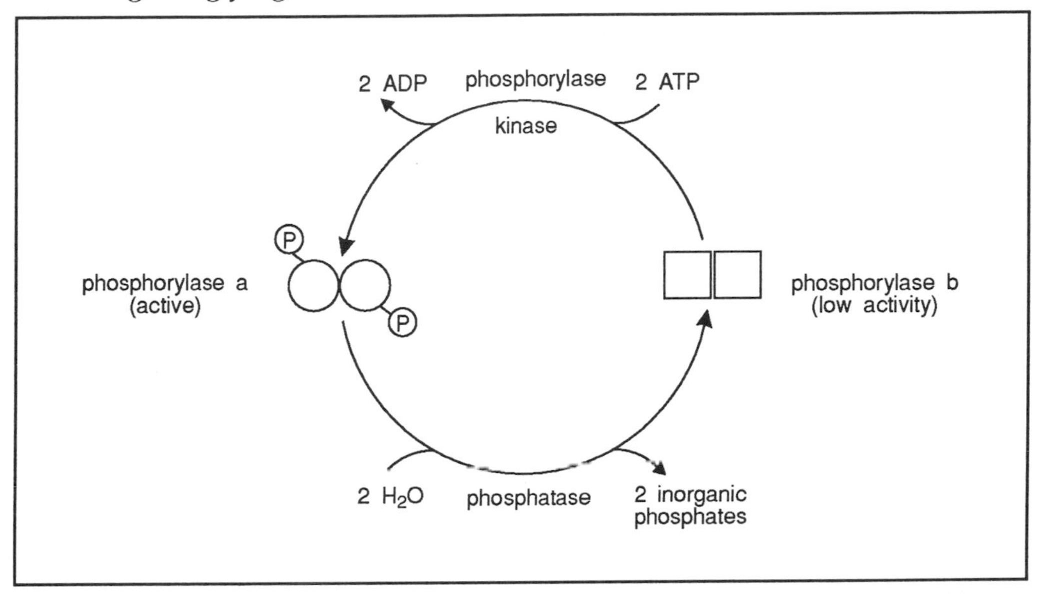

Figure 4.23 Enzyme activation by covalent modification. Phosphorylase is converted from a low activity form (phosphorylase b) to a high activity form (phosphorylase a) by phosphorylation.

4.7.4 Isoenzymes in metabolism and diagnosis

In the context of regulation of enzymes and metabolic pathways, the existence of isoenzymes (or isozymes for short) should be mentioned. These are different protein molecules within a cell or tissue showing the same enzymatic activity.

significance of isoenzymes

We have already discussed one case of isoenzymes when we reviewed the roles of hexokinase and glucokinase. If you think back to this, you will recall that the two enzymes catalyse the same reaction but have different kinetic properties. This allows each enzyme to be suited for different circumstances in the metabolism of a cell or organ. Isoenzymes are widespread and commonly arise through the occurrence of alternative subunits in an enzyme possessing a quaternary structure. Thus lactate dehydrogenase (LDH) in mammals consists of a tetramer, in which the subunits may be of two types (Figure 4.24). In this case the subunits are designated H (for heart type) or M (for muscle type); the H_4 tetramer is the heart-type form whilst M_4 is characteristic of skeletal muscle. Intermediate forms occur, 5 potential forms exist (on the basis of 4 subunits, of two alternative types) and different tissues display characteristic ratios of each of the 5 isoenzymes. The different subunits have different kinetic characteristics, which are reflected in the properties of the 5 isoenzymes. The particular isoenzyme complement of each tissue or cell is presumed to reflect the different metabolic circumstances or needs of different cells, these are discussed in greater detail elsewhere.

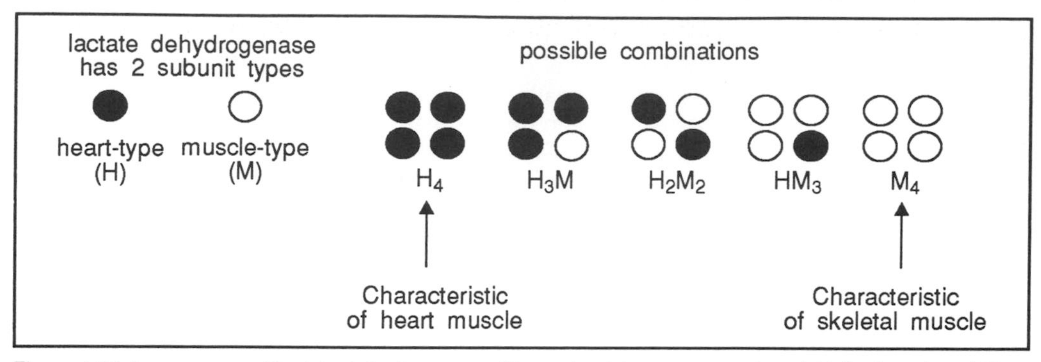

Figure 4.24 Isoenzymes of lactate dehydrogenase. Two subunit types are produced, H (for heart) and M (for muscle). These differ in their amino acid composition (and hence charge) and in their kinetic properties. Since the enzyme exists as a tetramer, 5 possible combinations exist. They may be separated by electrophoresis.

isoenzymes have diagnostic use

The occurrence of different isoenzymes in different tissues has lead to a diagnostic use of isoenzymes. As an example, we have noted that with mammalian lactate dehydrogenase (LDH), the H_4 isoenzyme (designated LD1) is characteristic of heart muscle, other tissues also have particular proportions of each isoenzyme. If damage occurs to a tissue, for example a heart attack (myocardial infarction), then the isoenzymes (and other enzymes) contained in that tissue may leak into the bloodstream. The serum pattern of particular isoenzymes is then a useful diagnostic aid for clinicians. In the case of a myocardial infarction, elevated levels of the LDH isoenzymes characteristic of heart muscle are observed.

SAQ 4.9

From the list which follows, select the most appropriate regulatory mechanism for the situations 1-4. Each mechanism should be used once!

1) There are two alternative fates for a compound, which are used to differing extents, depending on its concentration.

2) The end products of two different pathways are required in equal amounts.

3) A prokaryote is transferred to a different carbon source.

4) An enzyme which is involved in the release of utilisable carbohydrate from an energy store in a mammal. The store is used, when necessary, to maintain a steady glucose concentration in the bloodstream.

Mechanisms:

a) Allosteric regulation, involving activation and inhibition.

b) Phosphorylation and dephosphorylation.

c) Isoenzymes.

d) Induction of enzyme(s).

Summary and objectives

In this chapter we have examined the feature of enzymes. These are in effect the workers of cells which speed up the multitude of chemical reactions within cells. This function of catalysts is essential for metabolism to take place and understanding their activities and properties is vital to understanding metabolism.

We have examined how enzymes achieve their catalytic role, and the factors which influence their activities. We particularly focussed attention on the effects of inhibitors and allosteric regulators and the role of allosteric regulators of metabolic pathways.

Now that you have completed this chapter you should be able to:

- explain how enzymes achieve catalysis by reducing the activation energy required for the reaction to take place;
- explain why enzymes show specificity both in terms of the substrate they will interact with and in terms of the nature of the reactions they will catalyse;
- use data to calculate K_M values and to explain the meaning of K_M in terms of affinity of enzymes for their substrates;
- identify the types of inhibition from supplied data;
- describe how enzyme activity is influenced by chemical and physical factors;
- describe how enzyme activity may be regulated, especially by allostery, phosphorylation and through the occurrence of isoenzymes.

Coenzymes and cofactors

Coenzymes and cofactors

5.1 Introduction

cofactors are tightly bound

This chapter is about cofactors and coenzymes. These are non-proteinaceous compounds which are essential to enable some enzymes (not all) to function. Cofactors and coenzymes are distinguished in functional terms. Cofactors are usually thought of as being tightly bound to the enzyme and may be organic or inorganic. If organic, they may also be described as prosthetic groups. Metal ions which are tightly bound and essential for activity are also considered to be cofactors.

holoenzyme

Since cofactors are tightly bound (sometimes covalently) to the enzyme, they are normally present after isolation of the enzyme and are usually not removed by dialysis (see Chapter 4). Unless covalently bound, they are usually lost when protein denaturation occurs. Enzymes which require a cofactor for activity are sometimes called 'holoenzymes' when in their complete (active) form. Removal of the cofactor (perhaps by changing the pH) causes loss of activity.

apo-enzyme

The remaining, catalytically inactive, protein moiety is then known as an apo-enzyme (Figure 5.1). Activity of an apo-enzyme may often be restored by addition of the cofactor, this has for many years formed the basis for assay of several cofactors.

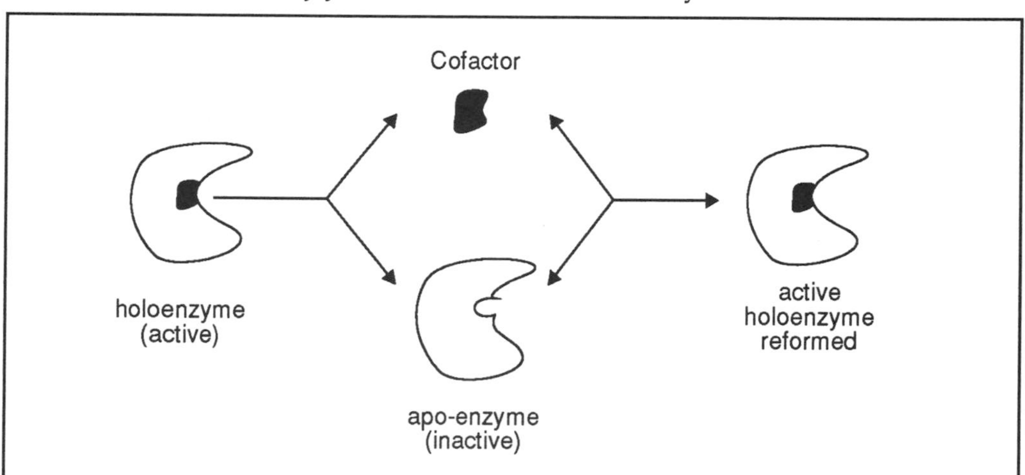

Figure 5.1 Removal of an essential cofactor converts active holoenzyme to an inactive form. The activity of this inactive apo-enzyme may be restored by binding of the cofactor, whereupon the holoenzyme is reformed.

co-enzymes are not tightly bound

Coenzymes, in contrast, are organic compounds which are easily removed from the enzyme, and frequently are only transiently bound to the enzyme in much the same way that substrates are. Some authorities view coenzymes as a type of cofactor, shown as follows:

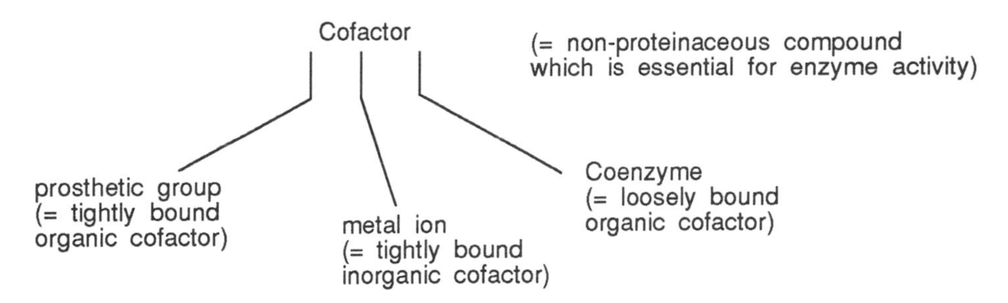

coenzymes present in small amounts

It really is irrelevant whether coenzymes are thought of as a type of cofactor or are categorised separately. We shall use the term cofactor for tightly-bound non-proteinaceous components of enzymes and coenzyme for easily dissociable compounds.

As we have noted, coenzymes readily dissociate from an enzyme. Many, such as the nicotinamide adenine dinucleotide (NAD^+ and its reduced form NADH, also the structurally similar $NADP^+$ and NADPH) used by dehydrogenases, may be viewed as though they are a second substrate of the enzyme. The reason that they are called coenzymes is partly historical, in that they were originally detected as thermally stable compounds which restored enzyme activity when added back to dialysed enzyme solutions (hence 'co'-enzyme). Coenzyme is still a useful term because such compounds (eg NAD^+) act in the same way in many reactions. In the case of NAD^+, it accepts reducing equivalents from a variety of compounds, their identity depends on the particular enzyme (for example lactate with lactate dehydrogenase, malate with malate dehydrogenase, etc). Unlike the other individual products of the various enzymes (pyruvate for lactate dehydrogenase and oxaloacetate with malate dehydrogenase), which are further metabolised, the coenzymes are converted back to their starting form. There is routinely only a small amount of each coenzyme in the cell.

If you glance through this chapter you are bound to notice that there are a large number of diverse chemical structures. Do not worry - you do not have to remember them in detail. What is important is that you gain an appreciation, in general terms, of their constituents, what they are derived from (often vitamins) and their roles ie. what sort of reactions they are involved in. You will then appreciate their importance more fully when you encounter them in later chapters on particular aspects of metabolism. The in text activities and self-assessment questions will provide guidance regarding what you should learn from this chapter.

5.2 Vitamins

5.2.1 General aspects

many vitamins are cofactors or coenzymes

Vitamins are organic compounds which many animals require in small amounts in their diets in order to remain healthy. They are needed because the animals fails to synthesise adequate quantities itself. This is usually because the enzymic pathway necessary for synthesis of the vitamin has been lost. In many cases, the requirement for the vitamin is because it, or a derivative, is a cofactor or a coenzyme. We shall concentrate on those vitamins which are needed as part of a cofactor or coenzyme, although other vitamins will be mentioned. Vitamin deficiency often leads to characteristic symptoms, which in some cases can be traced to reduction of a particular enzyme activity. Although a cell may synthesise the polypeptide part of an enzyme in adequate amounts, if it requires a cofactor for activity, then the polypeptide part will remain non-functional.

water soluble and fat soluble vitamins

Vitamins are normally categorised as either water-soluble or fat-soluble. We shall summarise them according to these categories. As you will see, the water-soluble vitamins are frequently required to fulfil a role as a cofactor or coenzyme. Fat-soluble vitamins are not usually used as cofactors or coenzymes, their mode of action, where known, will only be briefly referred to.

5.2.2 Water-soluble vitamins

Table 5.1 lists the most important water-soluble vitamins, the predominant forms in which they are used and the main processes that they are involved in.

This is quite a large table so take your time in reading through it carefully. Some of the names you may already be familiar with although you may not as yet be familiar with their biochemical functions. By the time you have completed your studies of metabolism you will have learnt a lot about the roles of many of these vitamins in metabolism.

results of vitamin deficiency

Water-soluble vitamins which are required for cofactors or coenzymes are indicated in Table 5.1 and these will be discussed further (section 5.3). In some cases, vitamin deficiency leads to particular, well-defined effects (for example, vitamin C deficiency results in scurvy, thiamine deficiency leads to beri-beri, niacin deficiency causes pellagra). In other cases, effects of deficiency are less clearly defined or identified (for example, riboflavin and biotin). In some cases, such as folic acid, it is extremely difficult to study vitamin deficiency, since the bacteria present in the gut produce adequate quantities. Studies on other animals may, or may not, be of any relevance to human nutrition and biochemistry and thus extrapolation of results gained with laboratory animals to humans must be approached with considerable caution. Effects of vitamin deficiency are described in more detail in texts on nutritions.

The one water-soluble vitamins which will not be discussed in section 5.3 is vitamin C, ascorbic acid.

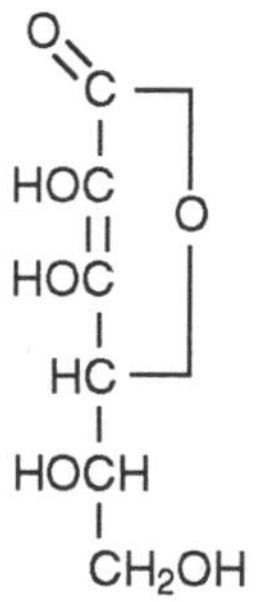

```
O
 \\
   C ──┐
   |   |
 HOC   O
   ||  |
 HOC   |
   |   |
  HC ──┘
   |
 HOCH
   |
   CH2OH
```

Amongst animals, only primates and guinea pigs require this vitamin, having lost an enzyme required for its synthesis. Fresh vegetables and fruit, especially citrus fruits, are good sources of vitamin C.

vitamin	common name	required for a cofactor or coenzyme?	form in which used	typical roles
one of the B_2 group	niacin	yes	nicotinamide adenine (NAD^+) nicotinamide adenine dinucleotide phosphate ($NADP^+$)	oxidation and reduction reactions
one of the B_2 group	riboflavin	yes	flavin adenine dinucleotide (FAD) flavin mononucleotide (FMN)	oxidation and reduction reactions
H	biotin	yes	biotin	carboxylation reactions
B_6	pyridoxal and related compounds	yes	pyridoxal 5'-phosphate (PLP)	various reactions of amino acids
B_1	thiamine	yes	thiamine pyrophosphate (TPP)	various reactions including decarboxylations and keto group transfer
one of the B_2 group	folic acid	yes	tetrahydrofolate (THF)	1-carbon transfers
B_{12}	cyanocobalamin	yes	coenzyme B_{12}	
one of the B_2 group	pantothenic acid	Yes	coenzyme A	transfer of groups eg acetyl groups
C	ascorbic acid	no		hydroxylations reactions probably also a water-soluble antioxidant

Table 5.1 Water-soluble vitamins.

Vitamin C deficiency causes abnormal collagen

Deficiency causes scurvy. Vitamin C is a strong reducing agent and is implicated in hydroxylation reactions in cells. One consequence of deficiency is abnormal collagen, vitamin C being required for conversion of proline residues in collagen to hydroxyproline. This occurs after incorporation of proline into the collagen polypeptide.

5.2.3 Fat-soluble vitamins

Hydrophobic vitamins are fat-soluble

The vitamins we have just identified are structurally very diverse but they share the property of being water-soluble and are thus found in the aqueous parts of cells and tissues. The reason that they are water-soluble is because they either possess charged groups or those displaying partial charge separation ie. polar groups. Compounds which do not possess such charge separations tend to be hydrophobic (literally, water-hating). They display very limited solubility in water but tend to 'dissolve' in hydrophobic lipid materials. By this we mean that they will be concentrated in, or partitioned into, lipid materials (membranes, storage lipid depots containing for example, neutral triglyceride). Such vitamins are called fat-soluble vitamins and are obtained from fats and oils in the diet. The most important ones are summarised in Table 5.2.

vitamin	common name	comments
A	retinol	derived from carotene. Required as retinol and retinal in the retina of eyes. Deficiency leads (initially) to night blindness.
D	calciferol and related compounds	various compounds may be converted to vitamin D. Involved in regulation of cellular levels of calcium binding protein. Deficiency causes rickets, which is characterised by bone brittleness.
E	tocopherols	these are isoprenoids, ie derived from isoprene, and are probably involved in the maintenance of functional mitochondria membranes acting as anti-oxidants.
K	menaquinone and related compounds	important in blood-clotting. Deficiency of vitamin K leads to a reduction in clotting capacity of blood.

Table 5.2 Fat-soluble vitamins.

fat-soluble vitamins do not act as cofactors or coenzymes

These vitamins do not, as far as we know, have roles as cofactors or coenzymes. As a result, we shall not discuss them further, although brief comments on their roles are given in Table 5.2

Π Use the material in the preceding sections to identify: 1) a vitamin needed as a coenzyme in redox (ie reduction- oxidation) reactions; 2) a vitamin commonly used as a cofactor in enzymes involved in amino acid metabolism; 3) a coenzyme involved in transfer of 1-carbon units, also identifying the vitamin from which it is derived.

The answer to 1) is niacin. Three vitamins are mentioned in Table 5.1 as being involved in redox reactions. Vitamin C was described as a strong reducing agent. However, it is not implicated as a coenzyme, although following loss of reducing equivalents (as a result of which dehydroascorbate is formed), ascorbic acid may be regenerated by other reductants, such as glutathione. This leaves niacin (used in NAD^+ and $NADP^+$ and their reduced forms) and riboflavin (present in FAD and FMN) which are both involved in redox reactions. The question specified use as a coenzyme. NAD^+/NADP have (in the text) been identified as coenzymes and thus niacin is the correct answer. From the information given, so far, you could not definitely eliminate riboflavin (ie. it might be used as either a cofactor or a coenzyme). In fact, FAD and FMN are usually tightly bound and should be regarded as cofactors.

The correct answer to 2) is pyridoxal (vitamin B_6). As will be described in Section 5.3.4, the cofactor derived from this vitamin (pyridoxal 5′-phosphate) is required by many enzymes involved in amino acid metabolism, including transaminases (amino transferase) and decarboxylases.

The answer to 3) is the, coenzyme tetrahydrofolate (THF), which is derived from folic acid. This coenzyme is widely used to transfer 1-carbon units, such as formate and formaldehyde. Other coenzymes may be involved in the transfer of larger units, for example, acetyl by coenzyme A.

SAQ 5.1

Which of the following apply to water-soluble and which to fat-soluble vitamins?

1) are hydrophobic molecules;
2) are hydrophilic molecules;
3) many act as coenzymes and cofactors;
4) are found in aqueous parts of the cell;
5) are often found associated with membranes.

5.3 Structure and function of vitamins and the cofactors and coenzymes derived from them.

In this section we will briefly describe the structures of the most important coenzymes and cofactors, together with the vitamins from which they are derived. We have not attempted to be encyclopedic, indeed, many alternative forms of some vitamins and many details of the roles of cofactors and coenzymes have deliberately been omitted. The purpose of this section is to give some idea of the structures and roles of these essential compounds of metabolism. Thus, when they are referred to in subsequent chapters on metabolism, you should have a general understanding of their structures and reactions.

5.3.1 Nicotinamide adenine dinucleotides

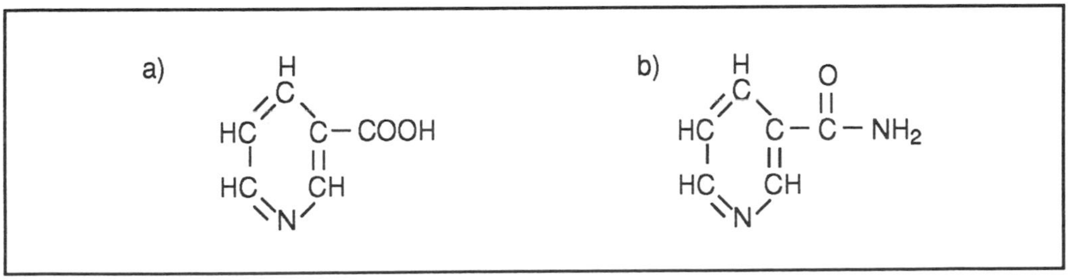

a)
```
        H
        C
HC          C - COOH
|           ||
HC          CH
        N
```

b)
```
        H        O
        C        ||
HC          C  - C - NH2
|           ||
HC          CH
        N
```

Figure 5.2 The structure of a) niacin, (also known as nicotinic acid) and b) nicotinamide.

Niacin (Figure 5.2, also known as a nicotinic acid) is part of the vitamin B_2 group. It is required for synthesis of nicotinamide adenine dinucleotide (NAD^+, Figure 5.3 a)) and the closely related nicotinamide adenine dinucleotide, phosphate, $NADP^+$. NAD^+ and $NADP^+$ are dinucleotides (see Figure 5.3 for structure).

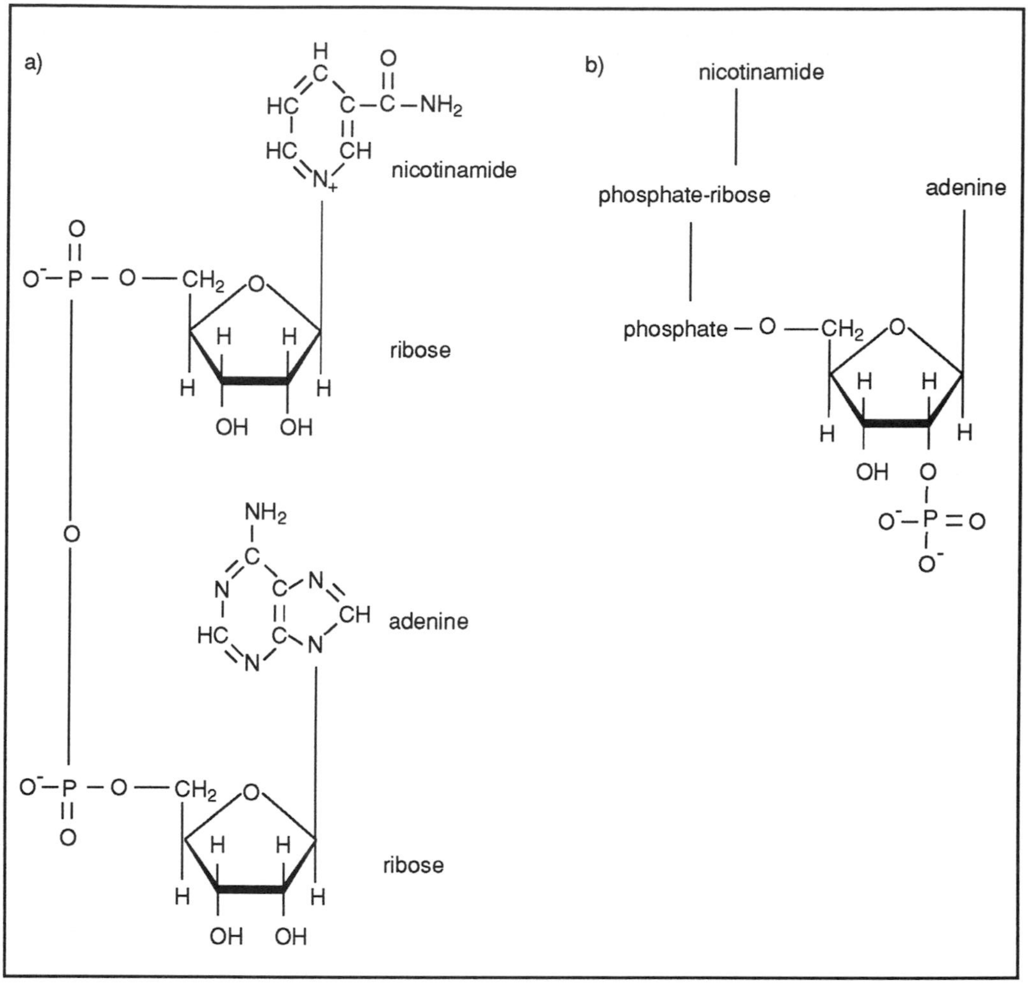

Figure 5.3 a) The structure of nicotinamide adenine dinucleotide NAD^+. b) $NADP^+$ differs only from NAD^+ in the additional phosphate group attached to the adenosine group.

A nucleotide consists of a nitrogenous base, such as adenine or cytosine, linked to a 5-carbon sugar (for example, ribose) to which are also attached one or more phosphates. This can be summarised as:

base $\rightarrow$ ribose $\rightarrow$ phosphate

NAD^+ can therefore be thought of as a dinucleotide, since it consists of two nucleotides linked through the phosphate groups, ie:

base (adenine) —— ribose —— phosphate
|
base (nicotinamide) —— ribose —— phosphate

Note that in these coenzymes the nicotinic acid (niacin) moiety is present as an amide (nicotinamide). The only difference between NAD^+ and $NADP^+$ is in the additional phosphate group attached to the ribose of the adenosine group of $NADP^+$ (Figure 5.4 b)); the significance of this will be considered shortly.

NAD^+ + $NADP^+$ involved in redox reactions

The importance of NAD^+ and $NADP^+$ lies in the reversible reductions which may occur at the nicotinamide group. This is summarised in Figure 5.4. Two reducing equivalents are involved, although the NAD^+ molecule only binds one proton and two electrons (ie $H^+ + 2e^- = H^-$, which is known as a hydride ion). Thus the overall reaction, as far as the coenzyme is concerned, is:

$$\underset{(\text{or } NADP^+)}{NAD^+} + H^- \rightleftharpoons \underset{(NADPH)}{NADH}$$

This should be more clear through consideration of an actual enzymic reaction. A typical redox reaction involving NAD^+ and NADH is that catalysed by lactate dehydrogenase. The reaction is as follows:

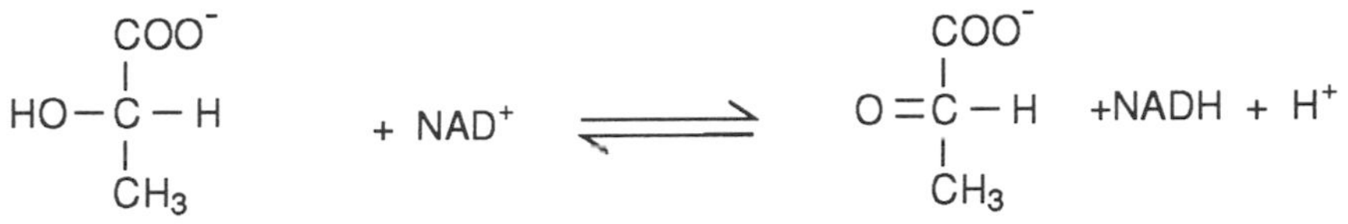

$$\underset{\text{L-lactate}}{HO-\overset{\overset{COO^-}{|}}{\underset{\underset{CH_3}{|}}{C}}-H} + NAD^+ \rightleftharpoons \underset{\text{pyruvate}}{O=\overset{\overset{COO^-}{|}}{\underset{\underset{CH_3}{|}}{C}}-H} + NADH + H^+$$

The substrate which acts as the reducing agent (L-lactate) loses two reducing equivalents (2H), of which NAD^+ accepts a hydride ion (H^-), with a free proton being left over. The reaction is thus balanced, stoichiometrically. Note that this reaction is potentially reversible. Thus reactions involving NAD^+/NADH may primarily involve NAD^+ reduction (for example, the malate dehydrogenase reactions, as part of the TCA cycle) or operate in either direction, depending on the metabolic circumstances (as in the case of the reaction shown above involving lactate dehydrogenase).

H
O
C – NH2
N+
ribose
etc
NAD(P)+
+ H-
H H
O
C – NH2
N
ribose
etc
NAD(P)H

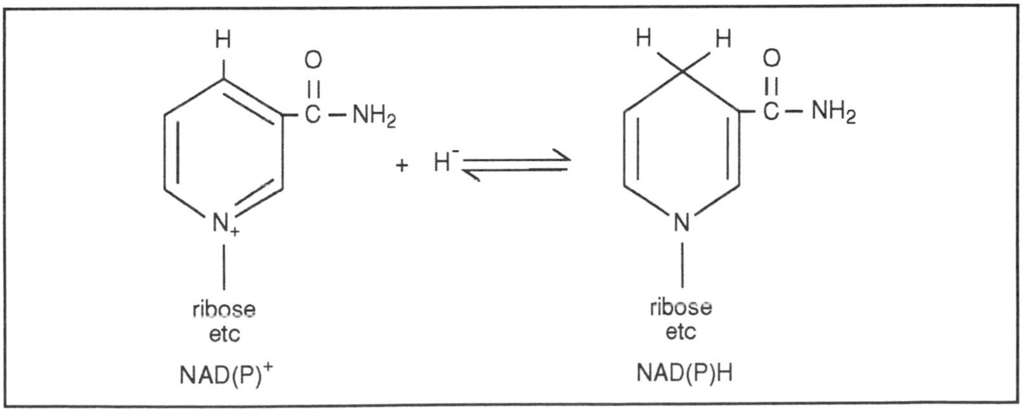

Figure 5.4 Reductions of NAD^+ and $NADP^+$ occur at the nicotinamide group. The oxidised nicotinamide ring accepts a hydride (H^-) ions, equivalent to a proton (H^+) and two electrons. For simplicity we have not labelled all the carbons of the nicotinamide rings.

pH affects redox reactions

Note also that, depending on the direction of the reaction, a proton is produced or consumed. pH will thus influence the equilibrium position of the reaction (as well as influencing the activity of the enzyme, but this is a separate issue, see Chapter 4).

Oxidation-reduction reactions involving $NAD^+/NADH$ are very common, you will encounter several reactions of this type in the central pathways of metabolism such as glycolysis and the TCA cycle.

We have already identified the structural difference between NAD^+ and $NADP^+$; let us now examine the reason for the occurrence of the two forms. $NADP^+$ can be reduced, in a similar manner to NAD^+. A typical reaction involving $NADP^+$ as reducing equivalent acceptor is that catalysed by glucose-6-phosphate dehydrogenase:

$$\text{glucose–6–phosphate } NADP^+ \rightleftarrows \text{6–phosphogluconate} + NADPH + H^+$$

This reaction represents the first step in the pentose phosphate pathway. The reaction on the nicotinamide ring is identical to that with NAD^+ - dependent dehydrogenases, yet enzymes are usually found to be specific for either NAD^+ or $NADP^+$; glucose-6-phosphate dehydrogenase will only reduce $NADP^+$. What is the explanation for the specificity?

We have already seen (Chapter 4) that enzymes are routinely capable of distinguishing between closely related compounds. Thus an enzyme acting on $NAD^+/NADH$, or on $NADP^+/NADPH$, is readily explained in terms of the precise shape of the coenzyme binding site of the enzymes. This does not address why such specificity occurs. The reason is as follows:

NAD^+ and $NADP^+$ form independent pools

NADH and NADPH are used by cells for different purposes and therefore cells maintain separate pools of $NAD^+/NADH$ and $NADP^+/NADPH$. This enables these redox agents to be, in principle, maintained in different ratios (ie. what proportion of the total NAD^+ + NADH is present as NAD^+), and it also facilitates the simultaneous operation of different processes.

NADH used in catabolic reactions

NADH is primarily an intermediate in energy generation (the formation of ATP) in which relatively reduced nutrients (fatty acids, sugars) donate reducing equivalents, during their metabolism by the main pathways of intermediary metabolism to NAD^+. The resulting NADH supplies reducing equivalents to the respiratory chain in mitochondria. The overall outcome of this is that ATP is synthesised (and water is formed, $2H + \frac{1}{2}O_2 \rightarrow H_2O$). This is described in detail in Chapter 8 but may be summarised as:

$$NADH + H^+ + 3ADP + 3Pi + \tfrac{1}{2}O_2 \longrightarrow NAD^+ + 3ATP + H_2O$$

Reactions involving $NAD^+/NADH$ may, as a generalisation, be described as catabolic, where the principal objective is the generation of useable cellular energy (ATP). Thus NAD^+ is the coenzyme which is used in the catabolism of monosaccharides and fatty acids, and is employed in reactions within mitochondria.

$NADP^+$ used in anabolic reactions

In contrast, NADPH is not routinely used to generate ATP and it does not donate reducing equivalents to the electron transport chain. Instead, it is used for reduction reactions during biosynthesis of cellular constituents, such as purines, pyrimidines and fats. Such reductive biosynthesis routinely use NADPH to provide the reducing equivalents. Examples are shown in Figure 5.5.

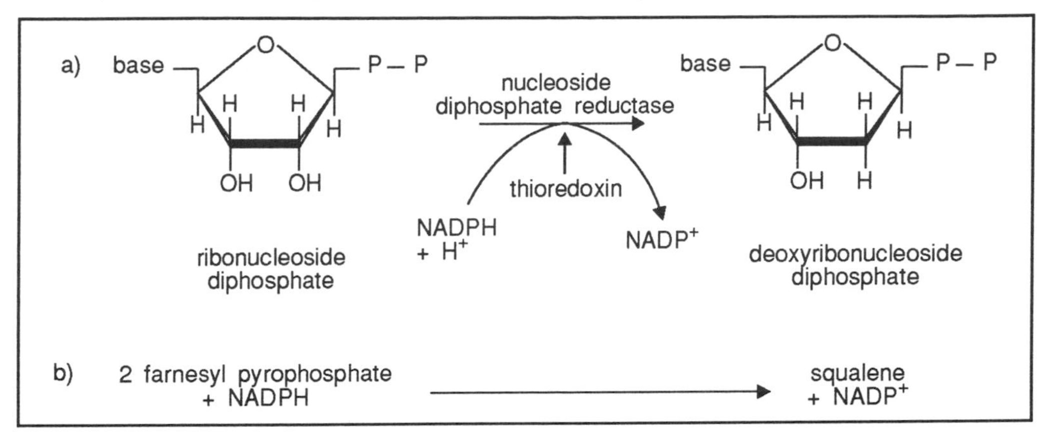

Figure 5.5 Use of NADPH is reductive biosynthesis. a) conversion of ribonucleoside diphosphates to deoxyribonucleoside diphosphate; b) conversion of farnesyl pyrophosphate to squalene.

The first is the formation of deoxyribonucleotides (required for DNA synthesis) from ribonucleotides; the second example is a preliminary reaction in the synthesis of complex fats, such as steroids. One of the roles of the pentose phosphate pathway (Chapter 6) is to produce NADPH required for reductive biosynthesis. You should note that the use of NAD^+/NADH for catabolic reactions and that of $NADP^+$/NADPH in anabolic pathways is a striking feature of metabolism.

Π Consider the following reactions, and the circumstances in which they occur, and decide whether NAD^+/NADH or $NADP^+$/NADPH would be used as coenzyme.

1) in mitochondria: isocitrate is oxidised by isocitrate dehydrogenase, as part of the TCA cycle;

2) in the conversion of acetyl CoA to fatty acids (ie biosynthesis of fatty acids), an acetoacetyl moiety (attached to an acyl carrier protein, ACP) is reduced to a β-hydroxybutyryl group.

Our responses are:

1) is part of the catabolic pathway by which carbohydrates and fats are used to generate useable energy in the form of ATP, here NAD^+ is the acceptor for reducing equivalents;

2) is a reductive biosynthesis which occurs in the cytoplasm. It is typical of such reactions in that it utilises NADPH to provide the reducing equivalents.

We have only discussed examples of reactions where NAD^+/NADH and $NADP^+$/NADPH are used as coenzymes. They are also used in other ways, notably to reduce the flavin-containing cofactors of respiratory chain components and to reduce glutathione reductase. NAD^+ is also involved in catalysis by some DNA ligases, which are important in genetic engineering.

5.3.2 Flavin-containing cofactors

FAD + FMN

Riboflavin (vitamin B_2) is required for synthesis of two cofactors used in redox reactions, flavin adenine dinucleotide (FAD) and flavin mononucleotide, FMN. Proteins containing these cofactors are known as flavoproteins. These cofactors are tightly bound to the protein, so that they are not lost during protein purification: drastic treatment is required to remove them, usually causing protein denaturation. Riboflavin is produced by plants and many micro-organisms and is obtained in the diet from vegetables, as well as animal tissues which are rich in the cofactors (eg liver).

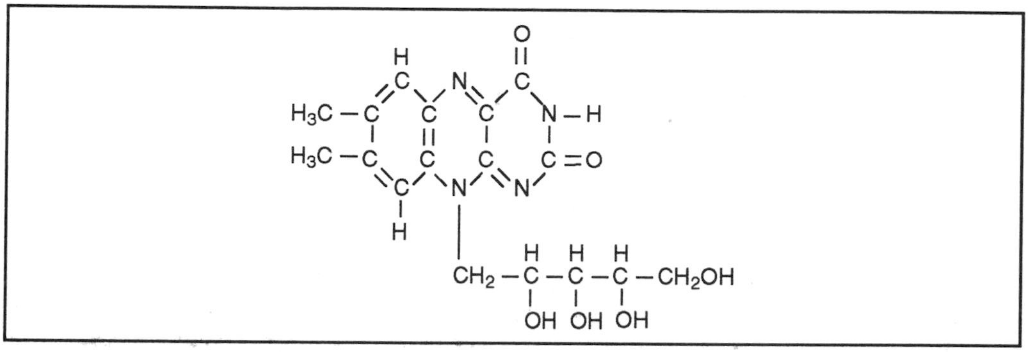

Figure 5.6 The structure of riboflavin.

Riboflavin (Figure 5.6) consists of an isoalloxazine ring attached to D-ribitol. If this is phosphorylated on the ribitol, flavin mononucleotide (Figure 5.7 a)) is formed. The association of a nitrogen containing ring structure with a sugar derivative (ribitol is a sugar alcohol) and a phosphate leads to the cofactor being considered to be a nucleotide, hence FMN.

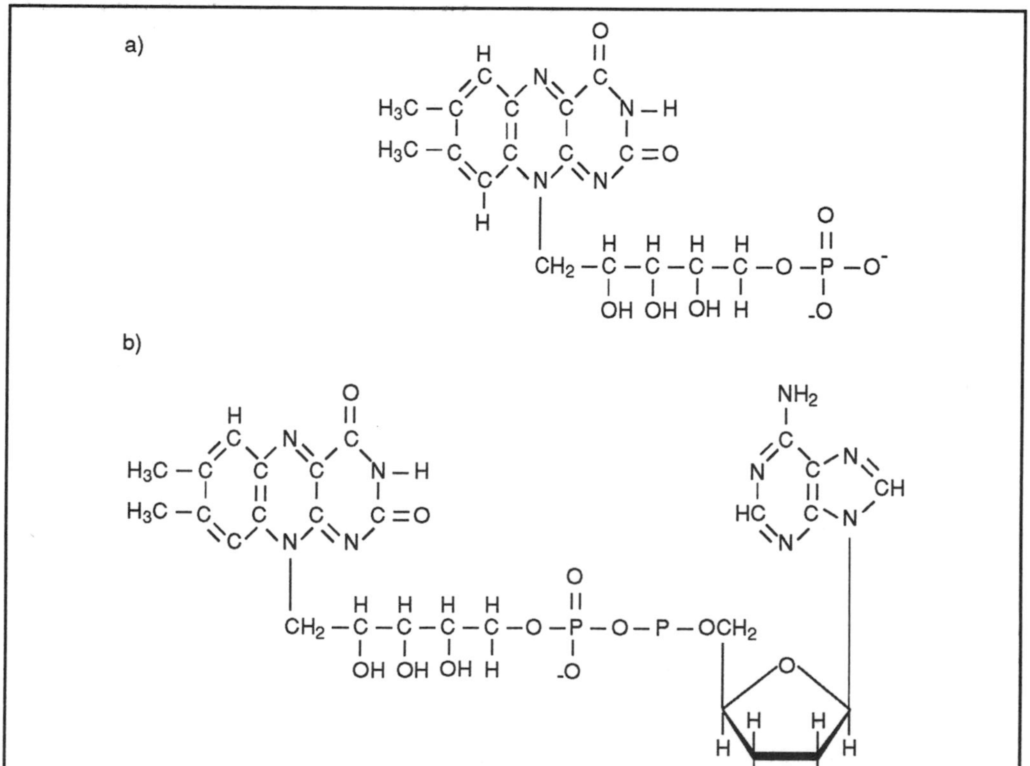

Figure 5.7 a) Flavin mononucleotide (FMN); b) Flavin adenine dinucleotide (FAD).

Flavin adenine dinucleotide (FAD, Figure 5.7 b)) consists of FMN linked to adenosine monophosphate (AMP) through the phosphate group, in a similar manner to that in $NAD(P)^+$.

yellow vitamins

In their oxidised forms, flavin nucleotides are yellow: thus enzymes containing them (for example, glucose oxidase, and the first flavoprotein to be isolated, NADH dehydrogenase, known as 'old yellow enzyme') are yellow. As we have noted, they are involved in redox reactions, in which two reducing equivalents (2H) are taken up by the isoalloxazine ring, as shown in Figure 5.8.

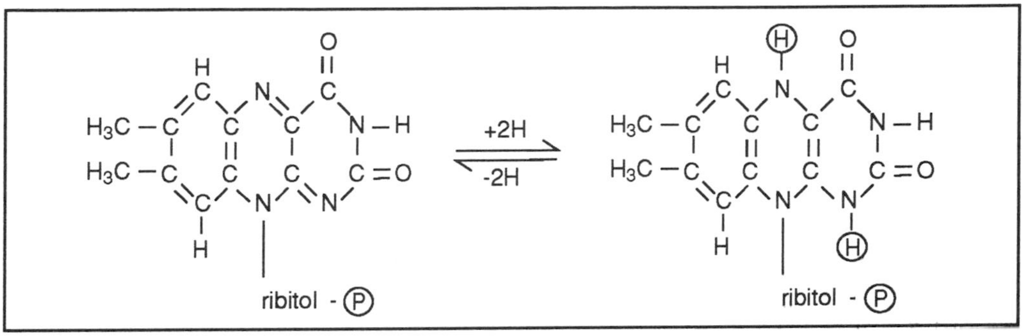

+2H ⇌ -2H

ribitol - Ⓟ

ribitol - Ⓟ

Figure 5.8 Reduction of FMN and FAD occurs within the isoalloxazine ring and involves binding of two reducing equivalents (2H). The strong yellow colour of FMN and FAD is lost when they are reduced.

Enzymes containing FAD or FMN are of widespread occurrence. We will cover three examples.

NADH dehydrogenase - this is the first step of the respiratory chain, where NADH supplies reducing equivalents. The reaction may be summarised as:

$$NADH + H^+ + FMN \rightarrow NAD^+ + FMNH_2$$

although in reality it is more complicated, in that mitochondrial NADH dehydrogenases are also associated with non-haem iron (NHI) proteins. Reduced $FMNH_2$ is re-oxidised by transfer of reducing equivalents to the next member of the respiratory chain, **coenzyme Q**.

Succinate dehydrogenase (SDH) - this enzyme is involved in the TCA cycle, catalysing the dehydrogenation of succinate in the following way:

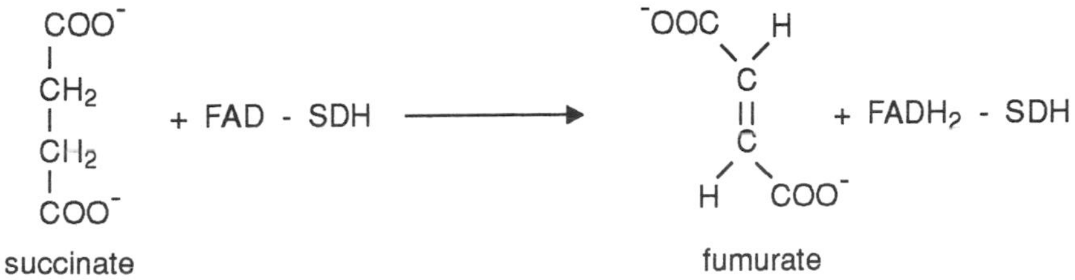

$$^-OOC-CH_2-CH_2-COO^- + FAD\text{-}SDH \longrightarrow {}^-OOC-CH=CH-COO^- + FADH_2\text{-}SDH$$

succinate fumurate

FAD is covalently bound to its enzyme

In this case the FAD is covalently bound to the enzyme. Oxidised FAD-SDH is regenerated by transfer of reducing equivalents from $FADH_2$ to coenzyme Q (see Chapter 8).

Glucose oxidase - this is typical of a large number of oxidases, these use oxygen as the acceptor for the reducing equivalents. The overall reaction is:

$$\text{glucose} + O_2 \rightarrow \text{gluconate} + H_2O_2$$

However, it proceeds as two distinct half reactions:

1) $\text{glucose} + \text{FAD} - \text{Enz} \rightarrow \text{gluconate} + \text{FADH}_2 - \text{Enz}$

2) $O_2 + \text{FADH}_2 - \text{Enz} \rightarrow H_2O_2 + \text{FAD} - \text{Enz}$

Other oxidases are active on L-amino acids, resulting in the release of ammonia and formations of oxo (keto) acids.

5.3.3 Biotin

Biotin (Figure 5.9) is widespread in nature. It is normally found covalently linked to a lysine residue of a protein, through the ε-amino (side chain) group of the lysine. Since it is covalently bound, it is not removed by dialysis.

Biotin is used as a cofactor in carboxylation reactions. These reactions may be thought of as two sequential reactions:

- synthesis of a carboxylated biotin-enzyme;
- transfer of the carboxyl group to an acceptor.

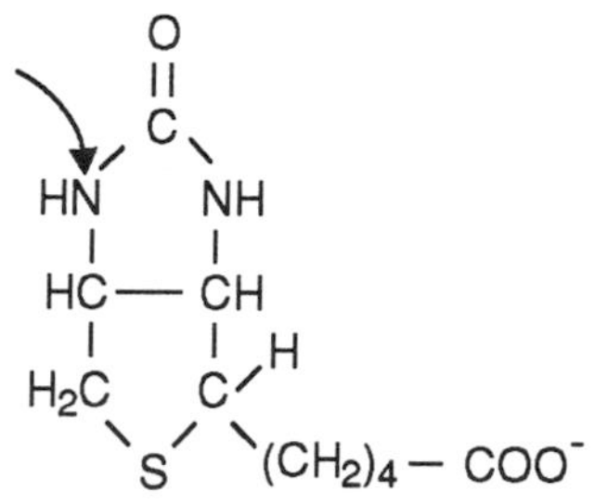

O
C
HN NH
HC — CH
H
H$_2$C C
S (CH$_2$)$_4$ — COO$^-$

Figure 5.9 The structure of biotin. Biotin acts as carboxyl carrier (see text), the carboxyl group is attached at the position arrowed.

anaplerotic reactions

The carboxylation of pyruvate to oxaloacetate (Figure 5.10) is a typical carboxylation. This is an important anaplerotic reaction, which serves to replenish TCA cycle intermediates. Since these intermediates may be used for biosynthesis, for example of amino acids, it is important that they can be replenished. Such topping-up reactions are called anaplerotic reactions. You will encounter other carboxylase reactions, notably in fatty acid metabolism, where the carboxyl acceptor is an acyl CoA (for example acetyl CoA or propenyl CoA).

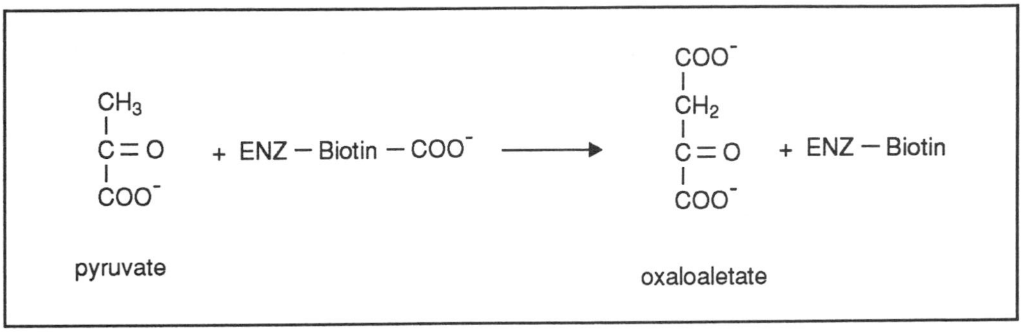

$CH_3-C(=O)-COO^-$ + ENZ — Biotin — COO^- ⟶ $COO^--CH_2-C(=O)-COO^-$ + ENZ — Biotin

pyruvate oxaloaletate

Figure 5.10 Carboxylation of pyruvate to oxaloacetate, catalysed by pyruvate carboxylase. This is an important anaplerotic reaction.

5.3.4 Pyridoxal and pyridoxal 5'-phosphate

pyridoxal
pyridoxine
pyridoxamine

Pyridoxal (vitamin B_6, Figure 5.11 a)) is widely available in plants, it is also found naturally as pyridoxine, where the aldehyde (-CHO) group of pyridoxal has been reduced to - CH_2OH. A further form of vitamin B_6 is pyridoxamine, in which $-CH_2NH_2$ replaces the aldehyde group.

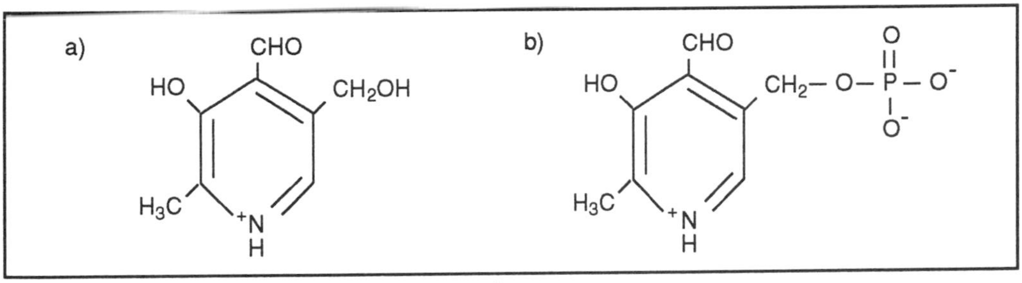

a) CHO, HO, CH_2OH, H_3C, ^+N, H

b) CHO, HO, $CH_2-O-P(=O)(-O^-)-O^-$, H_3C, ^+N, H

Figure 5.11 a) Pyridoxal, vitamin B_6.; b) Pyridoxal 5'-phosphate (PLP), which is the cofactor in transaminases and some other enzymes acting on amino acids. Note that we have not displayed all the carbons of the rings.

The form in which this vitamin is used as a cofactor, is pyridoxal 5'-phosphate, known as PLP (Figure 5.11 b)). It is an essential cofactor for a range of reactions, particularly those involving amino acids. Examples of the most important reactions are transaminations, decarboxylations and dehydrations. We will discuss each of these in turn.

transaminase reactions

Transaminations: These involve transfer of an α-amino group from an amino acid to an oxo acid, thereby generating different amino and oxo acids (Figure 5.12 a)). The enzymes involved are known as transaminases or amino transferases.

The generalised reaction is shown in Figure 5.12 b), in which it can be seen more clearly that the amino transfer occurs between two 'pairs' of compounds (amino acid 1 becomes oxo acid 1 and oxo acid 2 becomes amino acid 2). Glutamate transaminase (whose reaction is shown in Figure 5.12 a)) is specific for L-glutamate and α-oxoglutarate (amino acid 1 and oxo acid 1), and is highly active with oxoglutarate and L-aspartate as the second substrate pair (hence glutamate: oxaloacetate transaminase, GOT). However, transaminations with other pairs of L-amino acids and their corresponding oxo acids do occur in cells.

These transamination reactions provide a way of reversibly converting most of the amino acids to their corresponding oxo acids. Note that these reactions are easily reversed (the equilibrium constants are close to 1) and thus provide a highly effective system for interconverting L-amino acids.

a)

```
        COO⁻              COO⁻                  COO⁻            COO⁻
        |                 |                     |               |
        CH2               CH2                   CH2             CH2
        |           +     |        ⇌            |        +      |
        CH2               C=O                   CH2        H-C-⁺NH3
        |                 |                     |               |
   H-C-⁺NH3               COO⁻                  C=O             COO⁻
        |                                       |
        COO⁻                                    COO⁻

  L-glutamate        oxaloacetate         α-oxoglutarate     L-aspartate
```

b)

```
        R1                R2                    R1              R2
        |                 |                     |               |
   H-C-⁺NH3      +        C=O      ⇌            C=O      +  H-C-⁺NH3
        |                 |                     |               |
        COO⁻              COO⁻                  COO⁻            COO⁻

  amino group        amino group             oxo acid 1        amino
     donor             acceptor                                acid 2
  amino acid 1        oxo acid 2
```

Figure 5.12 a) Glutamate: oxaloacetate transaminase catalyses the reaction shown; b) The generalised reaction scheme for transamination. (Note that in older texts oxo acids are referred to as keto acids).

Π As the name suggests, glutamate: pyruvate transaminase is active with these substrates. Work out the structures of their products and identify them (the structures of L-glutamate and pyruvate have already been given in this chapter, if you need help).

The reactants (in the direction described) are L-glutamate and pyruvate, whose structures are:

```
          COO⁻              COO⁻
          |                 |
   H3⁺N-C-H       and       C=O
          |                 |
          CH2               CH3
          |
          CH2
          |
          COO⁻
```

The products will therefore be an oxo acid derived from glutamate, and an amino acid derived from pyruvate. The overall reaction is:

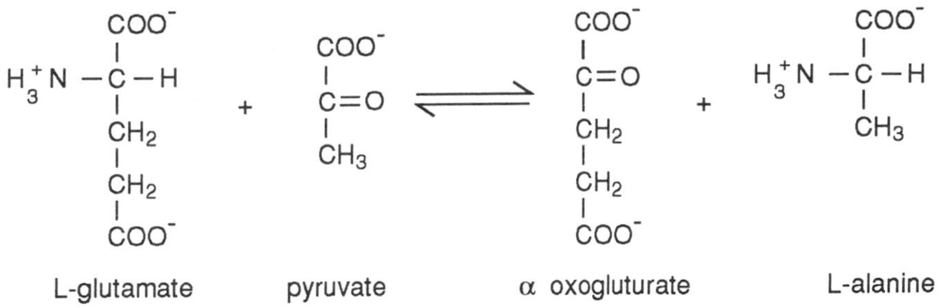

$$\begin{array}{ccccccc}
COO^- & & & & COO^- & & COO^- \\
| & & COO^- & & | & & | \\
H_3^+N-C-H & & | & & C=O & & H_3^+N-C-H \\
| & + & C=O & \rightleftharpoons & | & + & | \\
CH_2 & & | & & CH_2 & & CH_3 \\
| & & CH_3 & & | & & \\
CH_2 & & & & CH_2 & & \\
| & & & & | & & \\
COO^- & & & & COO^- & &
\end{array}$$

L-glutamate pyruvate α oxogluturate L-alanine

and the products are α-oxoglutarate and L-alanine. As a check, you could make sure that no atoms have been lost or gained, there should be the same numbers of each type of atom on both sides of the reaction.

decarboxylations

Amino acid decarboxylation: Another group of PLP-dependant enzymes which act on amino acids catalyse decarboxylations. These reactions are widespread and physiologically important. An example is shown in Figure 5.13 a).

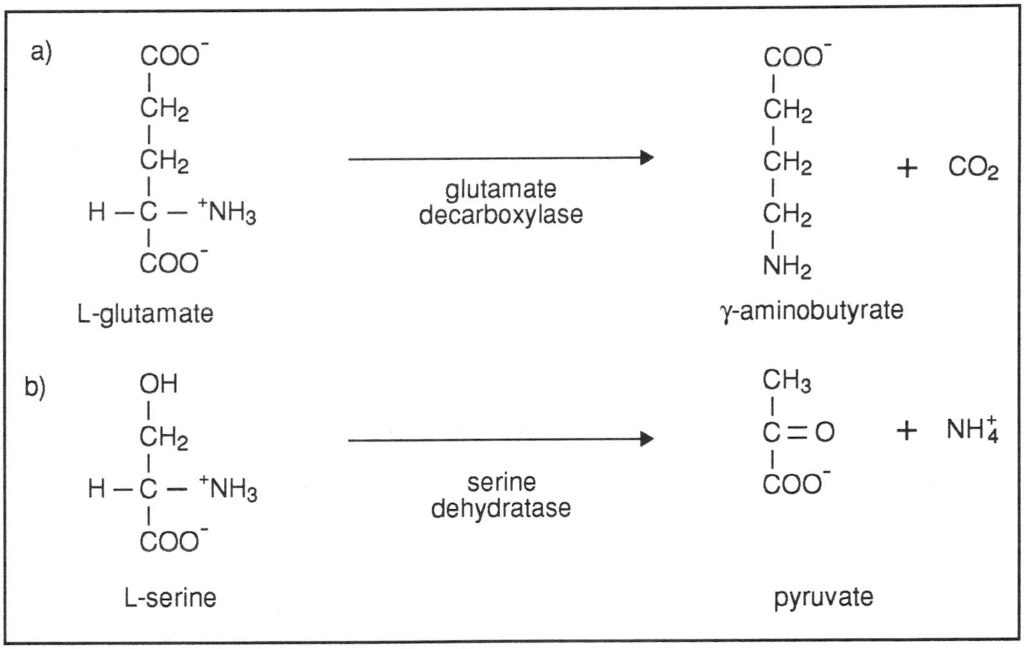

a)

$$\begin{array}{ccc}
COO^- & & COO^- \\
| & & | \\
CH_2 & & CH_2 \\
| & & | \\
CH_2 & \xrightarrow[\text{decarboxylase}]{\text{glutamate}} & CH_2 \quad + \quad CO_2 \\
| & & | \\
H-C-{}^+NH_3 & & CH_2 \\
| & & | \\
COO^- & & NH_2
\end{array}$$

L-glutamate γ-aminobutyrate

b)

$$\begin{array}{ccc}
OH & & CH_3 \\
| & & | \\
CH_2 & \xrightarrow[\text{dehydratase}]{\text{serine}} & C=O \quad + \quad NH_4^+ \\
| & & | \\
H-C-{}^+NH_3 & & COO^- \\
| & & \\
COO^- & &
\end{array}$$

L-serine pyruvate

Figure 5.13 a) Amino acid decarboxylases require PLP. Glutamate decarboxylase removes carbon dioxide from L-glutamate, producing γ-aminobutyrate; b) The reaction catalysed by serine dehydratase, a PLP-dependant enzyme.

Dehydrations: A third set of reactions involving PLP-containing enzymes is that catalysed by dehydratases, for example, serine dehydratase (Figure 5.13 b)). Here a non-oxidative deamination occurs, resulting in release of ammonia and formation of pyruvate.

formation of Schiff's base

How does pyridoxal 5′-phosphate enable these diverse reactions to be accomplished? In the absence of substrate, the PLP group is attached to the enzyme as a result of the formation of a Schiff's base with the ε-amino group of a lysine residue (Figure 5.14 a)).

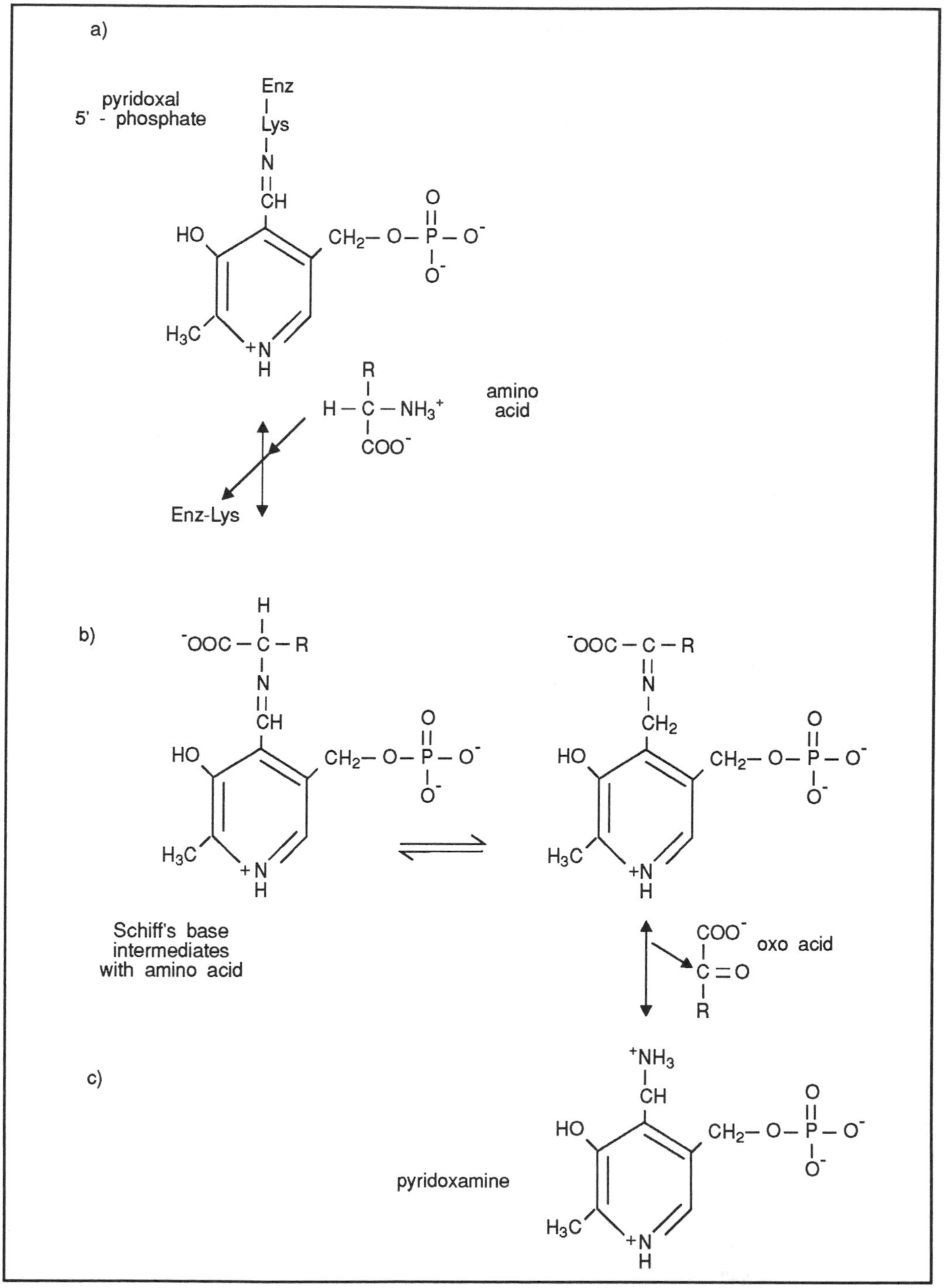

Figure 5.14 a) Mechanism of action of PLP-dependant enzymes, as exemplified by transamination. The PLP group is attached to the ε-amino group of a lysine residue of the enzyme; b) This is displaced by the α-amino group of the amino-donor amino acid, Schiff's base intermediates are formed; c) The oxo acid producer leaves, with pyridoxamine present within the enzyme. See text for full details.

The amino acid displaces the protein-lysine amino group, forming a Schiff's base with the PLP group (Figure 5.14 b)).

This undergoes rearrangement to form pyridoxamine phosphate (Figure 5.14 c)) and the 'product' oxo acid is released. The amino acid product is formed by reversal of these reactions, starting with the arrival of the oxo acid to act as an amino acceptor.

5.3.5 Thiamine and thiamine pyrophosphate

TPP involved in transketolase activity

The structure of thiamine (vitamin B_1) is shown in Figure 5.15 a). The form in which it is used is thiamine pyrophosphate (TPP, Figure 5.15 b)). TPP acts as a cofactor for several enzymes including transketolase, this is an important enzyme for carbohydrate interconversions, since reactions with several alternative substrates are possible.

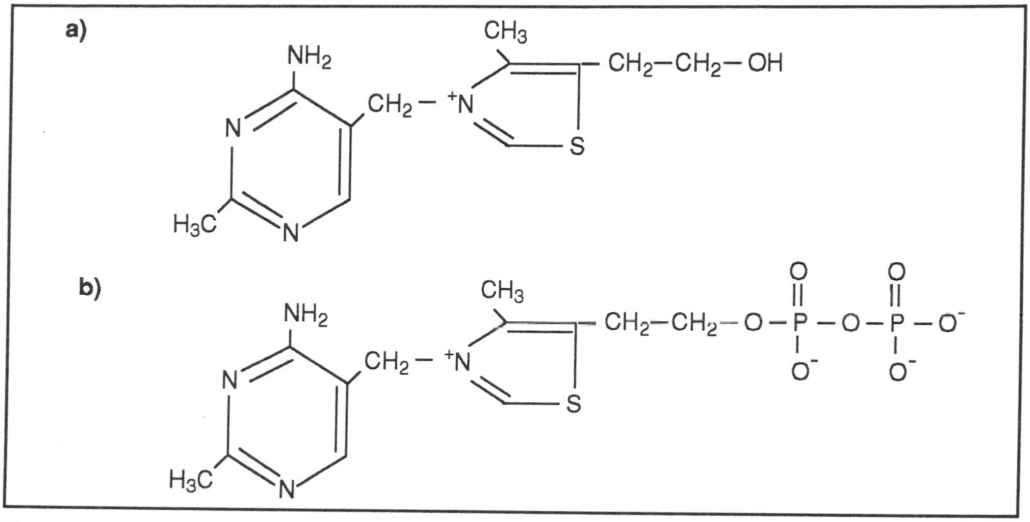

Figure 5.15 a) The structure of thiamine. b) Thiamine pyrophosphate (TPP), the 'active' form of this cofactor. Note we have not displayed all the carbons of the rings.

The generalised reaction scheme is shown in Figure 5.16, together with one specific reaction. The presence of transketolase and transaldolase (which does not require a cofactor) enables interconversion of many phosphorylated sugars, these enzymes are involved in the pentose phosphate pathway, which is discussed in the next chapter.

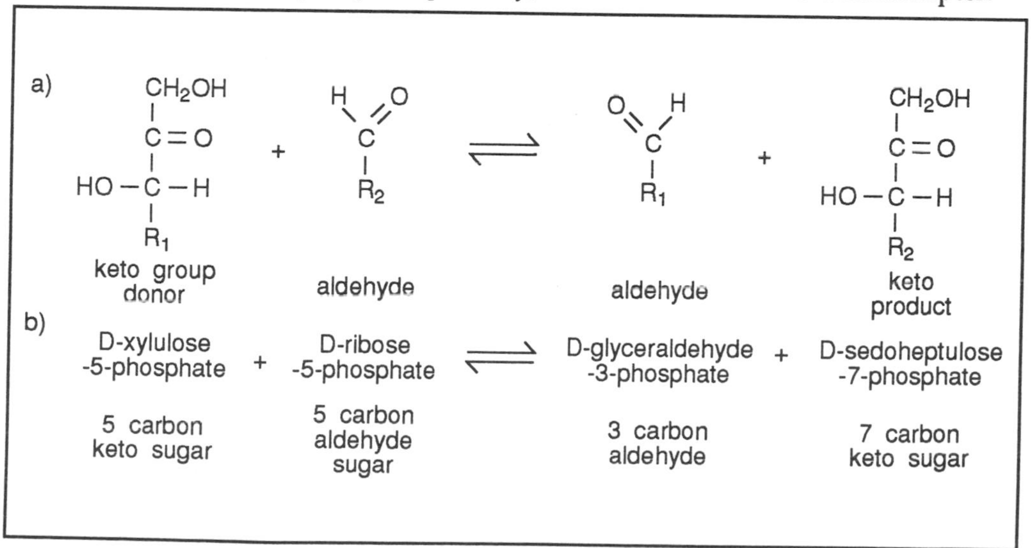

a) keto group donor + aldehyde ⇌ aldehyde + keto product

b) D-xylulose-5-phosphate (5 carbon keto sugar) + D-ribose-5-phosphate (5 carbon aldehyde sugar) ⇌ D-glyceraldehyde-3-phosphate (3 carbon aldehyde) + D-sedoheptulose-7-phosphate (7 carbon keto sugar)

Figure 5.16 a) The generalised reaction catalysed by transketolase, a TPP-dependent enzyme. A specific reaction is also shown b).

5.3.6 Pantothenic acid and coenzyme A

coenzyme A

Pantothenic acid (Figure 5.17) is required as a vitamin to enable the important coenzyme known as coenzyme A to be synthesised. It is also involved in the so-called acyl carrier protein (ACP), which we have briefly encountered whilst considering the role of NADPH in fatty acid biosynthesis. Since pantetheine (formed from cystamine and pantothenic acid) fulfils an analogous function in ACP as it does within coenzyme A, we shall not discuss it further.

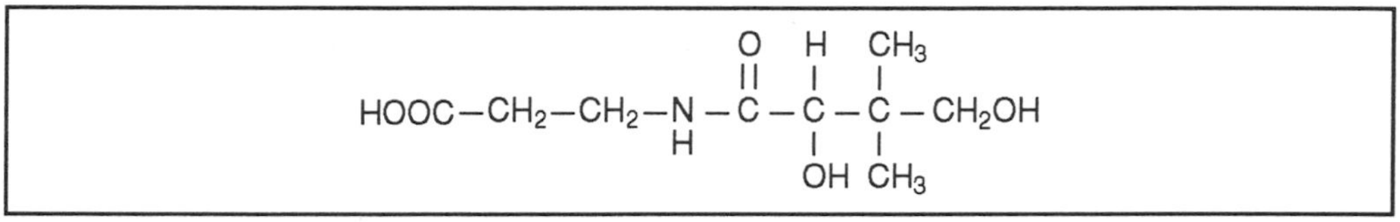

$$HOOC-CH_2-CH_2-\underset{H}{N}-\overset{O}{\overset{\|}{C}}-\underset{OH}{\overset{H}{C}}-\underset{CH_3}{\overset{CH_3}{C}}-CH_2OH$$

Figure 5.17 Pantothenic acid.

Coenzyme A consists of phosphorylated adenosine linked to pantothenic acid, to which cystamine is also attached (Figure 5.18). Coenzyme A is able to form energy-rich thioesters with carboxylic acids via the -SH group of the cystamine moiety, in the following way:

$$R-COO^- + HS-CoA \longrightarrow R-\overset{O}{\overset{\|}{C}}-S-CoA + OH^-$$

formation of thioesters

A thioester is analogous to a conventional ester, formed by reaction between a carboxyl group and an alcohol. The structural basis of the energy-rich nature of thioesters has been discussed earlier (Chapter 2). The most important thioester is acetyl CoA ($R = CH_3$ in the scheme above), formed from pyruvate during the catabolism of fatty acids and carbohydrates. Other acyl-S-CoA molecules are also involved in fatty acid catabolism.

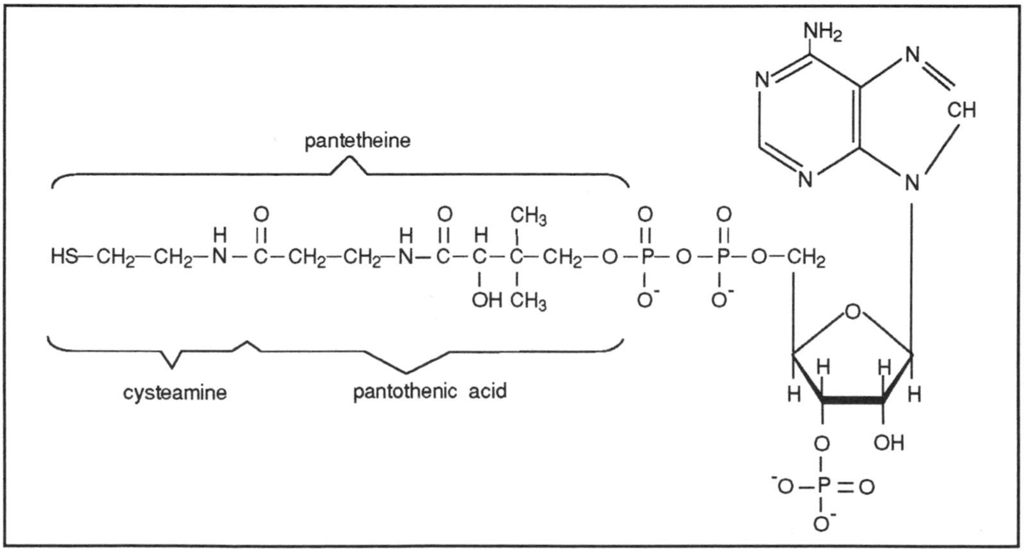

$$HS-CH_2-CH_2-\overset{H}{N}-\overset{O}{\overset{\|}{C}}-CH_2-CH_2-\overset{H}{N}-\overset{O}{\overset{\|}{C}}-\underset{OH}{\overset{H}{C}}-\underset{CH_3}{\overset{CH_3}{C}}-CH_2-O-\underset{O^-}{\overset{O}{\overset{\|}{P}}}-O-\underset{O^-}{\overset{O}{\overset{\|}{P}}}-O-CH_2$$

Figure 5.18 Coenzyme A.

5.3.7 Folic acid and tetrahydrofolate

Folic acid (shown in Figure 5.19 as folate, as it would occur at physiological pH) is required in order for the coenzyme tetrahydrofolate (THF) to be synthesised. This is accomplished by reduction by a nicotinamide-containing coenzyme.

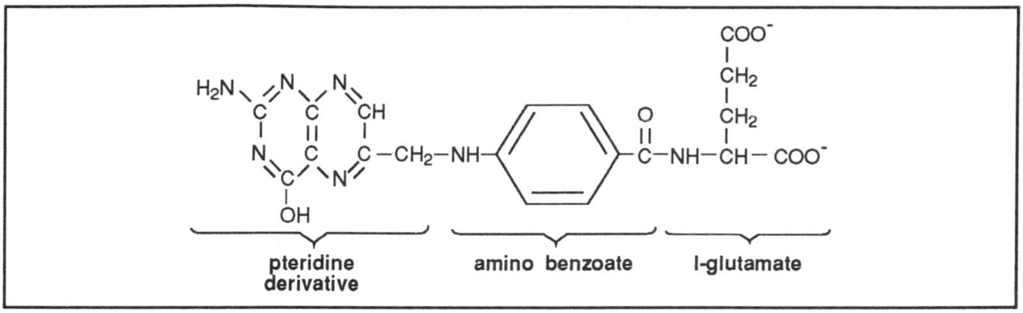

pteridine derivative | amino benzoate | l-glutamate

Figure 5.19 The structure of folate.

Π Which coenzyme do you think is used to reduce folate? (NADH or NADPH).

The answer is NADPH, this is another example of NADPH being used for a reductive biosynthesis. If in doubt, re-read section 5.3.1.

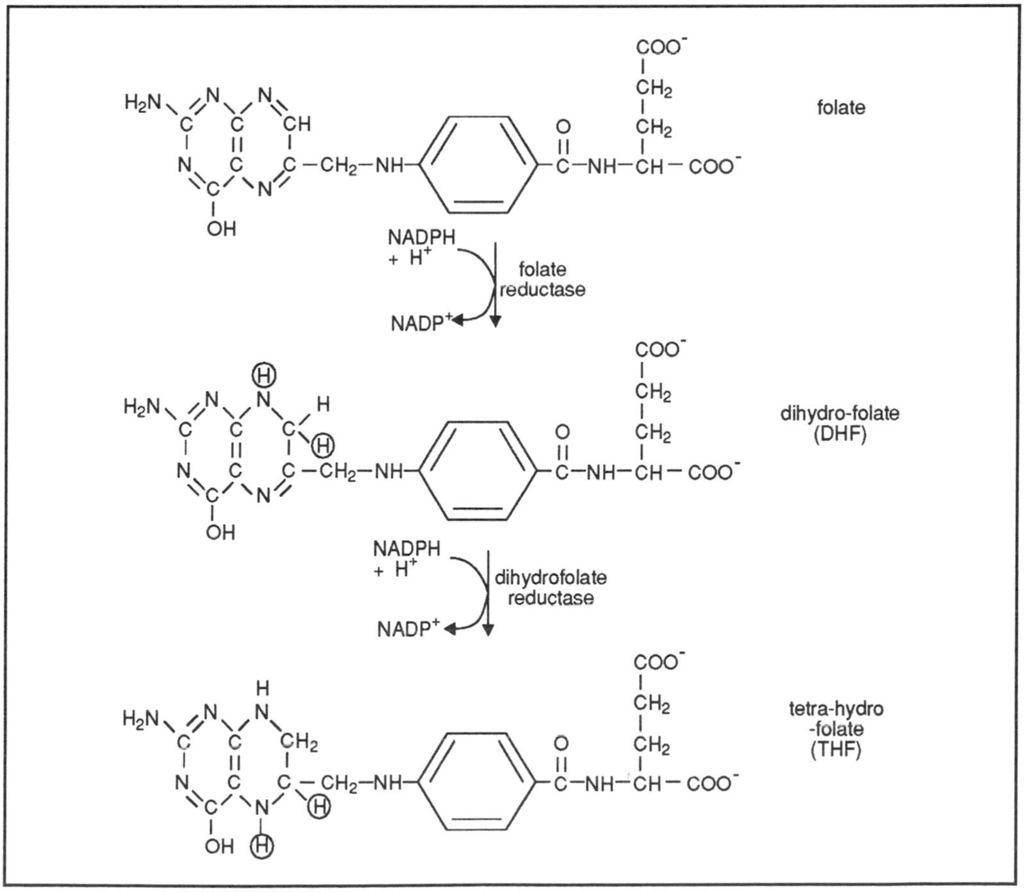

folate

NADPH + H^+ → $NADP^+$ (folate reductase)

dihydro-folate (DHF)

NADPH + H^+ → $NADP^+$ (dihydrofolate reductase)

tetra-hydro -folate (THF)

Figure 5.20 Conversion of folate to tetrahydrofolate (THF) which is the effective form of this coenzyme.

Reduction of folate is accomplished by NADPH, in a two stage process (Figure 5.20) involving folate reductase and then dihydrofolate reductase. The importance of THF is that it serves as a carrier of 1-carbon units, as either formate or formaldehyde. These groups are often referred to as 'active formate' and 'active formaldehyde'. One-carbon fragments may originate from serine, glycine, histidine and choline, as well as from formate and formaldehyde. Methyl transfer usually involves an additional compound, 5-adenosylmethionine. We will not explore the biochemistry of this any further at this stage.

5.3.8 Cyanocobalamine

methyl transfers

Vitamin B_{12}, normally isolated as cyanocobalamine (Figure 5.21), is needed to act as a coenzyme. The functional coenzyme has a 5-deoxyadenosyl group occupying the position occupied by cyanide in cyanocobalamine. This coenzyme is involved in several reactions, including methyl transfers.

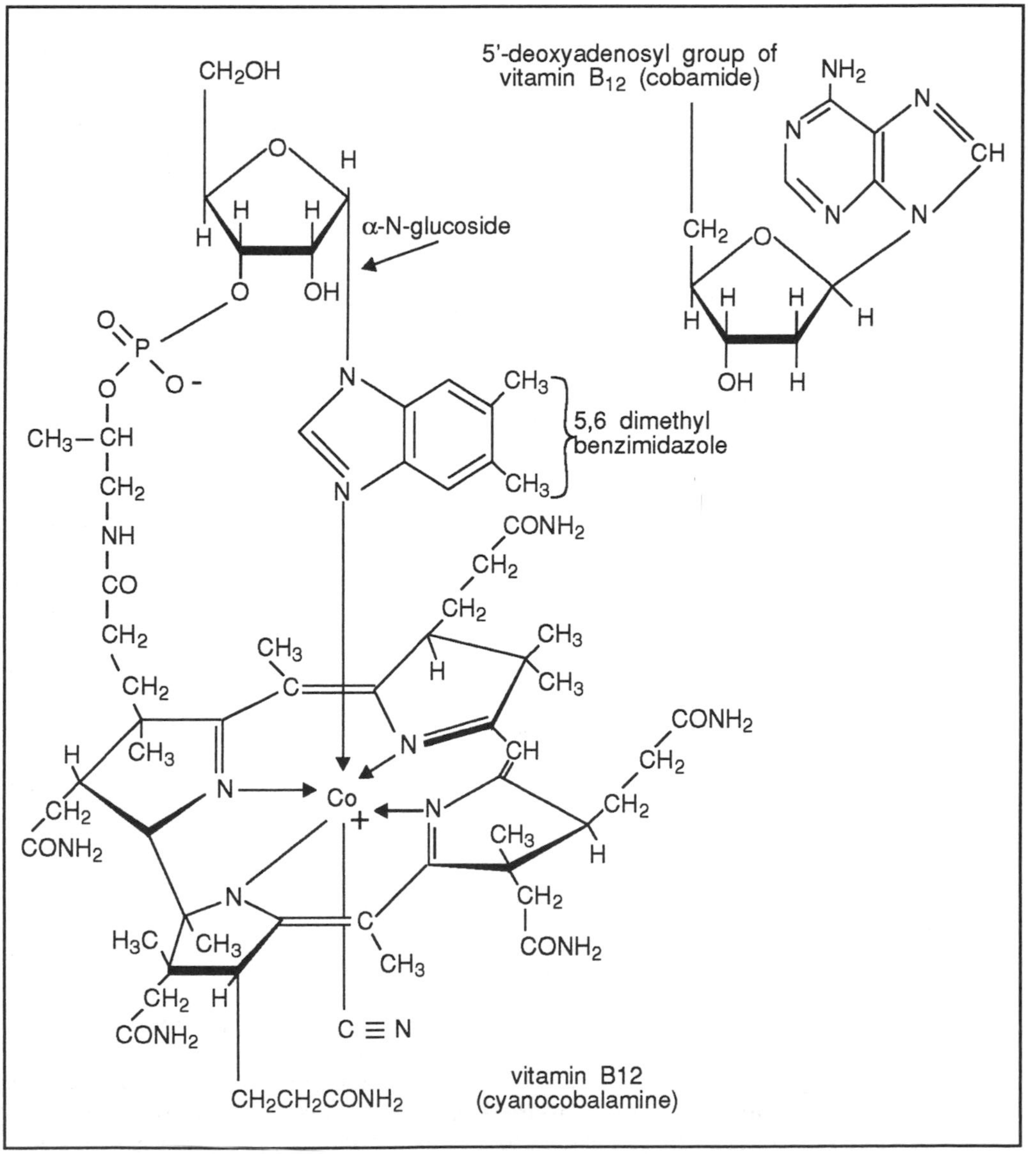

Figure 5.21 Cyanocobalamine. Note that cyanocobalamine is the usual form in which the vitamin is isolated. It is the cobamide form which acts as a coenzyme. In this form the cyanide group is replaced by a 5'deoxyadenosyl group.

SAQ 5.2

From the vitamins and cofactors/coenzymes which we have described, find examples of the following:

1) a cofactor containing a sugar alcohol;

2) a coenzyme in which thioesters are formed during coenzyme use;

3) a cofactor which forms Schiff's bases with both enzyme and substrate (but not at the same time);

4) a coenzyme involved in reductive biosynthesis;

5) a cofactor which changes colour on reduction (hint: being bleached is a change of colour!);

6) a cofactor involved in interconversions of 5-, 7- and 3-carbon phosphorylated sugars.

5.4 Other cofactors and coenzymes

We have already indicated that this chapter is not intended to be encyclopedic. Having discussed cofactors and coenzymes which are derived from vitamins, we must briefly refer to other, non-vitamins derived ones. We shall describe four types.

5.4.1 ATP and related nucleotides

coupled reactions

Numerous nucleotides are required by cells for growth, for example for incorporation into nucleic acids but here we are concerned with the role of ATP, in particular, as a coenzyme in metabolism. Any consideration of metabolic pathways reveals the importance of ATP as the energy currency of the cell. We have already seen, in earlier chapters, that the release of energy upon hydrolysis of the terminal phosphate of ATP enables energy-requiring reactions to occur. Such coupled reactions are widespread. Phosphorylation of glucose by hexokinase is a typical example. The individual components of the reaction may be thought of as:

a) $ATP + H_2O \rightarrow ADP + Pi$ $\qquad \Delta G^{o'} = -32.2 kJ\ mol^{-1}$

b) $\text{glucose} + Pi \rightarrow \text{glucose-6-phosphate} + H_2O$ $\qquad \Delta G^{o'} = +13.8\ kJ\ mol^{-1}$

The coupled reaction, catalysed by hexokinase, is:

c) $\text{glucose} + ATP \rightarrow \text{glucose-6-phosphate} + ADP$ $\qquad \Delta G^{o'} = -19.4 kJ\ mol^{-1}$

The phosphorylation of glucose by orthophosphate is energetically unfavourable (ie. it is endergonic). However, the energy made available by hydrolysis of ATP to ADP is more than sufficient to overcome this, such that the coupled reaction is now exergonic, and thus would be classified as a spontaneous reaction. Certainly, the equilibrium position for the coupled reaction (as written) is substantially in favour of the products.

catabolic pathways generate ATP

Understanding the role of ATP in the cells is vital. You will shortly realise that one of the principal objectives of the catabolic pathways described in succeeding chapters is the synthesis of ATP.

cyclic AMP/ protein kinases

ATP is not merely used for providing energy to enable energetically 'unfavourable' reactions to take place. It is also used to synthesise adenosine 3', 5' - monophosphate (cyclic AMP, also known as cAMP), which is an important intracellular messenger. Figure 5.22 gives the general structure of adenosyl nucleotides, including ATP, together with the structure of cAMP. ATP is also used to phosphorylate proteins. Such phosphorylations are usually catalysed by specific protein kinases, as we have seen in Chapter 4, phosphorylation of enzymes is widely used as a means of regulating their activity.

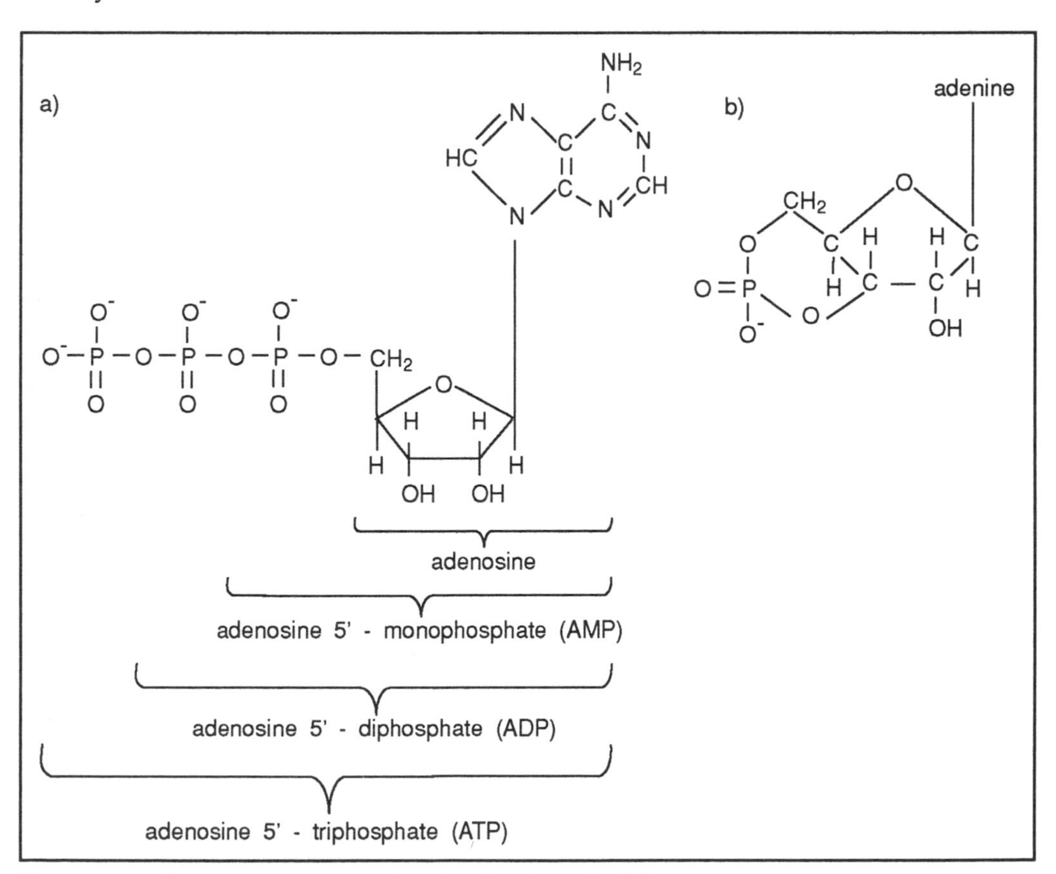

Figure 5.22 a) The structure of adenosyl nucleotides. b) cyclic AMP.

Guanosine triphosphate, GTP, is also used as a coenzyme, for example in the TCA cycle (synthesis of GTP from GDP and phosphate, using the energy released by cleavage of the thioester bond in succinyl CoA), and in protein synthesis (GTP is required for the initiation of protein synthesis on ribosomes). We have mentioned the cyclic nucleotide derived from ATP (ie cAMP). There is also a corresponding cGMP (whose structure is analogous to that of cAMP), which is derived from GTP and is involved in vision in which a light signal is converted to a nerve impulse.

5.4.2 Coenzyme Q

ubiquinone

The respiratory chain of mitochondria contains coenzyme Q. It was originally thought to be ubiquitous in biological systems, and has also been called ubiquinone. Coenzyme Q consists of a quinone moiety to which is attached a long isoprenoid chain (Figure 5.23 a)). The number of isoprene units varies, but the form shown in Figure 5.23 (Q_{10}, since it contains 10 isoprene units) is common in mammals.

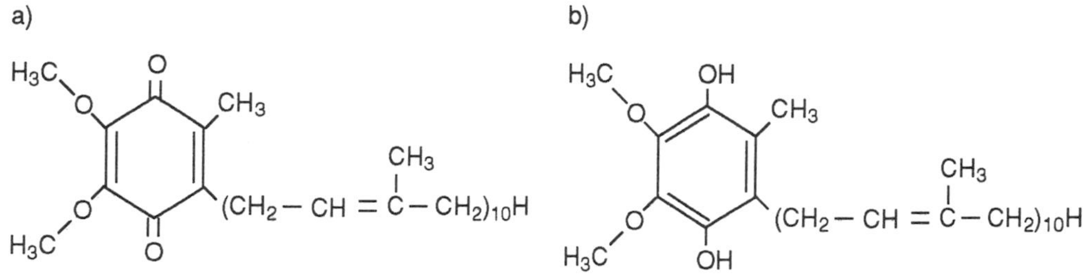

Figure 5.23 a) Coenzyme Q: various forms occur, depending on the length of the isoprenoid chains. The form shown, known as Q_{10}, is common in mammals. b) The reduced form of coenzyme Q.

As a consequence of the isoprenoid sidechain (which is pure hydrocarbon and hence non-polar) coenzyme Q is lipid soluble. It acts as a mobile electron carrier within the respiratory chain. We have already noted (Section 5.3.2) that it is the acceptor of reducing equivalents from flavoproteins, (two were mentioned, NADH dehydrogenase and the FAD-linked succinate dehydrogenase). Thus coenzyme Q is used as a link between reducing equivalents generated in TCA cycle reactions (NADH and $FADH_2$) and the cytochromes of the respiratory chain (Figure 5.24). The reduced form of coenzyme Q is shown in Figure 5.23 b), and results from transfer of 2H atoms from (in the cases we have mentioned) $FMNH_2$ of NADH dehydrogenase and $FADH_2$ of succinate dehydrogenase. Note that coenzyme Q is also used as reducing equivalent acceptor from other enzymes, notably from fatty acid degradation. We will describe the role of coenzyme Q in the respiratory chain in greater detail in Chapter 8.

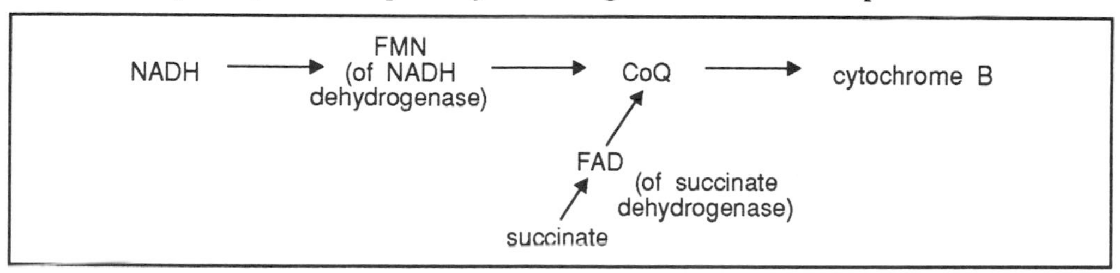

Figure 5.24 The role of coenzyme Q as a link between flavoproteins and cytochromes in the respiratory chain.

5.4.3 Haem

oxygen binding proteins

Neither haemoglobin nor the cytochromes are enzymes but they do require a cofactor, haem (Figure 5.25 a)), in order to function. Haem is a versatile compound composed of a porphyrin component containing an iron atom. The porphyrin component is derived from a tetrapyrrole compound, porphin (Figure 5.25 b)). The iron atom is chelated to the four pyrrole nitrogen atoms. Haem is found both in oxygen binding proteins (haemoglobin for oxygen transport, myoglobin for oxygen storage) and in the cytochromes of the respiratory chain and elsewhere.

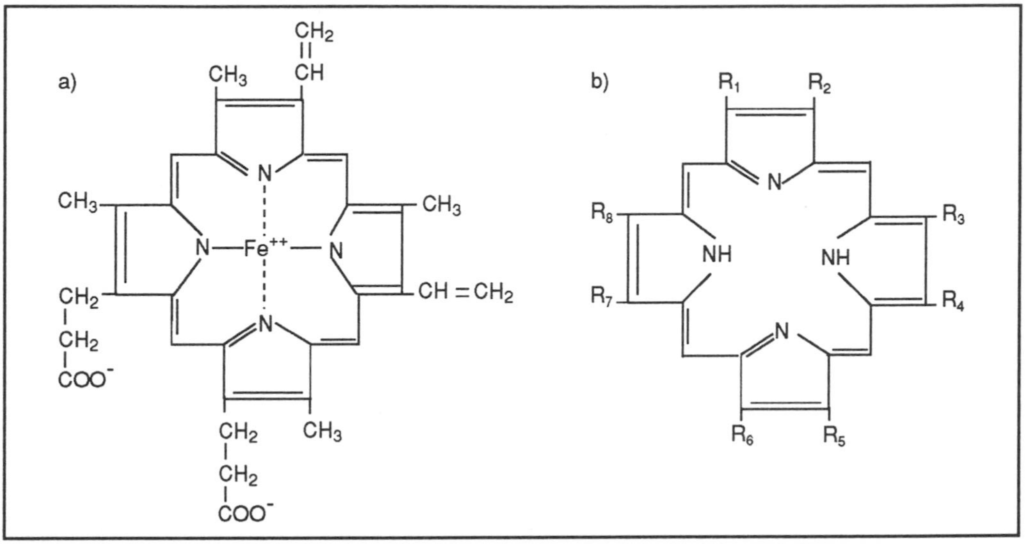

Figure 5.25 a) The structure of haem; b) Porphyrin. This tetrapyrole compound is the basis for synthesis of porphyrins (which differ in their subsituent groups. The same compound is the basis for chlorophyll, in which Mg^{++} is chelated to the nitrogen atoms of the four pyrrole rings.

It is the haem group which gives all of these proteins their characteristic red colour. The cytochromes are electron carriers, binding a single electron to the Fe.III (Fe^{3+}) atom, which is accordingly reversibly reduced to Fe.II (Fe^{2+}). Haem is non-covalently (but none-the-less tightly bound) in haemoglobin and myoglobin, but is covalently bound in cytochrome C. These proteins are the sites of action of well-known toxic compounds such as carbon monoxide which binds very tightly to the oxygen binding site of haemoglobin (thereby preventing oxygen transport) and cyanide (CN^-) which builds tightly to the terminal cytochrome of the respiratory chain.

5.4.4 Metal ions

Mg ATP

Numerous enzymes require metal ions in order to be catalytically active. These metal ions may be loosely or tightly bound. In some cases they are associated with a prosthetic group, for example Fe.II (Fe^{2+})/Fe.III (Fe^{3+}) in the haem group of the cytochromes. Otherwise the free metal ion is required for activity. Examples include Zn^{2+} in carboxypeptidase A and Cu^{2+} in cytochrome oxidase. Mg^{2+} is usually required for all reactions involving ATP. This is primarily because Mg^{2+} complexes to the negatively charged phosphate groups (see Figure 5.23) and the true coenzyme/substrate for reactions catalysed by kinases is Mg^{2+}-ATP. The presence of a catalytically essential metal ion can make enzymes susceptible to non-competitive inhibition by chelating agents, such as EDTA.

SAQ 5.3

Identify a common role for cofactors/coenzymes derived from riboflavin and niacin. Do any of the other cofactors and coenzymes which we have described also fulfil this role? If so, name them.

Summary and objectives

In this chapter we have reviewed the roles of cofactors and coenzymes in reactions catalysed by enzymes. We began by explaining the differences between cofactors and coenzymes and describing the role of water-soluble vitamins as sources of cofactors and coenzymes. We then examined the roles of the major cofactors. This discussion focussed on nicotinanide - containing coenzymes, flavin - containing cofactors, biotin, pyridoxal 5′-phosphate, thiamine pyrophosphate, pantothenic acid, folic acid and cyanocobalamine. The roles of ATP and related nucleotides, coenzyme Q, the haem moiety and metal ions were also described.

Now that you have completed this chapter, you should be able to:

- describe, in outline, the roles of the major cofactors and coenzymes in enzyme-catalysed reactions;
- outline the differences in water-soluble and fat-soluble vitamins;
- explain why two pools of nicotinamide - containing nucleotides are present in cells;
- draw, in outline, the structure of the major cofactors.

Primary pathways in the breakdown of carbohydrates to pyruvate

Primary pathways in the breakdown of carbohydrates to pyruvate

Overview

The earlier chapters have given an insight into both the meaning of energy in biological terms and the ways in which the biological catalysts (enzymes) function to carry out the chemical reactions. In this chapter we shall examine the ways in which polysaccharides, which are biopolymers of monosaccharides, are metabolised to glucose, the central hexose of metabolism. The main part of the chapter will be devoted to the consideration of the five pathways used by living organisms to degrade glucose to the central metabolite, pyruvate. We will also consider the relative importance of the pathways and their relationship to each other and their distribution in biology.

Introduction

Carbohydrates are the most common form of carbon and energy source for living cells and to be totally successful in this role they have to satisfy three major objectives. They must either directly or indirectly produce energy in the form of ATP, they must produce reducing power in the form of NADPH + H^+ for driving biosynthetic reactions and finally, they must produce twelve compounds referred to as 'precursor metabolites' from which all biosynthesis proceeds.

In this chapter we shall investigate the direct production of ATP in a process called substrate level phosphorylation and the way in which the cell produces the NADPH + H^+ required for biosynthetic reactions. Finally we will begin a study into how the precursor metabolites are produced. These metabolites are:

- glucose-6-phosphate;
- fructose-6-phosphate;
- glyceraldehyde-3-phosphate;
- 3-phosphoglycerate;
- phosphoenol pyruvate;
- pyruvate;
- ribose-5-phosphate;
- erythrose-4-phosphate;
- acetyl CoA;
- α-oxoglutarate;
- succinyl CoA;
- oxaloacetate.

When presented as a list they look rather daunting but you will see as you learn more about metabolism how their names and structures become familiar.

The production of the first eight compounds will be dealt with in this chapter. The last four will be covered in the following chapter when we discuss the tricarboxylic acid (or Krebs) cycle.

6.1 The occurrence of polysaccharides

Polysaccharides are carbohydrates and are polymers of monosaccharide units. They are biopolymers in that they are produced by living organisms and they fulfil a variety of structural and functional roles.

Inevitably an almost infinite variety of polysaccharides is possible because there are many different precursor molecules and the number of combinations of such precursors is theoretically unlimited. In addition the chain length of polysaccharides may vary from a small number of around ten monosaccharides (such compounds are called oligosaccharides) to long chains of several hundred units.

oligosaccharides

It will be clear that there is a wide variety of naturally occurring polysaccharides. In addition it is almost impossible to exaggerate the amount of polysaccharide production and turnover. It is an awe-inspiring fact that the most abundant organic compound in the biosphere is the polysaccharide cellulose; some sources claim that it represents around half of the world's polysaccharide carbon. Along with cellulose there are a few other polysaccharides such as chitin and hemicelluloses which are notable for their enormous presence in and on the earth's surface. More specialised polysaccharides will inevitably be found in smaller quantities, either in terms of the amounts per cell or in the restricted number of organisms which produce them. However we should remember that even these quantities will extrapolate to vast tonnages on a worldwide scale.

Table 6.1 lists some of the more important naturally-occurring polysaccharides with brief details of their chemistry and incidence in nature. The examples were chosen on the grounds of their large scale natural presence. Five examples contain only glucose and are termed homopolymers. Similarly chitin is a homopolymer because it contains only N-acetylglucosamine. The remaining three are termed heteropolymers as they contain more than one type of monosaccharide in their structure. The term complex polysaccharide is used to denote polysaccharides which are bound to non-carbohydrate materials, for example lipopolysaccharides of bacterial cell walls.

homo and heteropolymers

complex polysaccharide

Π Make a list of the homopolysaccharides in Table 6.1.

Your answer should be glycogen; starch; dextran; cellulose; laminarin; chitin. The other polysaccharides listed in Table 6.1 are heteropolysaccharides.

Note also that polysaccharides differ not only in the monosaccharides they contain but also in the ways in which these units are linked together. There may be differences in position (c.g $1 \rightarrow 4$, $1 \rightarrow 6$) but also in orientation (α and β). If you are unfamiliar with carbohydrate structure and chemistry we would recommend the BIOTOL text, 'The Molecular Fabric of Cells'. Because of the diversity and quantities involved it is not surprising that carbohydrates form the most important source of carbon and energy for most living organisms. In the carbon cycle, sugars and polysaccharides are produced from carbon dioxide by photosynthetic organisms for structural and energy storage purposes. Photosynthetic organisms are eaten by animals which, after death, decompose (with the aid of bacteria) and the carbon is recycled.

α and β links

Having established the variety and availability of polysaccharides and the fact that they are the principal carbon and energy source for non-photosynthetic organisms we can

polysaccharide	function	chemistry	distribution
glycogen	energy store	branched polymer of D-glucose linked α1-4 with some α1-6.	animals, bacteria
starch - amylose	energy store	unbranched polymer of D-glucose inked α1-4	plants
starch - amylopectin	energy store	as amylose but a few α1-6 linkages also	plants
dextran	energy store	branched polymer of D-glucose linked α1-4 with occasional α1-2, α1-3 and α1-6 linkages.	bacteria and yeasts
cellulose	structural	unbranched polymer of D-glucose linked β1-4	plants
laminarin	structural	unbranched polymer of D-glucose linked β1-3	marine algae
hemicelluloses	structural	mixtures of polysaccharides containing for example: β1-4 xylans, glucomannans, arabinogalactans	plants
pectins	structural	unbranched polymer of D-galacturonic acid linked α1-4 with possible presence of other sugars, for example: arabinose	plants
chitin	structural	unbranched polymer of N-acetyl-β-D-glucosamine linked β1-4	fungi and insects
agar	structural	unbranched polymer of β-linked galactose and galacturonic acid with substituted groups, for example: sulphate	marine algae

Table 6.1 Examples of important naturally-occurring polysaccharides; their function, chemistry and distribution.

now investigate the catabolism of polysaccharides to glucose or pentoses. This is important because virtually all cells can and even prefer to use glucose; a compound referred to as the 'universal substrate'.

6.2 Conversion of polysaccharides to monosaccharides

resistance of β-linked monosaccharides to catabolism

Considering the wide variety of polysaccharides and the large amount of potential energy which is available from their degradation, higher animals and plants have surprisingly limited abilities to convert them to central monosaccharides. If we consider the example of cellulose itself, higher animals, most eukaryotic micro-organisms and many bacteria cannot digest it because they do not produce cellulases, the enzymes which degrade the β1-4 glycosidic bonds. The property of producing significant amounts of cellulases is largely that of fungi (but not yeasts) and some bacteria. In particular, the aerobic soil bacteria of the genus *Cytophaga* degrade cellulose relatively quickly. This problem of breaking β–linkages is also reflected by the fact that few organisms can degrade chitin, hemicelluloses or agar. Inevitably catabolism will be slow because β-linked polysaccharides are relatively insoluble.

α and β amylases

Catabolism of α-linked glucose polysaccharides is a process which can be carried out by a much wider range of organisms. Enzymes degrading glycogen or the starches are called amylases. α-Amylases attack the reducing end of the polymer to yield glucose and the disaccharide maltose via oligosaccharides called dextrins. These enzymes are found in animals, plants and many bacteria. β-amylases attack the non-reducing end of the molecule and are found widely in plants but are rare in bacteria and absent in animals. It is beyond the scope of this chapter to list details of polysaccharides generally but you should by now be aware of two things: firstly that all known polysaccharides can be biologically degraded and secondly that it is largely bacteria and fungi which are responsible for these changes.

Π Where do you think that polysaccharide degradation occurs?

Your instinct may have been to write 'the soil' or in 'the alimentary tract of herbivores'. In principle you would be right but does it occur inside or outside of the cells? The answer is not a simple one. Extracellular polysaccharides are always degraded outside the cell membrane. This is because polysaccharides are largely insoluble and have a high molecular weight making them too large to pass through the membrane. Some degradative enzymes, therefore, function extracelluarly and are almost always hydrolases; that is they cleave the polysaccharide using water (hydrolysis) to yield free monosaccharides.

$$\text{polysaccharide}_{(\text{n residues})} + H_2O \rightarrow \text{polysaccharide}_{(\text{n-1})} + \text{monosaccharide}$$

hydrolysis and phosphorolysis

Polysaccharides produced intracellularly as energy stores are degraded inside the cell by phosphorolysis to yield monosaccharide phosphates, for example:

$$\text{glycogen}_{(\text{n residues})} + \text{Pi} \rightarrow \text{glycogen}_{(\text{n-1})} + \text{glucose-1-phosphate}$$

The enzyme which produces glucose 1-phosphate is glycogen phosphorylase and the reaction it catalyses is advantageous to living cells. Firstly the energy of the glycosidic bond is conserved in that it has been used to produce a sugar phosphate. This is essential because all intermediates of the pathways we shall study in a later section are phosphorylated. Secondly, the cell membrane is impermeable to the phosphorylated monosaccharide. The negatively charged phosphate group is highly polar and will not penetrate the lipid bilayer of the membrane.

facilitated diffusion

active transport

Following the degradation of extracellular polysaccharides there will be various mono-, di- and tri-saccharides on the periphery of the cell and they need to be brought across the cell membrane. The topic of uptake of nutrients is dealt with in the BIOTOL text, 'Biosynthesis and the Integration of Cell Metabolism,' but briefly we can say that hexoses are taken up by facilitated diffusion (in which specific membrane carriers are used in a non energy-requiring process) and disaccharides are generally taken up by active transport (like facilitated diffusion but the process is energy-requiring). Simple chemical diffusion is not important here.

glucose and xylulose are preferred substrates

A variety of hexoses and pentoses will be produced and most are acceptable carbon and/or energy sources for fungi and bacteria. However higher animals have a much more selective acceptance, particularly in terms of requirements for a carbon and energy source. In general terms the hexose of choice is glucose and the pentose of choice is xylulose. Figures 6.1 and 6.2 show generalised diagrams of pentose interconversions and hexose interconversions in a 'typical' cell. Not all living systems will either follow these pathways or have all of the required enzymes.

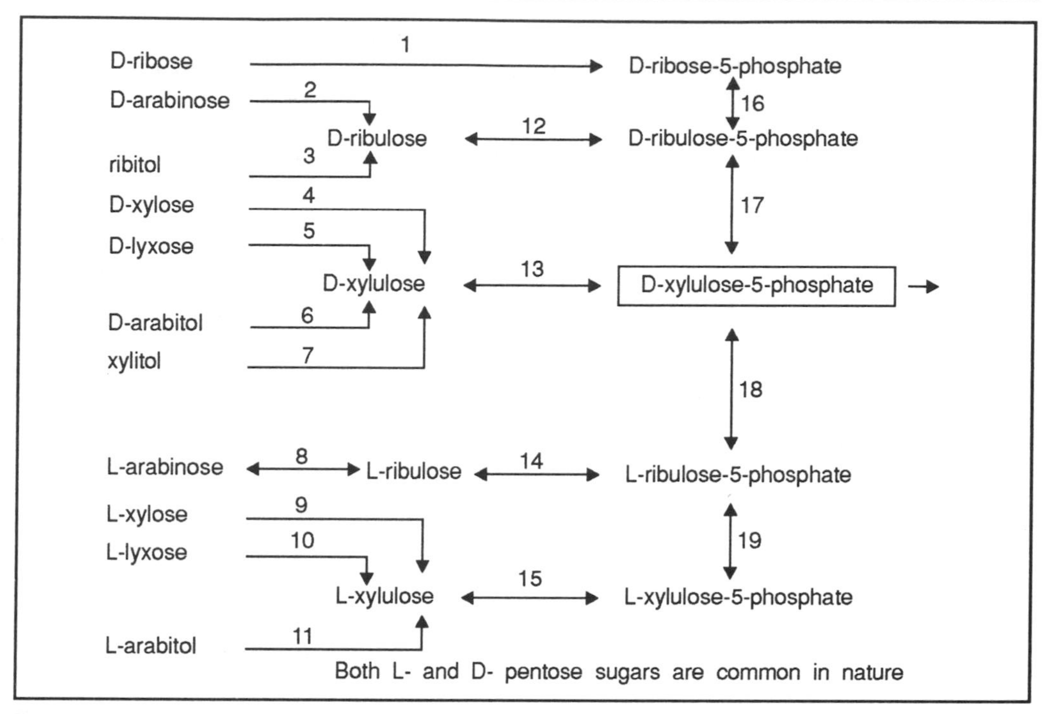

Figure 6.1 Interconversion of pentoses and pentitols to D-xylulose-5-phosphate.(Numbers are referred to within the text). Note that pentitols are polyols and are chemically related to the pentoses. Pentoses are converted to pentitols by reducing the keto or aldehyde groups to the corresponding alcohol. Note also the key position of D-xylulose-5-phosphate. This sugar phosphate can be passed into central metabolism.

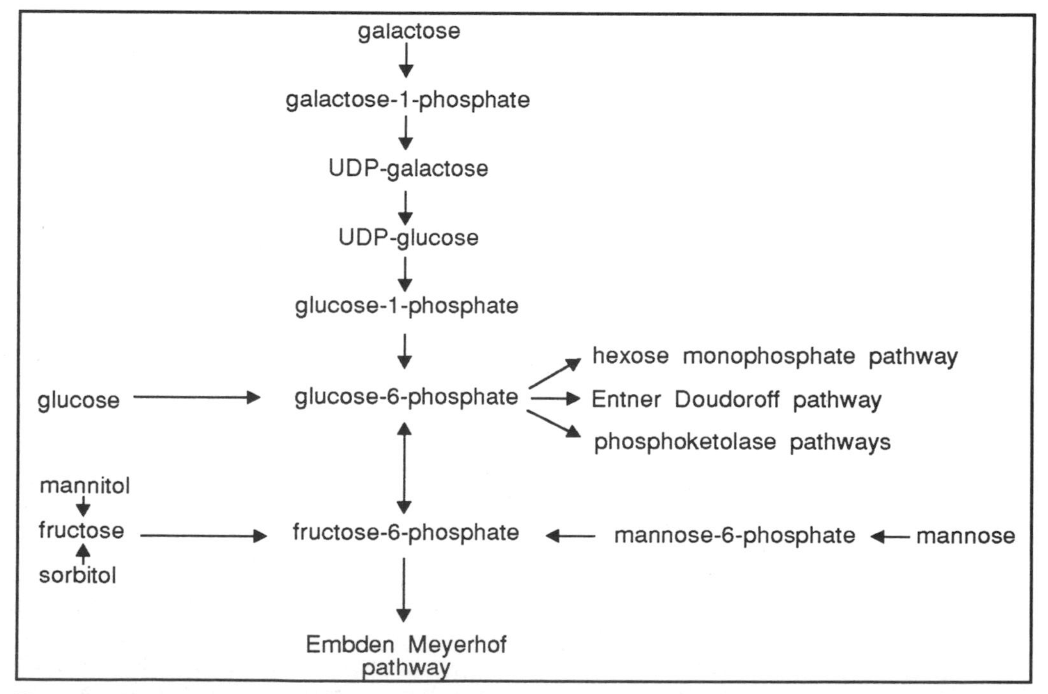

Figure 6.2 Interconversion of hexoses and hexitols for entry into central metabolic pathways. Hexitols are related to hexoses. Hexoses can be converted to the corresponding hexitols by reduction of their keto or aldehyde groups to form alcohols.

Π Indicate which reactions shown in Figure 6.1 are (a) isomerisations, (b) phosphorylations, (c) dehydrogenations.

Reactions 2, 4, 5, 8, 9, 10, 16, 17, 18, and 19 are isomerisations, since they only involved the reorganisation of the arrangement of the atoms present in the molecules.

Reactions 1, 12, 13, 14 and 15 are phosphorylations since they involve the addition of phosphate groups to the monosaccharides. The most common phosphorylating agent is ATP.

Reactions 3, 6, 7 and 11 are dehydrogenations since they involve the removal of hydrogen from the polyol to form the monosaccharides.

For these we can write the general formulae:

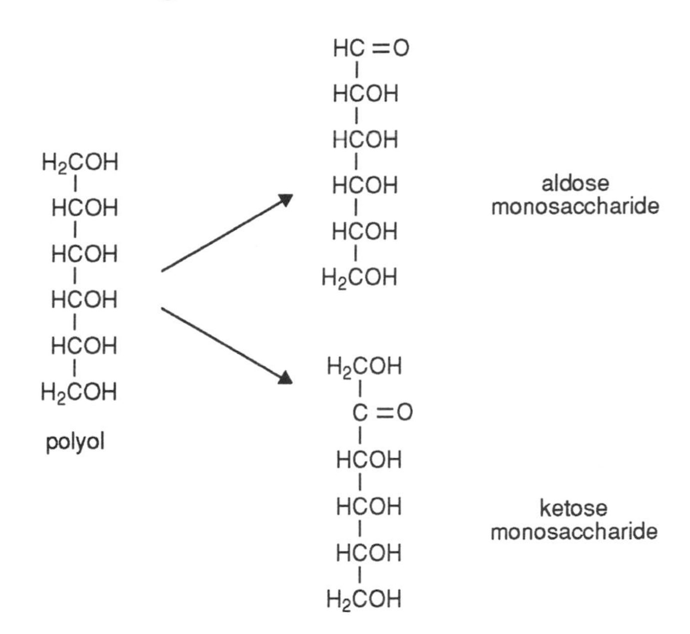

6.3 Commercially important polysaccharide degrading enzymes

In the previous section we saw that relatively few organisms degrade many of the naturally occurring polysaccharides. This means that individual organisms, when presented with a mixture of polysaccharides, may only be able to hydrolyse and benefit from some of these compounds. In an industrial situation this would be a problem because financial pressures would dictate that all of the substrate should be utilised. Addition of polysaccharide hydrolase(s) can often solve this problem. In addition, problems caused by the presence of unwanted polysaccharides can often be alleviated by the addition of enzymes. One of the best known use of enzymes is the employment of α-amylases in brewing. In Europe barley grains are used as the carbon source for beer production but yeasts cannot hydrolyse the starches present in barley. Thus the barley seeds, which initially contain little or no α-amylase activity, are allowed to undergo controlled germination, during which they produce large amounts of the enzyme to hydrolyse the starch to maltose and glucose. This can then be used by the yeast.

Further examples of the uses of α-amylases and other polysaccharide hydrolases are shown in Table 6.2.

enzyme	source	application
α-amylase	plant	in brewing, to degrade starches from barley for yeast to ferment to alcohol
	fungal	in bread production to hydrolyse starches in flour
	fungal	in the manufacture of glucose syrups for the confectionary industry
	fungal	as a digestive aid (particularly the acid resistant enzyme from *Aspergillus niger*)
	bacterial	cold swelling of starch in the laundry industry
pectinase	fungal	in the food and brewing industries to clarify fruit juices and beers
cellulases	fungal (particularly *Trichoderma spp.*)	as a digestive aid

Table 6.2 Examples of industrial uses of polysaccharide hydrolysing enzymes.

SAQ 6.1

Examine the following structure of a polysaccharide and answer the following questions.

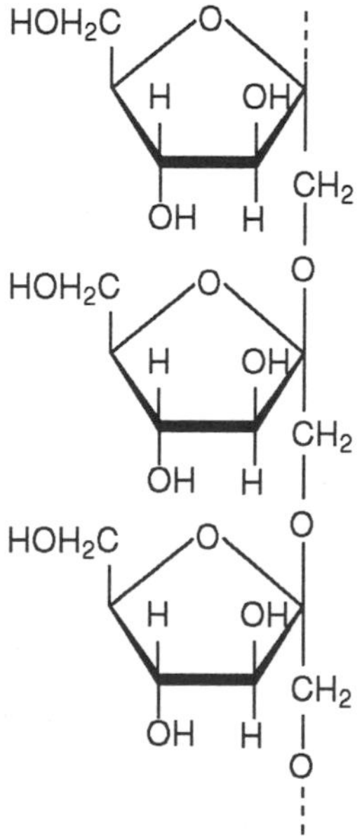

1). Is the polysaccharide a homopolysaccharide or a heteropolysaccharide?

2). Which of the following linkages best describe the linkages between the sugar residues: β1-4, α1-4, β1-2, α1-2?

3). If bacteria were supplied with a medium containing this polysaccharide describe the first stages in the catabolism of this polysaccharide.

6.4 Conversion of glucose to pyruvate

five main pathways

We have so far examined the ways in which polysaccharides, oligosaccharides and any unusual monosaccharides are usually converted to glucose or a similar central hexose or pentose. We have also noted the ways in which such compounds enter individual cells. The following rather large section will deal with the biodegradation of glucose. Five clearly defined pathways are known to occur in living systems and all eventually yield pyruvate as either the sole or one of the end-products. Not all pathways are found together in a single organism, in fact two functional pathways are the maximum at any one time. A survey of the distribution of these pathways throughout living systems will be found at the end of this chapter.

6.4.1 The Embden Meyerhof pathway

pyruvate and lactate production

In addition to the name above, this pathway is known by a variety of other names, for example the Embden-Meyerhof-Parnas pathway, or more commonly glycolysis. Glycolysis is a term literally meaning 'the dissolution of sugar' and is often used by mammalian biochemists for the process in muscle in which glucose is converted to two lactate molecules. This involves the use of the Embden Meyerhof pathway to yield two pyruvates from glucose; subsequently the pyruvate molecules are quantitatively reduced to lactate to maintain a redox balance. The production of lactate is also typical of the homofermentative lactic acid bacteria.

Of the five pathways to be considered in this section, the Embden Meyerhof pathway was the first one to be elucidated. Indeed when Büchner showed in 1897 that a cell free extract of yeast could rapidly ferment glucose to ethanol, a process mediated largely by the Embden Meyerhof pathway, the era of modern biochemical investigation really began.

Monosaccharides are aldoses or ketoses depending on whether or not they contain aldehyde groups, or ketone (oxo-) groups. We should also learn to recognise polyols.

a pentose (an aldose)	a pentose (a ketose)	pentol (a polyol)	hexose (an aldose)
			H O \\// C
H O \\// C	H_2COH	H_2COH	HCOH
HCOH	HC=O	HCOH	HCOH
HCOH	HCOH	HCOH	HCOH
HCOH	HCOH	HCOH	HCOH
H_2COH	H_2COH	H_2COH	H_2COH

The most common are pentoses or hexoses, i.e they contain five or six C atoms. Before studying the Embden Meyerhof pathway we should also learn to recognise the product of this pathway, the central metabolite, pyruvate.

COO^-
|
C=O
|
CH_3

One molecule of the six carbon aldohexose glucose quantitatively yields two molecules of pyruvate. We shall, after studying the pathway, follow the route of the individual carbons in glucose to examine their location in the pyruvate product. Carbon one for example always goes into the same position in the pyruvate molecule. This is indeed true of all of the carbons of glucose, there is never any deviation and the final positions are highly characteristic of this pathway.

Π What do you think is the first thing to happen to glucose as it enters the Embden Meyerhof pathway and why is it necessary?

glucose enters the pathway by activation

The first step, as we noted in Figure 6.2, is the phosphorylation of the sugar. The purpose of this is to activate the glucose and to prevent it leaking out of the cell. This process involves ATP and is possible at either carbon one or carbon six. In the Embden Meyerhof pathway it occurs at carbon six.

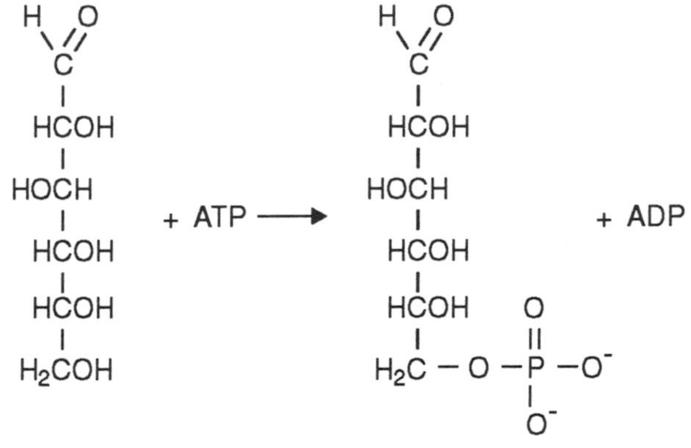

```
 H   O                      H   O
  \ //                       \ //
   C                          C
   |                          |
  HCOH                       HCOH
   |                          |
  HOCH                       HOCH
   |       + ATP ──►          |                + ADP
  HCOH                       HCOH
   |                          |
  HCOH                       HCOH     O
   |                          |       ||
  H2COH                      H2C — O — P — O⁻
                                      |
                                      O⁻

 glucose                   glucose-6-phosphate
```

You may find the naming of enzymes difficult. In this and several subsequent chapters simple explanatory guides will be included in order that you can learn to recognise types of reaction and thus more easily name the enzyme.

kinase

The enzyme in the previous reaction was a kinase. Generally enzymes catalysing reactions in which energy (in the form of ATP) is directly synthesized or, alternatively utilised are termed kinases. In this example the enzyme could be specifically called glucokinase. However the name usually quoted is hexokinase as the enzyme in most living organisms will phosphorylate other hexoses such as fructose, mannose and 2-deoxyglucose.

The next reaction is the conversion of glucose-6-phosphate to fructose-6-phosphate. This is an isomerization reaction and glucose and fructose are isomers of each other.

Π Can you remember what isomers are?

Isomers are compounds which have the same empirical formula. For example, glucose and fructose both contain six carbons, twelve hydrogens and six oxygens (empirical formula therefore is $C_6H_{12}O_6$); it is the arrangement of the atoms within the molecule which is different.

Π We have established that glucose is an aldohexose and that fructose is a ketohexose. The ketone group of hexoses usually occurs at carbon two. Can you draw the straight chain formulae for glucose and fructose for this isomerization reaction? Try it for yourself and check your drawing against ours.

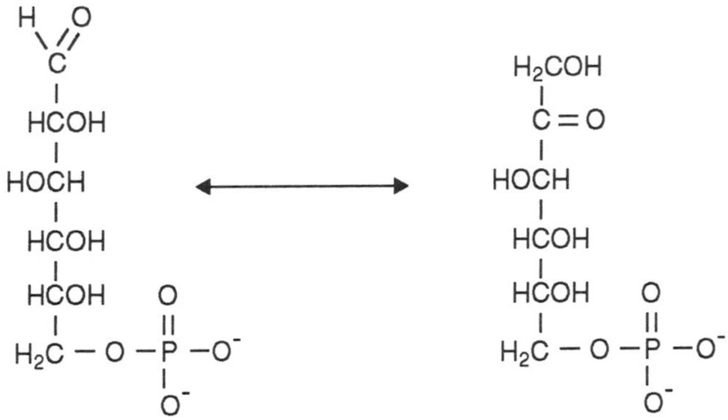

```
   H  O
    \//
     C                                   H2COH
     |                                     |
   HCOH                                   C=O
     |                                     |
  HOCH        <-------------->           HOCH
     |                                     |
   HCOH                                  HCOH
     |                                     |
   HCOH    O                             HCOH    O
     |     ||                              |     ||
  H2C - O -P -O-                        H2C - O -P -O-
           |                                     |
           O-                                    O-
glucose-6-phosphate                   fructose-6-phosphate
```

Notice that the substituents on carbons three to six are unchanged and the reaction, catalysed by hexose phosphate isomerase, is reversible.

further activation

In the next reaction a second phosphate molecule is added to the fructose-6-phosphate at the carbon one position, the reaction being irreversible.

Π Draw the formulae of fructose-6-phosphate and fructose-1, 6-bisphosphate and name the enzyme catalysing the reaction?

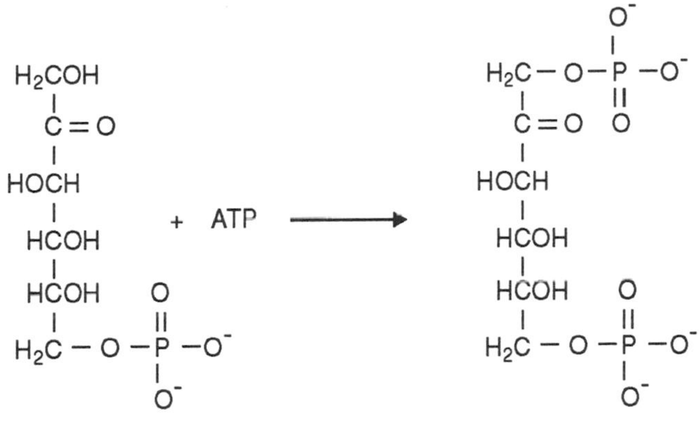

```
                                                  O-
                                                  |
  H2COH                                  H2C- O- P -O-
    |                                      |     ||
   C=O                                    C=O    O
    |                                      |
  HOCH                                   HOCH
    |         +  ATP   ------------>       |
   HCOH                                   HCOH
    |                                      |
   HCOH    O                              HCOH    O
    |      ||                              |      ||
  H2C - O -P -O-                         H2C - O -P -O-
           |                                      |
           O-                                     O-
fructose-6-phosphate                  fructose-1,6-bisphosphate
```

phospho-fructokinase is unique to the pathway

The name of the enzyme is phosphofructokinase; you would have been correct to suggest variations such as fructose phosphate kinase. This enzyme is very important as it is the only enzyme which is unique to the pathway.

In the preceding three reactions the strategy has been to activate the sugar and to convert it to a suitable molecule for cleavage into two usable triose phosphate intermediates. This cleavage occurs in the next reaction.

sugars form rings

Before proceeding there is one complication which must be mentioned. Glucose and fructose and their phosphorylated intermediates usually exist not as straight chains as we have indicated so far but as ring structures. Aldohexoses usually form a six membered ring involving carbons one to five plus an oxygen atom. This structure is similar to the compound pyran and hence the glucose is said to be in the pyranose ring form. Ketohexoses usually form a five membered ring involving carbons two to five and an oxygen atom. This is called a furanose ring structure due to its similarity to furan. These structures are technically correct but are usually more difficult for the student to follow. For completeness the first three reactions which we have studied are shown in Figure 6.3 as a flow diagram with the sugars in ring form.

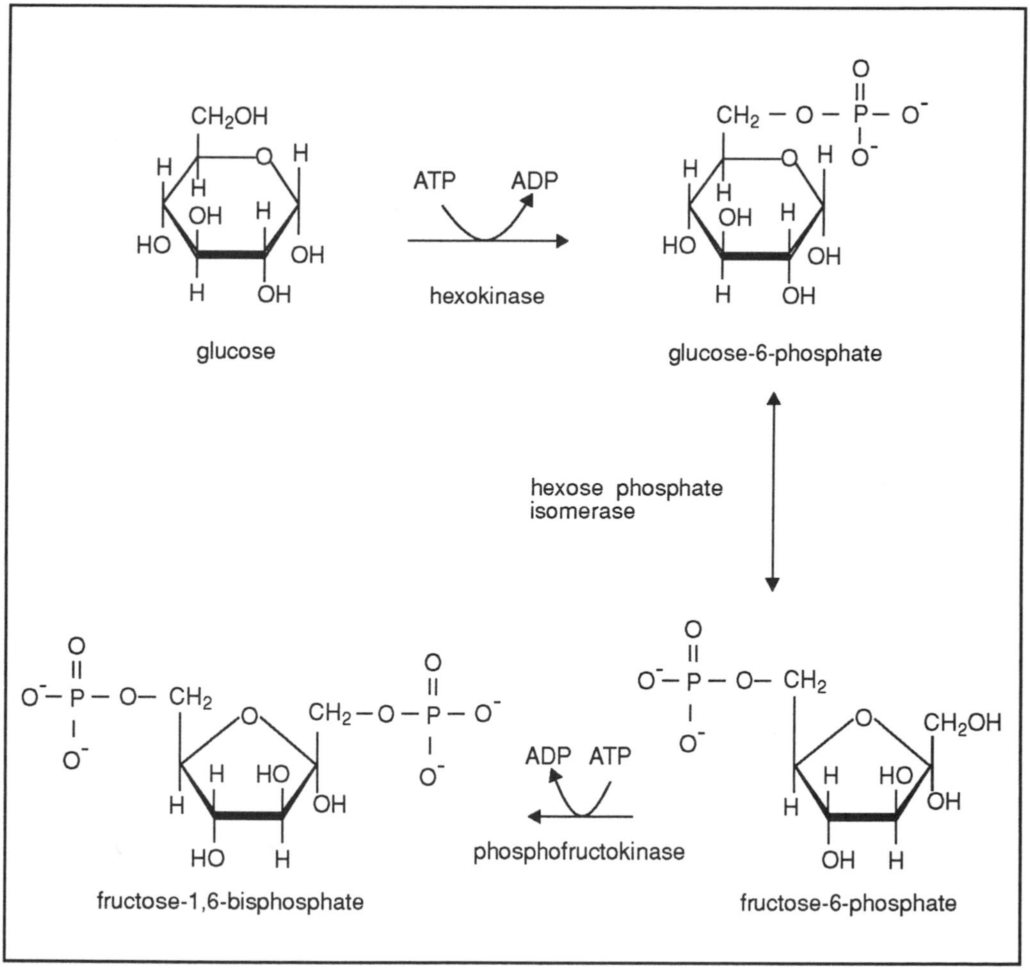

glucose — ATP, ADP, hexokinase → glucose-6-phosphate

glucose-6-phosphate ↔ fructose-6-phosphate (hexose phosphate isomerase)

fructose-6-phosphate — ATP, ADP, phosphofructokinase → fructose-1,6-bisphosphate

Figure 6.3 A flow diagram of the first three reactions of the Embden Meyerhof pathway showing the sugars in their ring form.

Π Copy out Figure 6.3 onto a fresh sheet of paper and then, as you learn about the remaining reactions of the pathway add them to your sheet. In this way you can produce a summary of the pathway which will help you learn the individual steps.

The fructose-1, 6-bisphosphate is now cleaved by fructose-1, 6-bisphosphate aldolase to yield glyceraldehyde-3-phosphate and dihydroxyacetone phosphate. The name aldolase is unusual but it is named for its ability to catalyse the reaction shown below from right to left. This is chemically an aldol condensation, between a hydroxyl group of a triose and an aldehyde group of a second triose to yield a hexose.

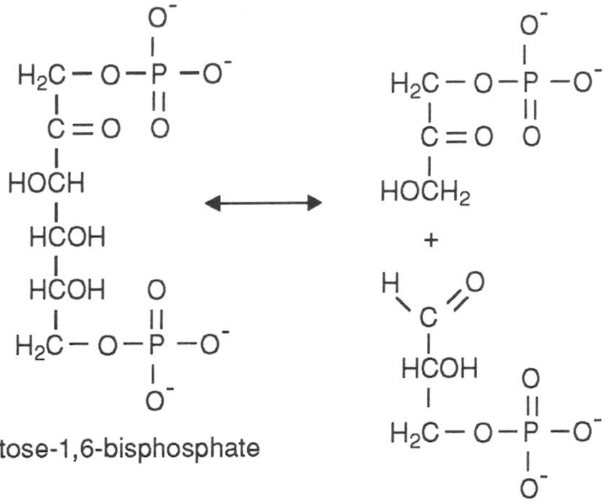

fructose-1,6-bisphosphate ⟷ dihydroxyacetone phosphate + glyceraldehyde-3-phosphate

In fact dihydroxyacetone phosphate is not part of the direct route of the Embden Meyerhof pathway but is something of a side arm. It can be converted to glyceraldehyde-3-phosphate by the action of an enzyme.

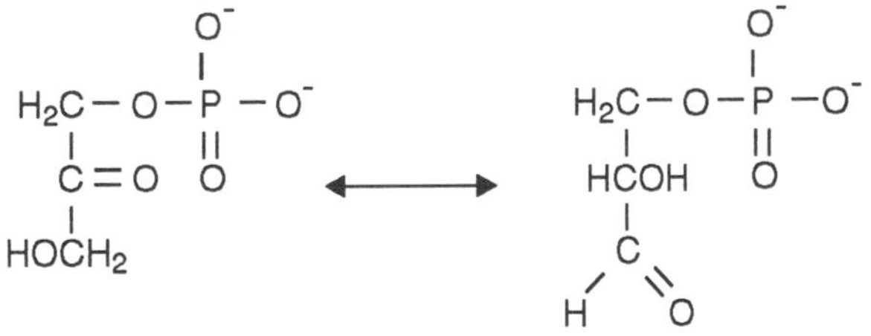

dihydroxyacetone phosphate ⟷ glyceraldehyde-3-phosphate

Π Can you name the enzyme or at least the type of reaction being catalysed?

The reaction is an isomerization and is carried out by an isomerase, the enzyme merely rearranging the hydrogen and oxygen atoms on two of the carbons. Since both compounds are triose phosphates the enzyme is called triose phosphate isomerase. As with other isomerases the reaction is reversible although the equilibrium lies over to the left, towards dihydroxyacetone phosphate, however, because of the continued removal of glyceraldehyde-3-phosphate in subsequent reactions the effective reaction in practice is left to right.

glucose forms two glyceraldehyde-3- phosphates

It is important to note that because of this the glucose molecule has been converted to two molecules of glyceraldehyde-3-phosphate. Thus each of the remaining reactions will occur twice during the conversion of glucose to two pyruvate.

Up to this point in the pathway two ATP molecules have been expended but no energy has been generated in the system. In the next reaction a high energy phosphate compound is produced at the same time as NAD^+ is reduced to $NADH + H^+$.

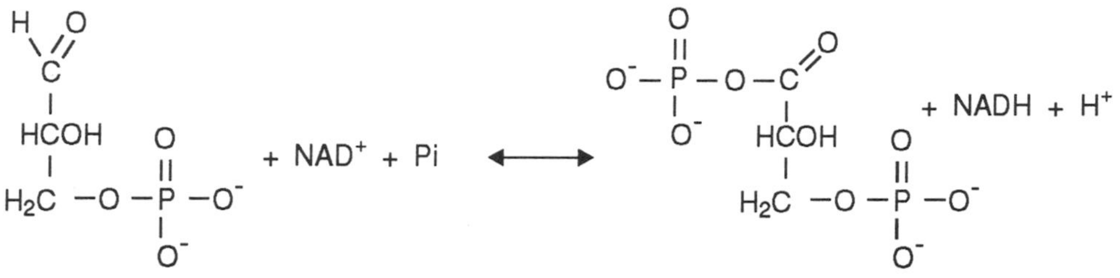

$$\text{glyceraldehyde-3-phosphate} + NAD^+ + Pi \longleftrightarrow \text{1,3-bisphosphoglycerate} + NADH + H^+$$

glyceraldehyde-3-phosphate 1,3-bisphosphoglycerate

first high energy compound made

The enzyme which carries out this reaction is a dehydrogenase, namely glyceraldehyde-3-phosphate dehydrogenase. Dehydrogenases are very important enzymes in metabolism and are generally easy to recognise. They usually catalyse reversible reactions using cofactors such as NAD^+, FAD or $NADP^+$.

substrate level phosphorylation

In the next reaction the production of ATP occurs for the first time in this pathway. This method of ATP production is called substrate level phosphorylation. It is important to distinguish it from ATP generated via the electron transport chain and oxidative phosphorylation. The stages at which substrate level phosphorylation occurs are important in many anaerobic organisms as it is generally the only way by which they obtain usable energy. The reaction is shown below:

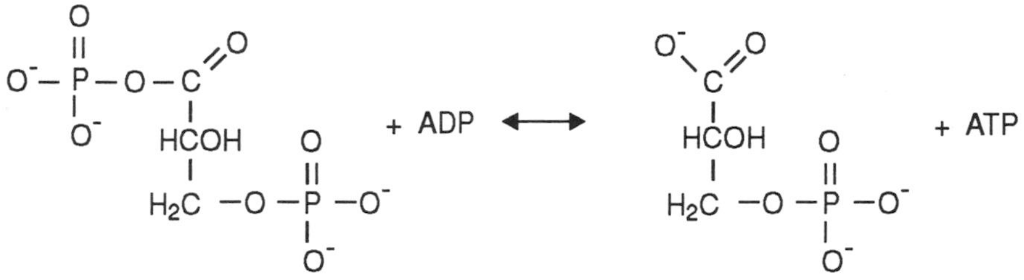

$$\text{1,3-bisphosphoglycerate} + ADP \longleftrightarrow \text{3-phosphoglycerate} + ATP$$

1,3-bisphosphoglycerate 3-phosphoglycerate

Π Can you name the enzyme catalysing the above reaction?

The clue lies in the fact that ATP synthesis occurs when the reaction moves to the right. The enzyme is therefore a kinase and its full name is phosphoglycerate kinase.

The next reaction is simply a rearrangement of the atoms within the molecule, specifically 3-phosphoglycerate is converted to 2-phosphoglycerate by moving the phosphate group from one carbon to another. The enzyme is called phosphoglycerate mutase.

$$\underset{\text{3-phosphoglycerate}}{^{-}OOC{-}HCOH{-}H_2C{-}O{-}PO_3^{2-}} \xrightleftharpoons[\text{mutase}]{\text{phosphoglycerate}} \underset{\text{2-phosphoglycerate}}{^{-}OOC{-}HC(O{-}PO_3^{2-}){-}H_2COH}$$

H2O removal increases phosphate transfer potential

Removal of a molecule of water from 2-phosphoglycerate follows producing an enol phosphate. Enol phosphate compounds have a high phosphate transfer potential and can be used to produce ATP.

$$\underset{\text{2-phosphoglycerate}}{^{-}OOC{-}HC(O{-}PO_3^{2-}){-}H_2COH} \xrightleftharpoons{\text{enolase}} \underset{\text{phosphoenol pyruvate}}{^{-}OOC{-}C(O{-}PO_3^{2-}){=}CH_2} + H_2O$$

In the final reaction of glycolysis the high energy phosphate group is transferred from phosphoenol pyruvate to ADP. The enzyme is pyruvate kinase. The equilibrium constant for this reaction favours the formation of pyruvate. In practice, this reaction can be regarded as virtually irreversible.

$$\underset{\text{phosphoenol pyruvate}}{^{-}OOC{-}C(O{-}PO_3^{2-}){=}CH_2} + ADP \xrightleftharpoons[\text{kinase}]{\text{pyruvate}} \underset{\text{pyruvate}}{^{-}OOC{-}C{=}O{-}CH_3} + ATP$$

Now we have reached pyruvate it is time to check your progress with the following SAQ.

SAQ 6.2

Draw a simplified flow diagram of the Embden Meyerhof pathway using names of compounds and not their formulae. Indicate energy and redox (reduction/oxidation) changes. For example your first two reactions will be:

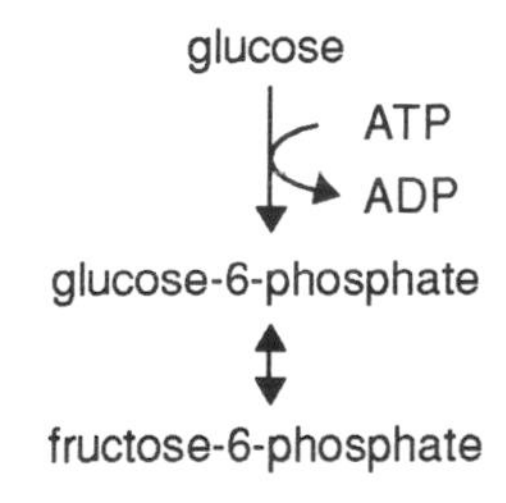

For each of the completed pathways you should always try to derive an overall equation. This is an equation which takes into account all of the changes which have occurred when comparing the starting materials to the final products. Overall equations are very useful because they give information not only about the metabolised compounds but also about energy and redox changes.

From the overall equation a net yield of ATP can be quoted, this is the number of molecules produced directly (that is by substrate level phosphorylation) from one molecule of glucose. It is worth noting that this does not have to be a whole number, for example if five ATP are produced from two glucose molecules, the net yield is 2.5.

SAQ 6.3

Derive the overall equation for the Embden Meyerhof pathway (using your diagram from SAQ 6.2) and calculate the net yield of ATP. In complicated pathways it is often beneficial to deal with the starting carbon compound and the carbon products on the left and right respectively of the equation and then to move on to redox and energy changes.

Π What is the general fate of each of the products of the pathway you established in SAQ 6.3? Remember each product will have a role and will not merely be left to accumulate in the cell.

Pyruvate as we shall see is the central metabolite and may be fed into the TCA (Krebs) cycle (Chapter 7) or be used for many biosynthetic reactions (dealt with in the BIOTOL text, 'Biosynthesis and the Integration of Cell Metabolism'). The ATP will of course be used to provide chemical energy for metabolic reactions. The NADH + H^+ must not be allowed to accumulate because, due to the fact that only small amounts of this cofactor are present, this would very quickly result in a deficiency of NAD^+. If you answered by writing that each NADH + H^+ is reoxidised by the electron transport chain in conjunction with the oxidative phosphorylation apparatus the answer is correct but needs qualifying. It is true only for aerobically-growing organisms. Remember many micro-organisms can grow anaerobically and do not have electron transport and oxidative phosphorylation facilities. Such organisms have a variety of ways to regenerate NAD^+ under anaerobic conditions and these will be dealt with in detail in Chapter 9. The principle of this, however, will be illustrated by a simple example at this point. A pathway possessed by some anaerobic bacteria and by mammalian muscle regenerates NAD^+ by reducing pyruvate to lactate:

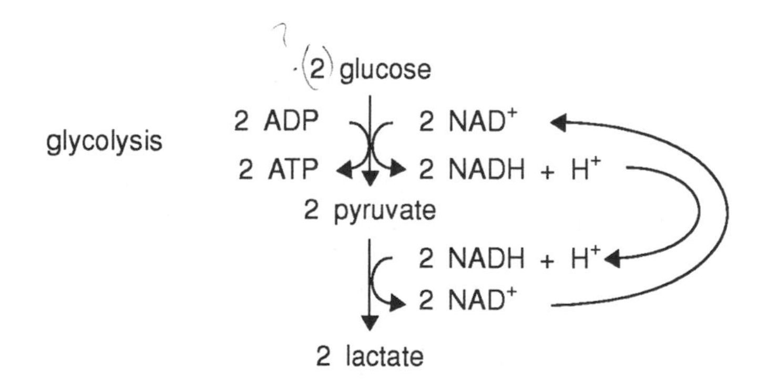

Thus the overall reaction for the above flow diagram is :

$$\text{glucose} + 2\text{ADP} \rightarrow 2 \text{ lactate} + 2\text{ATP}$$

reoxidation of NADH to NAD^+

It is important for us to remember that to aerobic organisms NADH + H^+ means more ATP. For anaerobic organisms the production of excess NADH + H^+ must be avoided at all costs.

The example above shows that the Embden Meyerhof pathway can occur in organisms growing totally anaerobically and for this reason the phrase 'the Embden Meyerhof pathway is an anaerobic pathway' is sometimes used. This is a phrase which, although correct, has to be treated with caution and explanation to avoid misunderstanding.

gluco-neogenosis versus glycolysis

To carry out a proper discussion of the control of this pathway, knowledge is required of a process called gluconeogenesis which converts pyruvate to glucose and occurs in all living cells. This process employs seven of the ten enzymes of the Embden Meyerhof pathway with additional ones where the Embden Meyerhof enzymes are not reversible. Effective control of both catabolic and anabolic pathways is therefore essential to establish both direction and flux of intermediates. In brief, control of the Embden Meyerhof pathway is exercised at the three stages catalysed by irreversible enzymes. This control centres largely on the energy charge or adenylate charge of the cell. As with the cofactor NAD^+/NADH + H^+ there is only a restricted, relatively-constant amount of ATP plus ADP plus AMP in cells. It is the ratio of ATP to ADP plus AMP which is important in control. If all of the compounds are present as ATP the energy charge is maximal; if all are present as AMP the energy charge is zero. The Embden Meyerhof pathway, in conjunction with the TCA cycle is the catabolic process by which glucose is degraded to yield energy. In contrast gluconeogenesis requires energy. Thus high energy charge favours gluconeogenesis and prevents glycolysis.

The complete Embden Meyerhof pathway was elucidated in 1940 and scientists tended to accept that this was the only pathway by which glucose was catabolised. However, certain metabolic essentials cannot be provided by this pathway (remember that glucose is often the only carbon and energy source). The missing essentials are pentoses and tetroses as well as the cofactor which supplies reducing power for biosynthesis, NADPH + H^+. The Embden Meyerhof pathway cleaves a six carbon compound to two three carbon compounds, bypassing pentoses and tetroses, and its single dehydrogenase enzyme reduces NAD^+ to NADH + H^+.

The experimental evidence that finally proved that a second pathway was in existence was that carbon dioxide evolved from glucose could, under certain conditions, arise from carbon-one of the glucose molecule. This is not possible if glucose is degraded via the Embden Meyerhof pathway as you should be able to demonstrate.

SAQ 6.4

Which carbon(s) of the glucose molecule would produce carbon dioxide if catabolism was via the Embden Meyerhof pathway?

To investigate this problem draw out the intermediates of the pathway using line forms and not ring structures. Follow the fate of the six carbons individually throughout the pathway. The one new piece of information which you may require is that any carbon dioxide arises from the carboxyl (COOH) of pyruvate.

This exercise demonstrates that carbon dioxide is derived from carbons three and four, during the operation of the Embden Meyerhof pathway. We will now examine the other pathways of glucose breakdown some of which result in the release of other carbons as carbon dioxide.

6.4.2 The hexose monophosphate pathway

Note that this pathway is also known as the phosphogluconate pathway or pentose phosphate pathway.

This pathway can be divided into two distinct parts, in the first part glucose is oxidised and decarboxylated to yield ribulose-5-phosphate and NADPH + H^+. In the second part a rearrangement of the five-carbon intermediates occurs to yield three-carbon, four-carbon and six-carbon intermediates. In this part of the pathway there are no energy changes, redox changes, carboxylations or decarboxylations.

As before, glucose begins by being activated by forming glucose-6-phosphate. In actual fact this occurs with all of the pathways described in this section. It is worth noting that the enzyme hexokinase is available to all these pathways; it is not the case that there is a separate one for each pathway.

In the second reaction glucose-6-phosphate is oxidised by the removal of two hydrogen atoms to yield 6-phosphogluconolactone.

```
 H  O                                       O
  \//                                       ||
   C                                        C ----
   |                                        |     |
 HCOH                                     HCOH    |
   |                                        |     |
 HOCH        +  NADP+   <------->        HOCH     O     +  NADPH + H+
   |                                        |     |
 HCOH                                     HCOH    |
   |                                        |     |
 HCOH    O                                 HC ----   O
   |     ||                                 |        ||
 H2C-O-P-O-                                H2C-O-P-O-
         |                                           |
         O-                                          O-
glucose-6-phosphate                     6-phosphogluconolactone
```

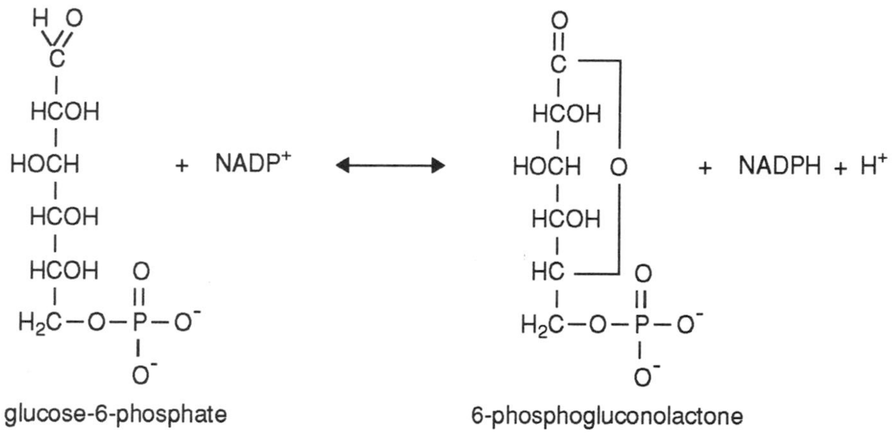

Π Can you name the enzyme catalysing the reaction?

NADPH + H^+ is formed for the first time

Look at the reaction for clues. It is a reversible reaction in which a metabolite yields two hydrogens to $NADP^+$ producing NADPH + H^+. Thus a dehydrogenation has occurred in which glucose-6-phosphate has lost two hydrogens. The name of the enzyme therefore is glucose-6-phosphate dehydrogenase. Note that this enzyme is specific for $NADP^+$ and does not use NAD^+.

The next reaction is carried out by a specific lactonase, 6-phosphogluconate lactonase, an enzyme which breaks open the lactone ring yielding 6-phosphogluconate.

6-phosphogluconolactone + H_2O ⟷ 6-phosphogluconate

In the final reaction of the first half of this pathway the 6-phosphogluconate is oxidatively decarboxylated, probably via the transient intermediate 3-keto-6-phosphogluconate. The enzyme catalysing the reaction(s) is 6-phosphogluconate dehydrogenase.

6-phosphogluconate + $NADP^+$ ⟷ 3-keto-6-phosphogluconate + NADPH + H^+

3-keto-6-phosphogluconate → carbon dioxide + ribulose-5-phosphate

Note that the carbon dioxide in this pathway can only arise from carbon one of glucose.

Π Can you write an overall reaction for the conversion of glucose to ribulose-5-phosphate?

Your answer should be:

glucose + $2NADP^+$ + ATP →

$$\text{ribulose-5-phosphate} + CO_2 + 2NADPH + 2H^+ + ADP \qquad (E - 6.2)$$

The way to work this out is to write the individual reactions out like this:

glucose + ATP → glucose-6-phosphate + ADP

glucose-6-phosphate + $NADP^+$ → 6-phosphogluconolactone + NADP + H^+

6-phosphogluconolactone + H_2O → 6-phosphogluconate

6-phosphogluconate + $NADP^+$ → 3 keto-6-phosphogluconate + NADPH + H^+

3 keto-6-phosphogluconate → ribulose-5-phosphate + CO_2

We can now cancel out molecules that are produced in one reaction, but are used in a subsequent reaction:

glucose + ATP → ~~glucose-6-phosphate~~ + ADP

~~glucose-6-phosphate~~ + $NADP^+$ → ~~6-phosphogluconolactone~~ + NADP + H^+

~~6-phosphogluconolactone~~ + H_2O → ~~6-phosphogluconate~~

~~6-phosphogluconate~~ + $NADP^+$ → ~~3 keto-6-phosphogluconate~~ + NADPH + H^+

~~3 keto-6-phosphogluconate~~ → ribulose-5-phosphate + CO_2

Adding the remainder together yields:

glucose + $2NADP^+$ + ATP + H_2O

→ ribose-5-phosphate + CO_2 + 2NADPH + $2H^+$ + ADP

The elucidation of this first part of the pathway remedies two of the inadequacies of the Embden Meyerhof pathway; namely it provides a source of NADPH + H^+ and there is production of a pentose sugar. As we saw earlier, once a pentose is produced, cells can convert it into other pentoses and the next step in the pathway is the conversion of ribulose-5-phosphate to either ribose-5-phosphate or xylulose-5-phosphate.

Two problems have still not been resolved however. Firstly, there is no production of tetroses yet and no obvious mechanism whereby the cell can produce varying ratios of NADPH + H^+ with respect to ribulose-5-phosphate. From the above overall equation they are produced in a strict (NADPH + H^+: ribulose-5-phosphate) ratio of 2:1.

transaldolase and transketolase

The second half of the pathway, which consists only of the interconversion of carbon skeletons, relies heavily on two key enzymes and is an extremely flexible and useful series of processes. The two enzymes are transaldolase and transketolase and they are found in all living cells. Transaldolase transfers a three carbon dihydroxyacetone group from a ketose sugar to an aldehyde recipient (see Figure 6.4). Transketolase transfers a two carbon 'active aldehyde group' from a ketose sugar to an aldehyde sugar. The sequence of events for this part of the pathway is shown in Figure 6.4 and to provide enough carbon skeletons to balance we will start with three molecules of ribulose-5-phosphate.

After isomerase 1) and epimerase 2) activities to produce ribose-5-phosphate and xylulose-5-phosphate respectively from ribulose-5-phosphate the first of the two key enzymes to act is transketolase 3). A two carbon fragment, carbons one and two of xylulose-5-phosphate, is transferred onto carbon one of ribose-5-phosphate. This results in the formation of glyceraldehyde-3-phosphate and sedoheptulose-7-phosphate.

These two compounds now react under the influence of transaldolase 4), the enzyme transferring a three carbon fragment from sedoheptulose-7-phosphate to glyceraldehyde-3-phosphate, yielding erythrose-4-phosphate and fructose-6-phosphate respectively. Erythrose-4-phosphate is the first tetrose we have seen.

Finally, the transketolase is again used to transfer a two-carbon fragment from xylulose-5-phosphate to the erythrose-4-phosphate to yield glyceraldehyde-3-phosphate and fructose-6-phosphate respectively, the end-product of this pathway. These last few paragraphs are quite complicated but all of the reactions are shown in Figure 6.4 and a careful study of them is strongly recommended before proceeding.

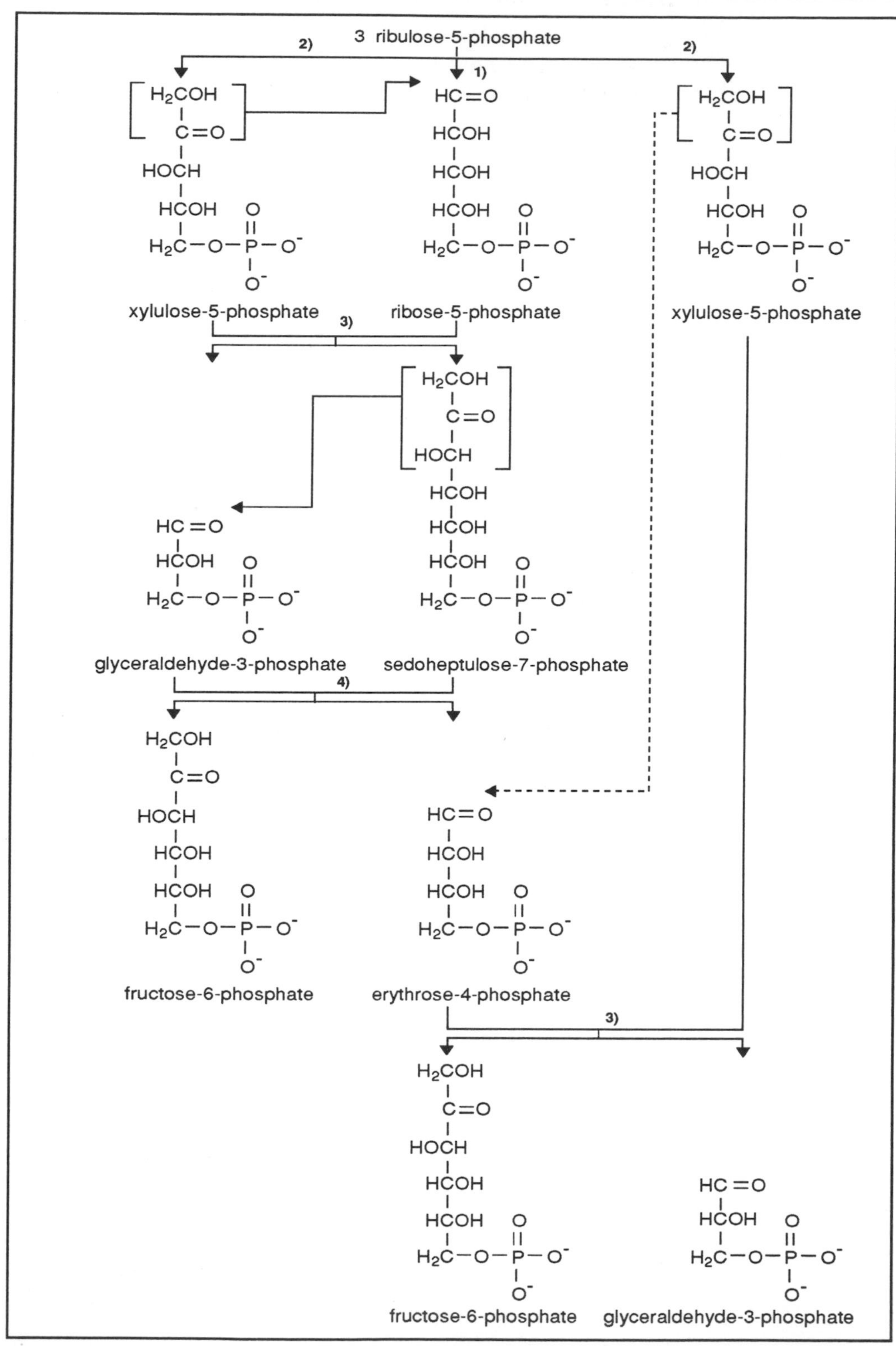

Figure 6.4 Flow diagram of the non-oxidative half of the hexose monophosphate pathway. (See text for details).

Π Can you derive an overall equation for this part of the hexose monophosphate pathway as shown in Figure 6.4?

Your result should be:

3 ribulose-5-phosphate →

2 fructose-6-phosphate + glyceraldehyde-3-phosphate (E - 6.3)

If you did not get this result, use the method we described for the conversion of glucose to ribulose-5-phosphate (i.e write out all the reactions and cancel the products that are made and subsequently removed). Note that no energy or redox changes are involved in the carbon rearrangement. As a useful check always add up the number of carbons on each side of an equation, in this case:

$(3 \times 5) \rightarrow (2 \times 6) + (1 \times 3)$

thus 15 = 15 and the equation is balanced.

Sedoheptulose-7-phosphate is an uncommon metabolite occurring only in this pathway. Glyceraldehyde-3-phosphate is an intermediate of the Embden Meyerhof pathway and could theoretically be converted to pyruvate.

Π Can you think of a reason why glyceraldehyde-3-phosphate is not removed from the hexose monophosphate pathway at this stage?

The problem lies not with the glyceraldehyde-3-phosphate but with the sedoheptulose-7-phosphate. These two compounds are required to react with each other for the pathway to proceed. If glyceraldehyde-3-phosphate was removed, sedoheptulose-7-phosphate would be wasted, because it does not feature in any other pathway. The glyceraldehyde-3-phosphate produced in the final reaction shown on Figure 6.4 could however be removed.

It must be noted that the reactions described in Figure 6.4 are reversible i.e the reaction sequence can also go in the opposite direction. Thus, two molecules of fructose-6-phosphate plus a glyceraldehyde-3-phosphate can be converted into three molecules of ribulose-5-phosphate. We will return to this point later.

Π Can you now write an overall equation for the whole of the hexose monophosphate pathway from glucose?

The result should be:

3 glucose + 6NADP$^+$ + 3 ATP →

$3CO_2$ + 2 fructose-6-phosphate + glyceraldehyde-3-phosphate + 6NADPH + 6H$^+$ + 3ADP (E - 6.4)

You may find it easier to produce this equation in stages. The overall equation of the first half of the pathway (equation 6.2) has to be multiplied throughout by three as three ribulose-5-phosphates are required for the second half of the pathway. Thus:

$$\text{glucose} + 2\text{NADP}^+ + \text{ATP} \rightarrow CO_2 + \text{ribulose-5-phosphate} + 2\text{NADPH} + 2H^+ + \text{ADP}$$

becomes:

$$3\ \text{glucose} + 6\text{NADP}^+ + 3\text{ATP} \rightarrow 3CO_2 + 3\ \text{ribulose-5-phosphate} + 6\text{NADPH} + 6H^+ + 3\text{ADP}$$

now add equation (E - 6.3) to it:

$$3\ \text{ribulose-5-phosphate} \rightarrow 2\ \text{fructose-6-phosphate} + \text{glyceraldehyde-3-phosphate} \quad \text{(E - 6.3)}$$

This gives:

$$3\ \text{glucose} + 6\text{NADP}^+ + 3\text{ATP} \rightarrow 3CO_2 + 2\ \text{fructose-6-phosphate} + \text{glyceraldehyde-3-phosphate} + 6\text{NADPH} + 6H^+ + 3\text{ADP} \quad \text{(E - 6.4)}$$

SAQ 6.5

We have seen that if the glyceraldehyde-3-phosphate formed as a result of the first transketolase reaction of Figure 6.4, were removed, sedoheptulose-7-phosphate would be wasted because it features in no other pathway. Would the removal of erythrose-4-phosphate also be a waste of carbon skeletons?

6.4.3 The use of the hexose monophosphate pathway and, where necessary, the Embden Meyerhof pathway to provide NADPH + H^+ or pentoses or both.

We have seen that the hexose monophosphate pathway provides both NADPH + H^+ and pentoses to satisfy cellular requirements. According to the overall equation for the first part of the pathway (Equation E 6.2) each glucose produces two NADPH + H^+ to one pentose; a rather rigid ratio which may not satisfy requirements.

Π Work out a way in which the pathway(s) are used if NADPH + H^+ but not pentoses are required?

A possible solution is that the first part of the hexose monophosphate pathway could be used to generate NADPH + H^+ and that the pentose (ribulose-5-phosphate) formed could be converted by transaldolase and transketolase into fructose-6-phosphate and glyceraldehyde-3-phosphate. These could be oxidised by the Embden Meyerhof pathway.

We can summarise the relationship between the pathways in the following way:

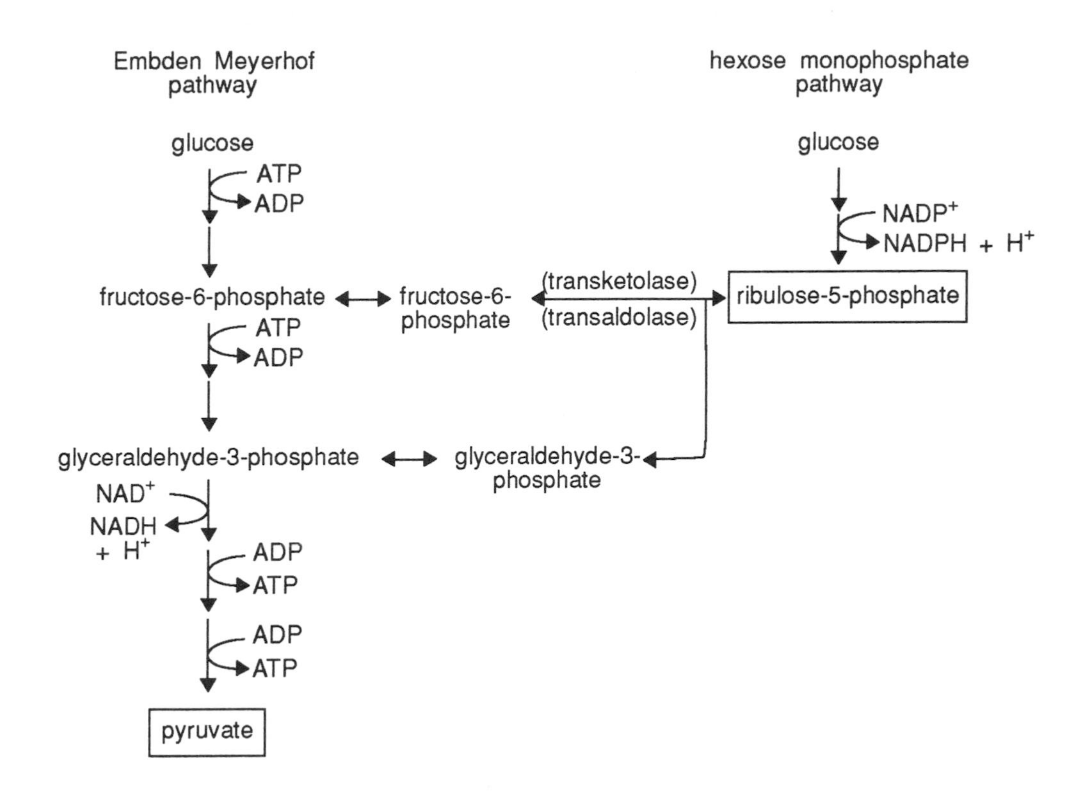

SAQ 6.6

Can you suggest a way in which the pathways operate to yield pentoses but not $NADPH + H^+$?

Before proceeding it is important that you understand the previous ITA and SAQ 6.6.

Before concluding this section we must remember that the Embden Meyerhof pathway produces $NADH + H^+$, which is required in aerobic cells for ATP production. The cell will therefore need to balance the rate of flow (flux) of intermediates through the Embden Meyerhof pathway with that through the hexose monophosphate pathway. An explanation of the control of this balance is described in the BIOTOL text, 'Biosynthesis and the Integration of Cell Metabolism'.

pathways are flexible

At this point it is worth emphasising one or two points which have arisen in Section 6.4 so far. The hexose monophosphate pathway is considered by many students to be unduly complicated. One of the reasons for this is that it is, in practice, extremely flexible. A common early misconception in student studies is that metabolic pathways always start with a particular substrate and slavishly proceed to the final product without pause or addition or removal of intermediates. This is far from the truth: metabolic pathways are generally very flexible. Intermediates can be inserted into the hexose monophosphate pathway at any point. One can start the pathway at any point and, with one exception, intermediates can be withdrawn as required.

Π Can you recall the one exception where an intermediate may not be removed?

The answer is the glyceraldehyde-3-phosphate formed along with the sedoheptulose-7-phosphate following transketolase activity. This is because metabolic processes would not allow such a wasteful situation as to combine two useful pentoses to yield a useful three carbon compound but an unusable seven carbon product.

6.4.4 The Entner Doudoroff pathway

micro-organisms can use different pathways

This pathway, considered to be an alternative to the Embden Meyerhof pathway, was discovered in 1952 by Entner and Doudoroff working on *Pseudomonas saccharophila*. The pathway has since been shown to be the major route for oxidative metabolism of sugars in *Pseudomonas spp.* and related organisms. All are obligate aerobes with the notable exception of *Zymomonas mobilis*, a strict anaerobe. Glucose fermentation by this organism will be discussed later in Chapter 9. More recently, though, unrelated organisms such as *E. coli* have been shown to be capable of producing the enzymes needed for this pathway if grown on substrates such as gluconic acid as sole source of carbon and energy. This is referred to as enzyme induction.

The first reaction is, as in previous pathways, the activation of glucose to glucose-6-phosphate. The next two reactions are in common with the hexose monophosphate pathway as discussed in section 6.4.2. These reactions are shown in Figure 6.5.

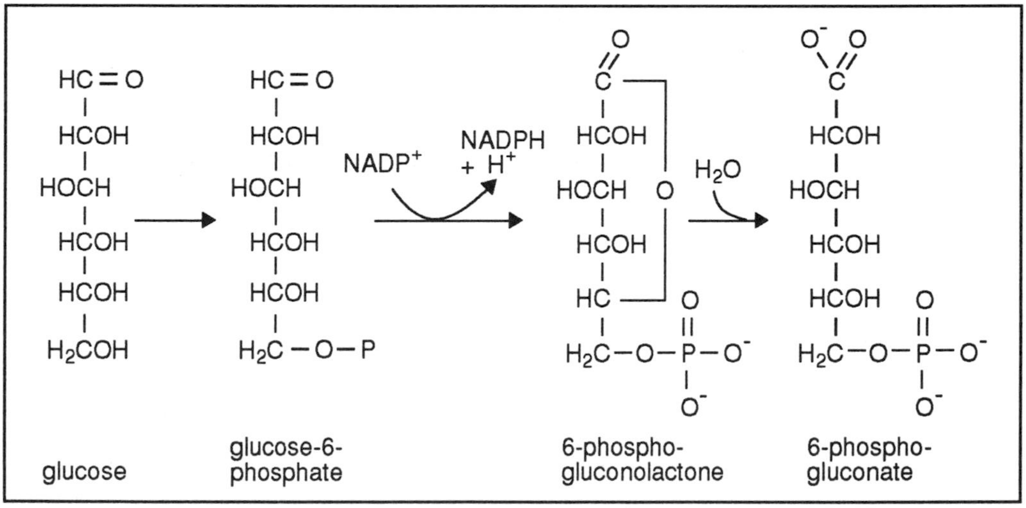

HC=O | HCOH | HOCH | HCOH | HCOH | H_2COH → HC=O | HCOH | HOCH | HCOH | HCOH | H_2C-O-P → ($NADP^+$ → NADPH + H^+) → 6-phosphogluconolactone → (H_2O) → 6-phosphogluconate

glucose; glucose-6-phosphate; 6-phospho-gluconolactone; 6-phospho-gluconate

Figure 6.5 The first three reactions of the Entner Doudoroff pathway.

Π Looking at Figure 6.5, can you remember or deduce the names of the three enzymes involved?

They are, in order: hexokinase, glucose-6-phosphate dehydrogenase and 6-phosphogluconolactone lactonase.

In the next stage the 6-phosphogluconate is dehydrated rather than dehydrogenated as in the hexose monophosphate pathway. This process forms an intermediate unique to this pathway, 2-keto-3-deoxy-6-phosphogluconate.

Π Using the formula for 6-phosphogluconate shown in Figure 6.5 try to work out the formula of 2-keto-3-deoxy-6-phosphogluconate. Do this on a piece of paper and then check it with the structure shown below.

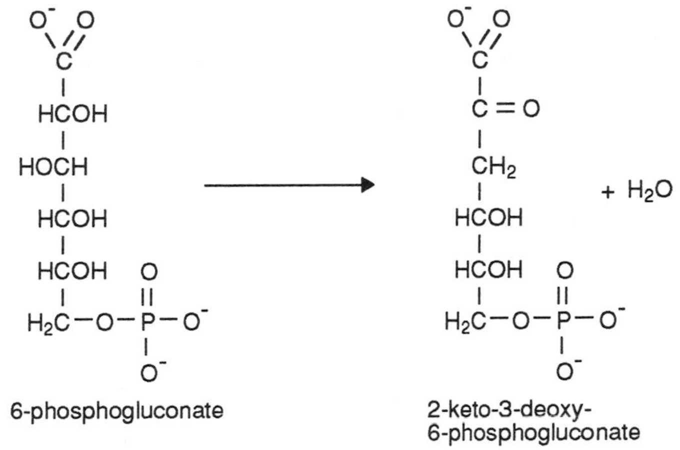

O^- O \ // C | HCOH | HOCH | HCOH | HCOH O | || $H_2C-O-P-O^-$ | O^-

6-phosphogluconate

→

O^- O \ // C | C = O | CH_2 | HCOH | HCOH O | || $H_2C-O-P-O^-$ | O^- + H_2O

2-keto-3-deoxy-
6-phosphogluconate

The enzyme catalysing the reaction is 6-phosphogluconate dehydrase. It is important not to be put off by these long chemical names: they are really guidelines and if you separate them down dealing with one step at a time you should find them far less daunting. Thus 2 keto means it has a keto (C = O) group at position 2; 3-deoxy (no oxygen at position 3) 6-phospho is a phosphate group at position 6, gluc means it is based on glucose and ate is the salt of a carboxylic acid (i.e C_1 is $-COO^-$).

In the next step the 2-keto-3-deoxy-6-phosphogluconate is cleaved to two three carbon compounds in a reaction similar to the one occurring in the Embden Meyerhof pathway.

Π Using the structure of the molecules as a guide try to decide what compounds these are.

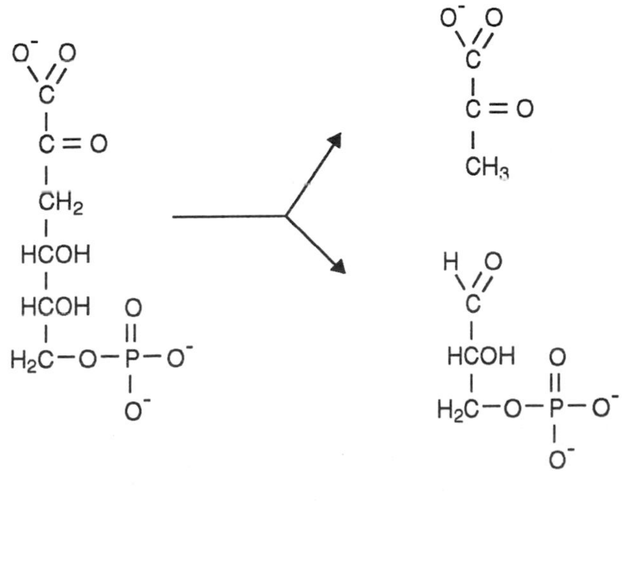

O^- O \ // C | C = O | CH_2 | HCOH | HCOH O | || $H_2C-O-P-O^-$ | O^-

→

O^- O \ // C | C = O | CH_3

H O \ // C | HCOH O | || $H_2C-O-P-O^-$ | O^-

a quick way to get to pyruvate

The enzyme catalysing the reaction is 2-keto-3-deoxy-6-phosphogluconate aldolase. The products formed are pyruvate and glyceraldehyde-3-phosphate.

The glyceraldehyde-3-phosphate is itself converted to pyruvate, using the last five enzymes of the Embden Meyerhof pathway, and this can be summarised by:

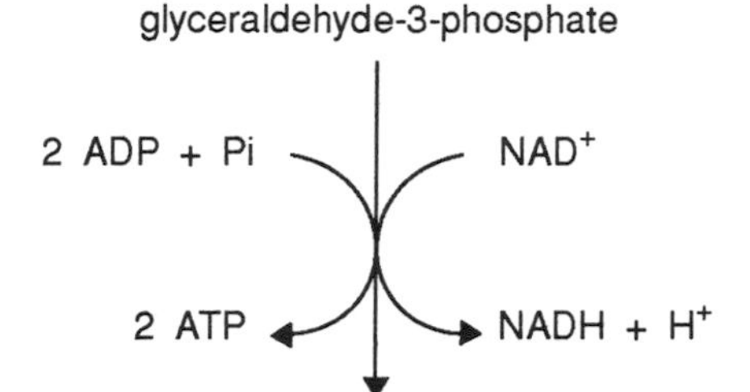

This type of summary diagram is very useful and we shall produce several more throughout this text. Their use is similar to writing out a series of linear equations but they have the advantage that we can interconnect summary diagrams and describe a complex process in a simple way.

Π You should now be able to write an overall reaction for the Entner Doudoroff pathway from glucose to pyruvate.

glucose + NAD^+ + $NADP^+$ + ADP + Pi →

2 pyruvate + NADH + H^+ + NADPH + H^+ + ATP (E - 6.5)

Note that in the equation the NAD^+ and $NADP^+$ are kept separate. This is important in that certain enzymes are specific for NAD(H) and others for NADP(H). In addition, the functions of the two carriers is generally different; NADH is used to produce ATP in aerobes and NADPH to supply reducing power for biosynthetic reactions in all organisms.

SAQ 6.7

Bearing in mind that the carboxyl groups of pyruvate yield carbon dioxide which glucose carbons yield carbon dioxide via the Entner Doudoroff pathway?

6.4.5 Comparison of the Embden Meyerhof and Entner Doudoroff pathways

We have now established the overall equation for each of these two pathways as follows:

glucose + 2 ADP + 2 Pi + 2 NAD^+ →

2 pyruvate + 2 ATP + 2 NADH + 2 H^+ (E - 6.1)

(see response to SAQ 6.3)

glucose + ADP + Pi + NAD^+ + $NADP^+$ →

2 pyruvate + ATP + NADH + H^+ + NADPH + H^+ (E - 6.5)

The first and obvious comment is that the two pathways are very similar to each other, both providing pyruvate, reducing power and energy. The main advantage of the Embden Meyerhof pathway is that the net yield of ATP is two compared to one in the Entner Doudoroff. This on the face of it is a doubling of energy yield, but, as we shall see in the next chapter, in the context of overall energy production from glucose by aerobic organisms, this difference of one ATP is not very significant. The major advantage of the Entner Doudoroff pathway is that both NADH and NADPH are produced. Neither pathway, however, includes pentose and tetrose production.

6.4.6 The pentose phosphoketolase pathway

aeroduric

This pathway is largely restricted to members of the genus *Leuconostoc* and some members of the genus *Lactobacillus*. These organisms are relatively common and they have no mechanism for using oxygen. They are, therefore, anaerobic or fermentative. In fact although they ferment glucose, they are unlike strict anaerobes in that they grow both in the presence and absence of oxygen. They are therefore termed aeroduric. In this chapter we shall confine our studies to the pathway itself: the metabolic problems arising from the fermentative mode will be discussed in chapter 9. The pathway was elucidated in 1952 following the discovery that these organisms did not contain fructose-1, 6-bisphosphate aldolase. This enzyme is an essential part of the Embden Meyerhof pathway which, therefore, can not function in its entirety.

The first part of the pathway is identical to the part of the hexose monophosphate pathway (section 6.4.2) in which glucose is converted to ribulose-5-phosphate and then to xylulose-5-phosphate. The next reaction is the key reaction of this pathway, the cleavage of xylulose-5-phosphate involving inorganic phosphate and the enzyme xylulose-5-phosphate phosphoketolase to yield glyceraldehyde-3-phosphate and acetyl phosphate. The enzyme is absolutely specific for this sugar.

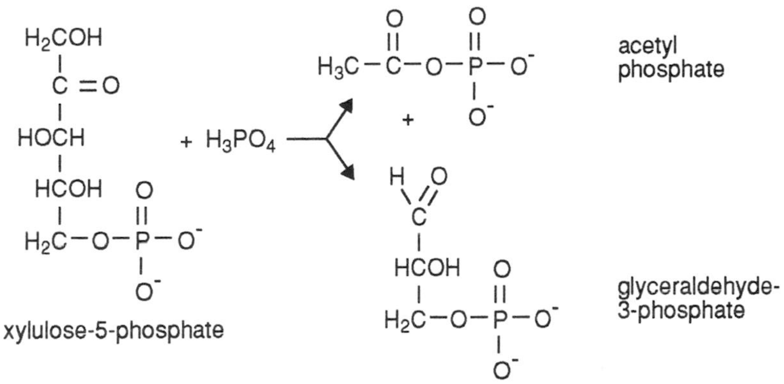

$$H_2COH-C(=O)-HOCH-HCOH-H_2C-O-P(=O)(O^-)-O^- + H_3PO_4 \rightarrow H_3C-C(=O)-O-P(=O)(O^-)-O^- + HC(=O)-HCOH-H_2C-O-P(=O)(O^-)-O^-$$

xylulose-5-phosphate

acetyl phosphate

glyceraldehyde-3-phosphate

appearance of acetyl phosphate

Acetyl phosphate is sometimes written as acetyl ~P, the squiggle indicating that the phosphate bond is a high energy bond which on hydrolysis will yield a similar amount of energy as the reaction ATP $\rightarrow$ ADP + P_i. In fact in certain organisms the following reaction occurs:

$$\text{acetyl} \sim P + ADP \rightarrow \text{acetate} + ATP$$

However in the organisms we are discussing acetyl ~P is reduced to ethanol without production of ATP. The reasons for this will be discussed in chapter 9.

As in all previous pathways the glyceraldehyde-3-phosphate which is formed is converted to pyruvate using the enzymes of the Embden Meyerhof pathway.

We must now work out an overall reaction for this pathway and because it is complicated we will build it up in stages as follows:

1) Equation (E - 6.2), the oxidative part of the hexose monophosphate pathway;

2) epimerisation of ribulose-5-phosphate to xylulose-5-phosphate;

3) the key reaction of this pathway;

4) the last five reactions of the Embden Meyerhof pathway;

5) the overall reaction.

Π Try to work out equations for each of these stages for yourself before reading on.

The equations for each of these processes are:

1) glucose + 2 $NADP^+$ + ATP → CO_2 + ribulose-5-phosphate + 2 NADPH + 2 H^+ + ADP

2) ribulose-5-phosphate ↔ xylulose-5-phosphate

3) xylulose-5-phosphate + Pi → acetyl-phosphate + glyceraldehyde-3-phosphate

4) glyceraldehyde-3-phosphate + NAD^+ + 2ADP + Pi →

pyruvate + NADH + H^+ + 2ATP

We can add these all together and, as before, we cancel everything which appears on both sides of the equation. The remainder gives us the overall reaction.

5) glucose + 2 $NADP^+$ + NAD^+ + ADP + Pi →

pyruvate + acetyl~phosphate + CO_2 + 2 NADPH + 2 H^+ + NADH + H^+ + ATP

The overall reaction is correct but technically needs further revision in that, by convention, overall reactions for anaerobic or fermentative organisms should be in redox balance. That is to say that there should be no reduced or oxidised carriers appearing in the equation. The details of this revision will be discussed in Chapter 9.

6.4.7 The hexose phosphoketolase pathway

We have so far looked at the Embden Meyerhof, the hexose monophosphate, the Entner Doudoroff and the pentose phosphoketolase pathways.

Π As a piece of revision, work out where each pathway would stop if for some reason the enzymes fructose-1,6-bisphosphate aldolase and glucose-6-phosphate dehydrogenase were missing.

This is actually quite simple because pathways would stop at the substrate for which the enzyme was missing. Still, the revision was useful. Thus, the Embden Meyerhof pathway would stop at fructose-1, 6-bisphosphate and the other three at glucose-6-phosphate.

If this were to happen to an organism it would soon die through lack of ATP. Strangely, a group of organisms is known which does not possess either of these two enzymes although they contain the other enzymes of the Embden Meyerhof and hexose monophosphate pathways. These are members of the genus *Bifidobacterium* and they, too, are aeroduric.

These organisms alleviate their problem by possessing a unique enzyme, fructose-6-phosphate phosphoketolase. It is specific for fructose-6-phosphate and is found only in the few organisms in this genus.

fructose-6-phosphate phosphoketolase

Fructose-6-phosphate is produced as in the Embden Meyerhof pathway.

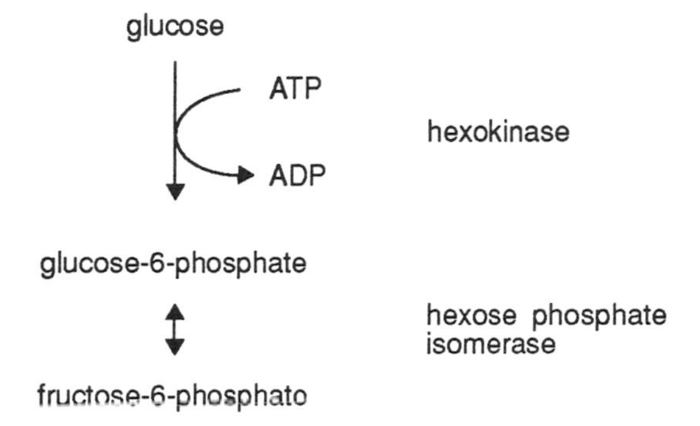

In the next reaction a cleavage of fructose-6-phosphate occurs in the same manner as in the cleavage of xylulose-5-phosphate described in Section 6.4.6. The products of this cleavage are acetyl phosphate and erythrose-4-phosphate.

$$\text{fructose-6-phosphate} + \text{Pi} \rightarrow \text{acetyl phosphate} + \text{erythrose-4-phosphate}$$

(Use Section 6.4.6 to help you draw the structures of these compounds).

Π We know that each of the four pathways studied to date are incomplete in these members of *Bifidobacterium* but we must find a way for the organisms to progress from this point. Can you think of a way forward?

If you are having trouble, look at the following clues one by one until you find the answer.

- only small amounts of erythrose-4-phosphate are required by the organism therefore we would expect this compound to participate in a reaction fairly soon;
- apparently the only available intermediates as well as erythrose-4-phosphate are glucose, glucose-6-phosphate and fructose-6-phosphate;
- think of the hexose monophosphate pathway;
- can you remember how transketolase works?

The answer is the use of transketolase and transaldolase catalysing the following:

$$\text{fructose-6-phosphate} + \text{erythrose-4-phosphate} \leftrightarrow \text{sedoheptulose-7-phosphate} + \text{glyceraldehyde-3-phosphate}$$

and

$$\text{sedoheptulose-7-phosphate} + \text{glyceraldehyde-3-phosphate} \leftrightarrow \text{xylulose-5-phosphate} + \text{ribose-5-phosphate}$$

no reduced cofactors required for pentose or textrose production

The *Bifidobacterium spp.* produce pentoses and tetroses without producing reduced co-factors such as NADH + H^+ or NADPH + H^+. We noted in the last section that an overproduction of NAD(P)H + H^+ must be avoided in anaerobic or fermentative organisms. The pathway described above is the only one in which redox reactions are not involved in hexose to pentose and tetrose conversion. This gives *Bifidobacterium* distinct advantages in terms of energy production as we shall see shortly.

The pathway to this stage however does have two problems to be solved. Firstly there is currently a net loss of one ATP and secondly there has been no obvious way to produce pyruvate or some of the other important precursor metabolites. However, *Bifidobacterium spp.* also have the enzyme xylulose-5-phosphate phosphoketolase which, you may remember, splits xylulose-5-phosphate into glyceraldehyde-3-phosphate and acetyl phosphate.

Further, *Bifidobacterium spp.* are able to generate ATP from acetyl phosphate:

$$\text{acetyl}\sim\text{P} + \text{ADP} \rightarrow \text{acetate} + \text{ATP}$$

We will now test your ability to complete a partial diagram describing the hexose phosphoketolase pathway.

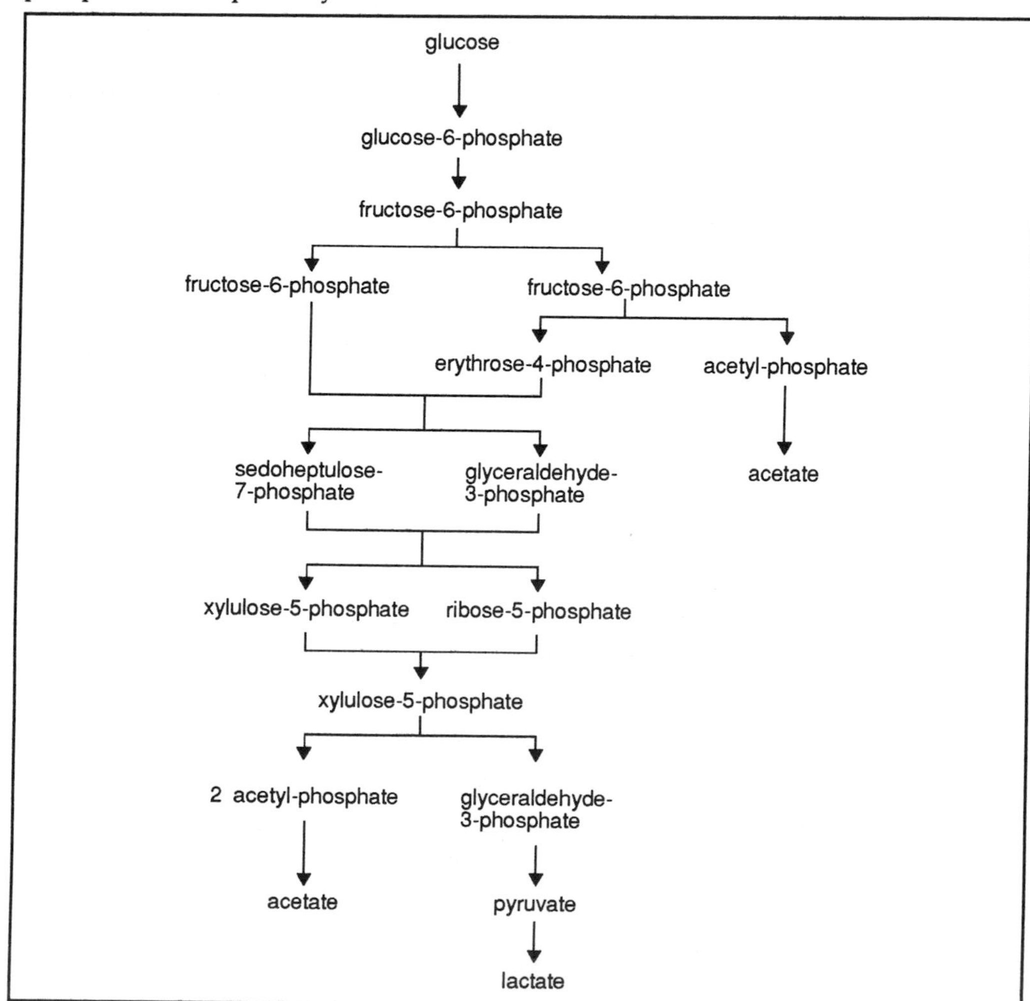

Figure 6.6 Summary of the hexose phosphoketolase pathway for SAQ 6.8.

SAQ 6.8

Consider the incomplete summary diagram in Figure 6.6. Complete the diagram by inserting ATP synthesis or degradation redox changes and addition of inorganic phosphate. In addition produce an overall reaction and work out the net yield of ATP per molecule of glucose utilised. (Hint - start with two molecules of glucose).

2.5 ATP per glucose

Thus *Bifidobacterium spp.* can produce pyruvate, tetroses and pentoses readily and the fact that they produce pentoses and tetroses without redox changes gives them a higher-than-average net yield of ATP (2.5) when compared with other anaerobes. Thus they are very well equipped to grow fermentatively.

6.5 The distribution and overall comparison of the five pathways

We can summarise the distribution of the five pathways as follows:

Embden Meyerhof pathway:	in all higher animals and plants and many bacteria; used in conjunction with the hexose monophosphate pathway
hexose monophosphate pathway:	in all higher animals and plants and most bacteria; used in conjunction with either the Embden Meyerhof or Entner Doudoroff pathways
Entner Doudoroff pathway:	in *Pseudomonas spp.* and related aerobes plus the anaerobic *Zymomonas mobilis* (it may be induced in certain organisms, for example, *E. coli.*); generally acts as an alternative to the Embden Meyerhof pathway
pentose phosphoketolase pathway:	in some of the lactic acid bacteria, namely all of the genus *Leuconostoc* and some members of the genus *Lactobacillus*; these are aeroduric organisms which ferment glucose to produce lactate and acetate
hexose phosphoketolase pathway:	only found in the bacterial genus *Bifidobacterium*; these organisms also require the pentose phosphoketolase pathway

A comparison of the pathways is very useful for several reasons. SAQ 6.9 invites you to initiate this comparison in diagrammatic form and there will follow a discussion of the important aspects.

SAQ 6.9

Consider the diagram in Figure 6.7. it looks complicated but is effectively a summary of the five pathways studied in Section 6. Study the diagram and then identify each pathway. Complete the diagram inserting coloured arrows for the enzymes for each pathway. For example the colours could be:

Embden Meyerhof pathway - red; hexose monophosphate pathway - green; Entner Doudoroff pathway - blue etc.

The first reaction, glucose to glucose-6-phosphate is common to all five pathways and therefore should have five arrows, one in each of the five colours representing the pathways.

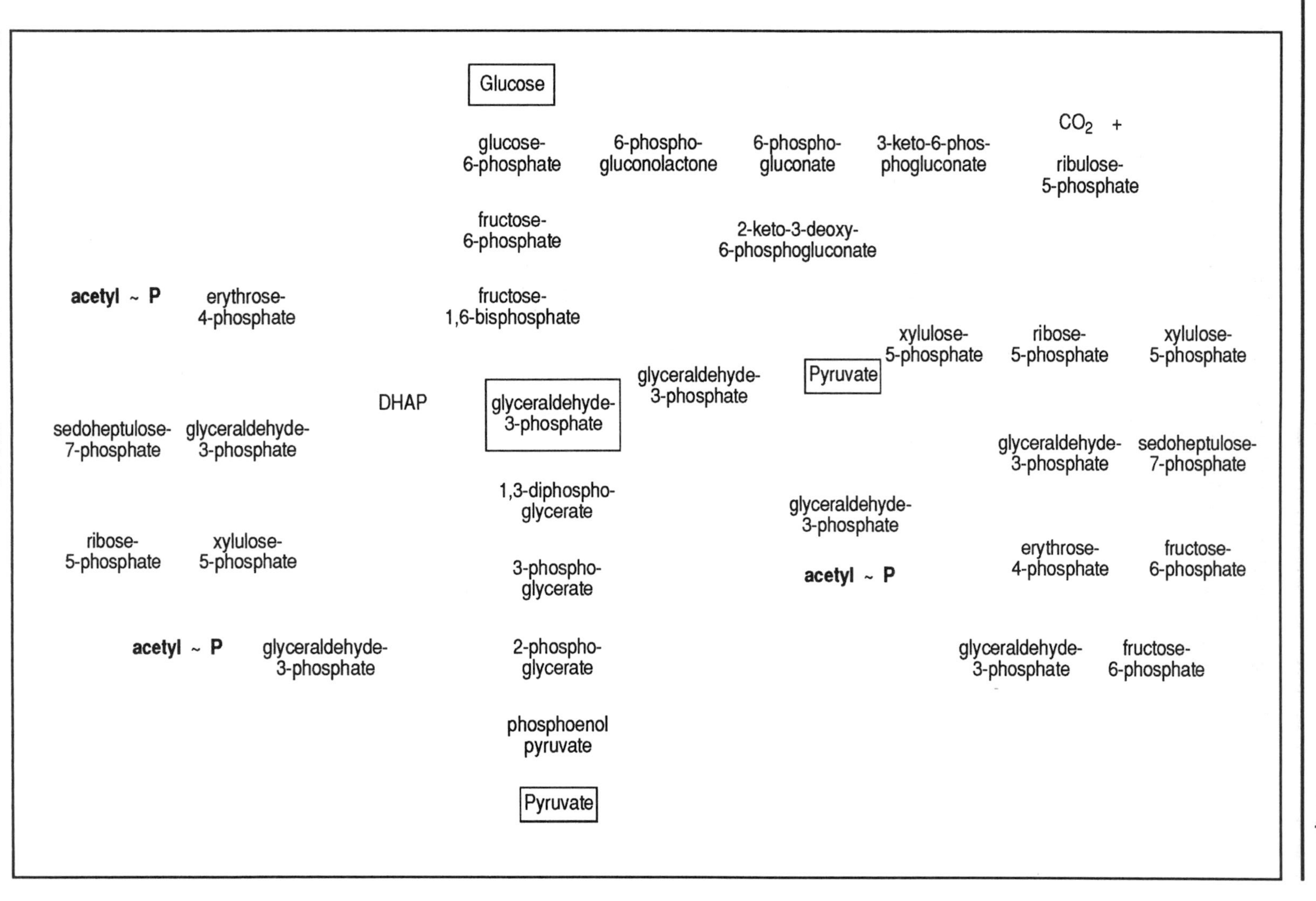

Figure 6.7 Summary diagram required to answer SAQ 6.9.

Completing and studying the pathway in Figure 6.7 has several benefits to us:

- completion should give you a sense of achievement;
- a series of pathways which, in isolation, seem to be very complicated are often made more understandable by looking at a composite diagram;
- a composite diagram gives an excellent overall impression of central carbohydrate metabolism;
- such a diagram demonstrates that although there are never more than two complete pathways in any one organism, there are some 'key reactions' common to all pathways;
- the composite diagram shows that the starting point is the same and that the products are very similar despite divergence in the middle;
- the diagram shows that there are one or two reactions for each pathway which are unique to that pathway. These are very important as potential control points because control of these would only affect the one pathway. This is important in situations where two or more pathways are present.

Π Re-examine your completed Figure 6.7 and identify unique reactions in the various pathways which may be useful for the control of these pathways?

The ones we anticipate you would spot are:

Embden Meyerhof pathway	phosphofructokinase
Entner Doudoroff	6-phosphogluconate dehydrase and 2-keto-3-deoxy-6-phosphogluconate aldolase
pentose phosphoketolase	xylulose-5-phosphate phosphoketolase
hexose phosphoketolase	fructose-6-phosphate phosphoketolase

Summary and objectives

Carbohydrates are the principal source of carbon and energy in most living systems.

Complex carbohydrates in the form of polysaccharides are broken down to monosaccharides, usually the hexose glucose or, if a pentose, xylulose.

Five pathways are recorded for the degradation of glucose to the central metabolite pyruvate. They are varied in their sequence of reactions but they all produce ATP and pyruvate.

The Embden Meyerhof pathway and Entner Doudoroff pathway are alternatives to each other and both work in combination with the hexose monophosphate pathway.

The remaining two pathways are restricted to specific aeroduric bacteria.

You should now be able to:

- discuss the usefulness of polysaccharides as carbon and energy sources and show how they can be converted into one or two key monosaccharide intermediates;

- list the reactions of the Embden-Meyerhof, hexose monophosphate, Entner Doudoroff, pentose phosphoketolase and hexose monophosphate pathways, derive their overall equations and discuss their interrelationships;

- explain the use of the Embden Meyerhof and hexose monophosphate pathways to yield different ratios of NADH, NADPH and pentoses and show how the second pathway can be used to support the growth of micro-organisms on pentoses as sole carbon and energy supplies;

- describe the unique feature of the hexose monophosphate pathway and discuss its superiority as a pathway for strictly anaerobic bacteria.

The tricarboxylic acid and glyoxylate cycles

The tricarboxylic acid and glyoxylate cycles

Overview

You have learnt from the previous chapter that carbohydrates, the principal carbon and energy source of most organisms, are metabolised to pyruvate. This occurred by one of a variety of pathways in organisms growing aerobically or anaerobically. In this chapter we shall follow the metabolism of pyruvate and consider its importance as a source of both energy and intermediates for biosynthesis.

The majority of this chapter will be devoted to a consideration of a cyclic series of reactions called the Krebs or tricarboxylic acid (TCA) cycle, a pathway also known as the citric acid cycle. The metabolic importance of this pathway cannot be overstressed; it is the central metabolic pathway and it is found in either complete or virtually complete form in all living systems.

central pathway

Introduction

Pyruvate does not enter the TCA cycle directly but first must be decarboxylated to yield an 'active acetyl' group. This active acetyl is then oxidised by a cyclic process to yield two carbon dioxide molecules. A simplified diagram of this process is shown in Figure 7.1.

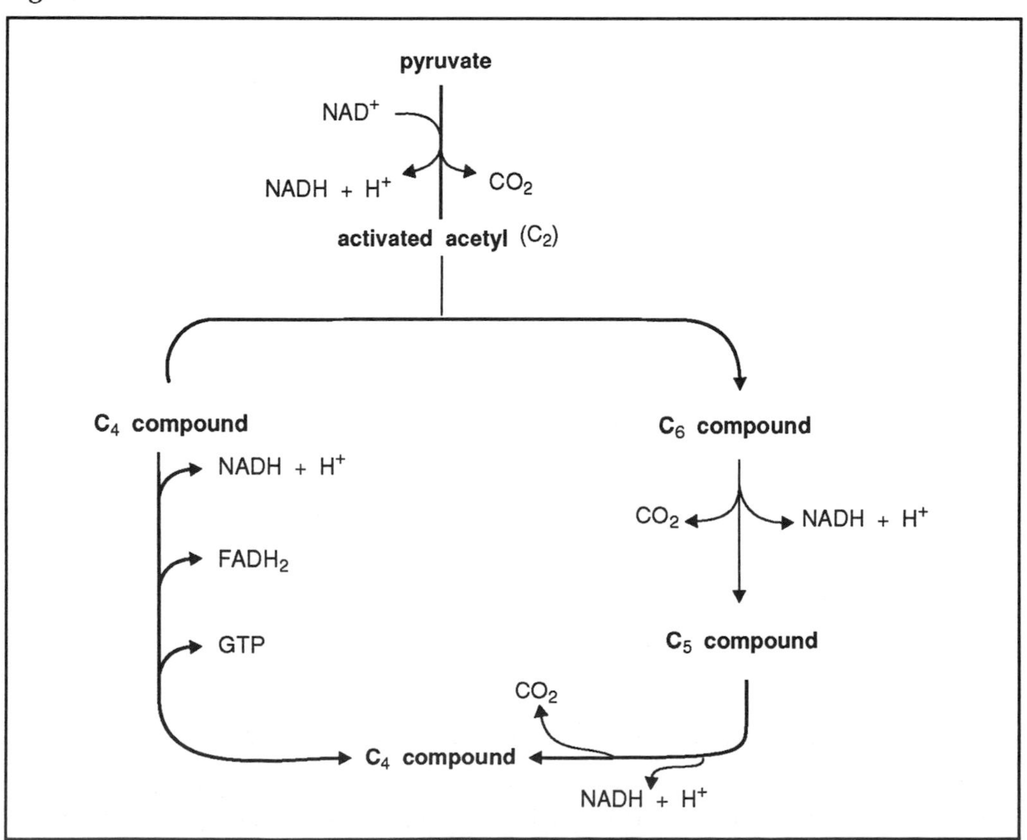

Figure 7.1 A simplified diagram of the TCA cycle.

This figure shows that during each turn of the cycle a four-carbon acceptor molecule binds to the acetyl moiety to give a six-carbon compound. This is oxidatively decarboxylated to a four-carbon compound via a five-carbon compound. The four-carbon compound is successively oxidised back to the original acceptor molecule such that there is evolution of a relatively large amount of reducing power in the form of hydrogen ions. These hydrogen ions will be linked to a redox carrier, generally NAD (nicotinamide adenine dinucleotide) and the reoxidation of NAD will be via an electron transport chain coupled to oxidative phosphorylation.

Π How many molecules of NAD^+ are reduced per turn of the cycle? How many FAD^+ are similarly reduced?

By following the cycle from pyruvate via acetyl CoA you can identify four recurrences of NAD^+ reduction and one of FAD.

Π Would you conclude that the type of metabolism being carried out by organisms using pyruvate as shown in Figure 7.1 is aerobic or anaerobic?

These cells must be growing aerobically, that is, they are growing in the presence of and are utilising molecular oxygen as their terminal electron acceptor. They would require the electron transfer chain (ETC) and oxidative phosphorylation for this process. These mechanisms are absent in anaerobic organisms. (Note, some organisms can use alternatives to oxygen such as nitrate and sulphate ions as terminal electron acceptors. These cells which are anaerobes may also have a fully functional tricarboxylic acid cycle. We will discuss this type of metabolism in Chapter 9).

amphibolic pathway

The oxidative degradation of pyruvate to yield energy is obviously the catabolic role of the TCA cycle and is restricted to aerobically growing organisms. However the TCA cycle has an anabolic role in all organisms, that of providing intermediates for biosynthesis. A pathway exhibiting both catabolic and anabolic properties is called an amphibolic pathway.

different roles of TCA cycle

In this chapter we will examine the operation of the TCA cycle from the following viewpoints:

- the catabolic role of the cycle in aerobes;
- the anabolic role of the cycle in aerobes and anaerobes;
- replenishing mechanisms of the cycle;
- its involvement in growth on two-carbon compounds;
- compartmentalisation and control of the cycle;
- manipulation of the cycle to yield industrially useful quantities of some intermediates.

7.1 The catabolic role of the TCA cycle in aerobic organisms

bridging reaction

The final product of glycolysis is pyruvate and, as mentioned, it has to be converted initially to an active acetyl group before it can be incorporated into the TCA cycle itself. The conversion of pyruvate to acetyl CoA is sometimes referred to as the bridging reaction as it bridges or links glycolysis to the TCA cycle.

The overall reaction is:

$$CH_3\overset{O}{\overset{||}{C}}-COO^- + CoASH + NAD^+ \longrightarrow CH_3\overset{O}{\overset{||}{C}}-SCoA + CO_2 + NADH + H^+$$

pyruvate → acetyl CoA

Π Name three basic reactions occurring within the above equation.

The equation shows that a decarboxylation, an oxidation and a transfer of a two-carbon compound to coenzyme A occur. It is obviously a complicated sequence of reactions and it is catalysed by the pyruvate dehydrogenase enzyme complex.

Coenzyme A is a molecule which acts as a cofactor in enzyme catalysed acylations. The acyl group is often acetyl - hence acetyl CoA - and this acyl group is always linked to a terminal -SH group, thus the free cofactor is often written as CoASH. The bond between acetyl and coenzyme A is a thioester bond.

three enzymes form a multi-enzyme complex

The pyruvate dehydrogenase complex from several organisms has been extensively studied and is known to be a large multi-enzyme system. That of *E. coli*, for example, has a molecular weight of 4.6 x 10^6 daltons and contains 60 polypeptide chains. These chains are of three enzymes, pyruvate dehydrogenase (24 chains), dihydrolipoyl transacetylase (24 chains) and dihydrolipoyl dehydrogenase (12 chains). The reaction it catalyses is more completely represented in Figure 7.2.

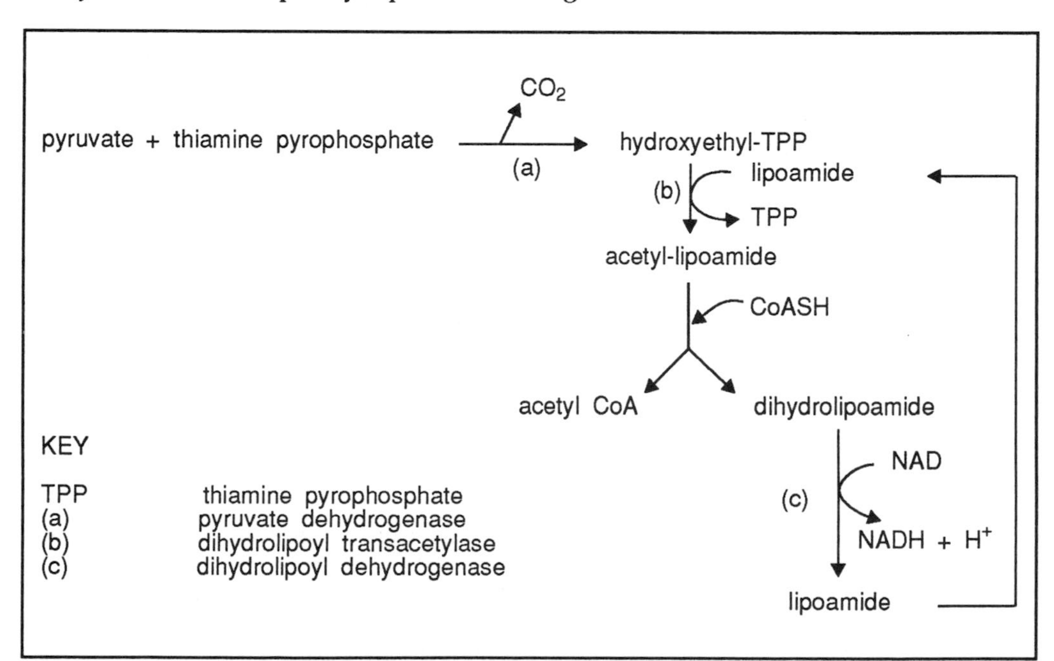

Figure 7.2 The reaction bridging glycolysis and the TCA cycle.

The acetyl moiety of the acetyl CoA now enters the TCA cycle by combining with the acceptor molecule, oxaloacetate, to form citrate.

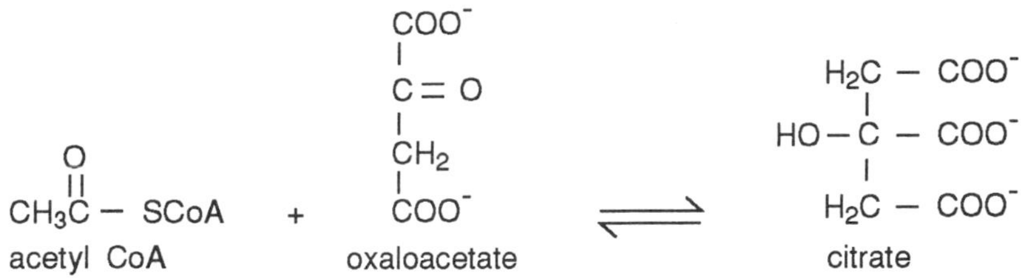

$$\underset{\text{acetyl CoA}}{CH_3\overset{O}{\overset{\|}{C}}-SCoA} + \underset{\text{oxaloacetate}}{\begin{array}{c}COO^-\\|\\C=O\\|\\CH_2\\|\\COO^-\end{array}} \rightleftharpoons \underset{\text{citrate}}{\begin{array}{c}H_2C-COO^-\\|\\HO-C-COO^-\\|\\H_2C-COO^-\end{array}}$$

citrate formed

This is a condensation type of reaction and the earlier trivial name of the enzyme was the condensing enzyme but it is more correctly known as citrate synthetase.

In the next step citrate is isomerised to isocitrate. Remember an isomerisation reaction is simply a rearrangement of atoms.

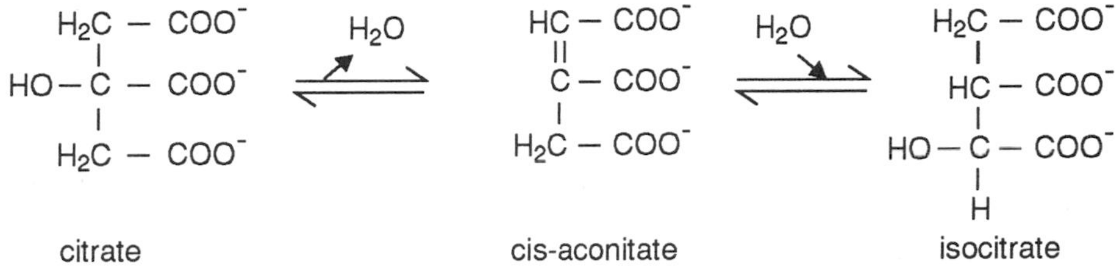

$$\underset{\text{citrate}}{\begin{array}{c}H_2C-COO^-\\|\\HO-C-COO^-\\|\\H_2C-COO^-\end{array}} \overset{H_2O}{\rightleftharpoons} \underset{\text{cis-aconitate}}{\begin{array}{c}HC-COO^-\\\|\\C-COO^-\\|\\H_2C-COO^-\end{array}} \overset{H_2O}{\rightleftharpoons} \underset{\text{isocitrate}}{\begin{array}{c}H_2C-COO^-\\|\\HC-COO^-\\|\\HO-C-COO^-\\|\\H\end{array}}$$

isocitrate formed

The enzyme catalysing this step is called aconitase because the compound, cis-aconitic acid, is considered to be an intermediate between citrate and isocitrate.

A single enzyme catalyses both reactions.

The next reaction is the first of two consecutive oxidative decarboxylation reactions, initially isocitrate being converted to α-oxoglutarate by the enzyme isocitrate dehydrogenase. As in the previous reaction, conversion is via an intermediate which is very unstable when bound to isocitrate dehydrogenase, the overall reaction being:

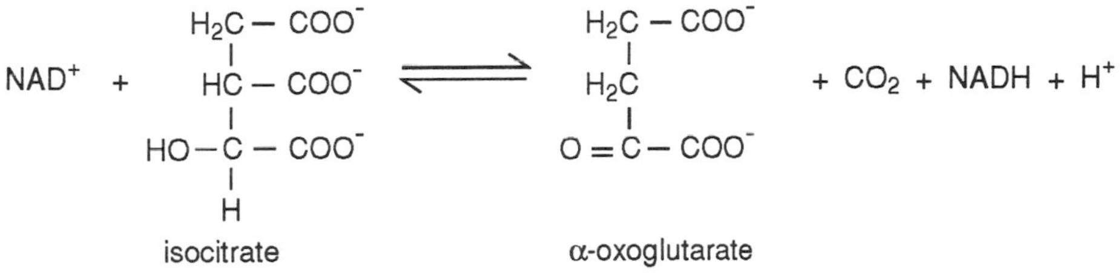

$$NAD^+ + \underset{\text{isocitrate}}{\begin{array}{c}H_2C-COO^-\\|\\HC-COO^-\\|\\HO-C-COO^-\\|\\H\end{array}} \rightleftharpoons \underset{\alpha\text{-oxoglutarate}}{\begin{array}{c}H_2C-COO^-\\|\\H_2C\\|\\O=C-COO^-\end{array}} + CO_2 + NADH + H^+$$

There is also an NADP-linked isocitrate dehydrogenase enzyme. We will see that in eukaryotic cells this enzyme has a different role to that of the NAD-linked enzyme. In prokaryotes the enzyme which is associated with the TCA cycle is NADP-linked.

α-oxoglutarate is now converted to succinyl CoA as follows:

$$\text{CoASH} + \text{NAD}^+ + \begin{array}{l} H_2C - COO^- \\ | \\ H_2C \\ | \\ O = C - COO^- \end{array} \rightleftharpoons \begin{array}{l} H_2C - COO^- \\ | \\ H_2C \\ | \\ O = C - SCoA \end{array} + CO_2 + \text{NADH} + H^+$$

α-oxoglutarate succinyl CoA

Π You should be able to name this enzyme and make comments about the reaction.

The reaction is catalysed by α-oxoglutarate dehydrogenase, a complex enzyme which is very similar to the pyruvate dehydrogenase complex discussed earlier in the chapter. The same cofactors are required in a similar series of reactions.

The thioester bond of succinyl CoA is an energy rich bond, the hydrolysis of which has a $\Delta G^{o\prime}$ of approximately -34 kJ mol^{-1}. Thus in reactions utilising succinyl CoA the phosphorylation of a nucleoside diphosphate is possible. This is what happens in the next reaction of the TCA cycle but phosphorylation of GDP (guanosine diphosphate) rather than ADP occurs. The reaction is catalysed by the enzyme succinate thiokinase (also known as succinyl CoA synthetase).

$$\text{GDP} + \text{Pi} + \begin{array}{l} H_2C - COO^- \\ | \\ H_2C \\ | \\ O = C - SCoA \end{array} \rightleftharpoons \begin{array}{l} H_2C - COO^- \\ | \\ H_2C \\ | \\ COO^- \end{array} + \text{GTP} + \text{CoASH}$$

succinyl CoA succinate

All cells contain nucleoside diphosphokinases which transfer phosphate groups thus:

$$\text{ADP} + \text{GTP} \rightleftharpoons \text{ATP} + \text{GDP}$$

formation of GTP

In the succinyl CoA: succinate reaction, energy (in the form of GTP) is produced directly. The production of a triphosphate occurs nowhere else within the TCA cycle.

Π What name is given to this type of reaction?

It is called a substrate level phosphorylation, you encountered this before in Chapter 6.

The cycle continues with the oxidation of succinate thus:

$$\text{E} - \text{FAD} + \begin{array}{l} H_2C - COO^- \\ | \\ H_2C \\ | \\ COO^- \end{array} \rightleftharpoons \begin{array}{l} HC - COO^- \\ \| \\ CH \\ | \\ COO^- \end{array} + \text{E} - \text{FADH}_2$$

succinate fumarate

FAD involved in formation of fumarate

The E - FAD denotes that the hydrogen acceptor, FAD, is tightly bound to the enzyme. FAD is used as the hydrogen acceptor because in the reaction the free energy change is insufficient to reduce NAD^+. We will return to this point later. The enzyme is called succinate dehydrogenase.

succinate dehydrogenase is membrane-bound

Succinate dehydrogenase is unusual in that it is the only membrane-bound enzyme of the TCA cycle; it is an integral part of the inner mitochondrial membrane (of eukaryotes) or cell membrane (of prokaryotes). In this sense it is different from the other TCA cycle enzymes which are dissolved in the mitochondrial matrix (eukaryotes) or cytoplasm (prokaryotes).

Fumarate is hydrated in the next reaction to form L-malate, the reaction being catalysed by fumarase.

$$H_2O + \underset{\text{fumarate}}{HC(COO^-){=}CH{-}COO^-} \rightleftharpoons \underset{\text{L-malate}}{HOC(H)(COO^-){-}H_2C{-}COO^-}$$

malate formed

The final reaction of the cycle, which converts L-malate to the original acceptor molecule, oxaloacetate, is catalysed by the enzyme malate dehydrogenase.

$$NAD^+ + \underset{\text{L-malate}}{HOC(H)(COO^-){-}H_2C{-}COO^-} \rightleftharpoons \underset{\text{oxaloacetate}}{O{=}C(COO^-){-}H_2C{-}COO^-} + NADH + H^+$$

starting compound reformed

dehydration and dehydrogenation

It is worth pausing for a few sentences to consider the reactions from succinate to oxaloacetate. You will notice that the only difference between these two compounds is that succinate has a $-CH_2$ at carbon 2 whereas oxaloacetate has a -C=0 group. This conversion is by a three step sequence of dehydrogenation - hydration - dehydrogenation (or hydrogenation - dehydration - hydrogenation in the opposite direction).

This sequence of three reactions occurs in several other metabolic pathways, for example in fatty acid synthesis and degradation. Recognition of the simple steps involved will help in future chapters.

We are now in a position to produce a more complete diagram than the one presented as Figure 7.1 and this is presented as Figure 7.3. You might like to draw this figure out for yourself and put on the chemical structures of the intermediates.

SAQ 7.1

Imagine that you have an organism which is being fed on uniformly ^{14}C labelled glucose during a period sufficiently long for all three of the carbon atoms of the resultant pyruvate to carry a ^{14}C label. Draw the TCA cycle as shown in Figure 7.3 but in the form of the structural formulae in order to follow the fate of each carbon from pyruvate. This will take some time but you should be able to manage it. When you have done this, indicate where these carbons would be at the end of a second turn of the cycle, assuming that the pyruvate used for the second turn was unlabelled.

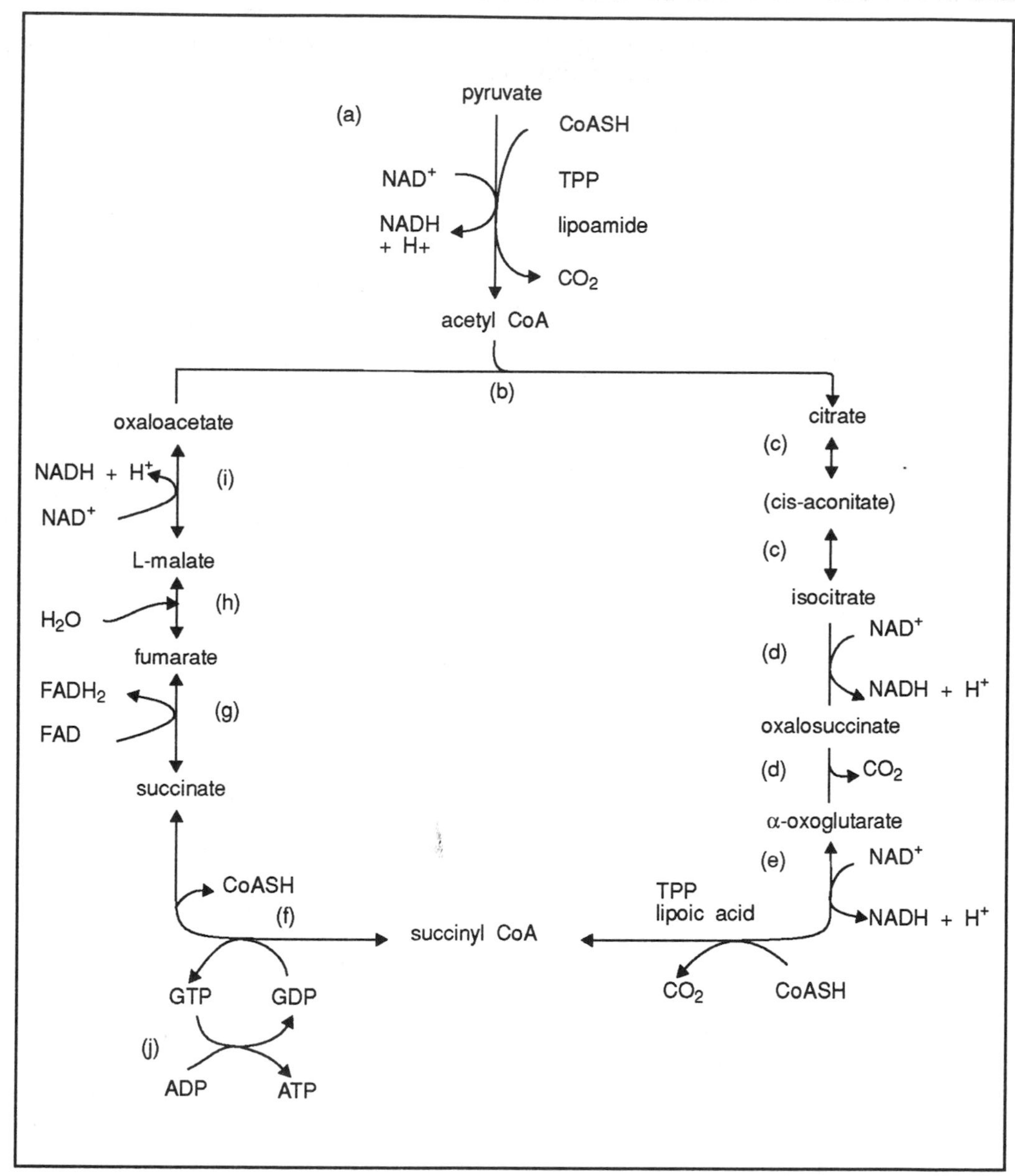

Figure 7.3 The tricarboxylic acid cycle. Key to enzymes of the TCA cycle: (a) pyruvate dehydrogenase complex, (b) citrate synthetase, (c) aconitase, (d) isocitrate dehydrogenase, (e) α-oxoglutarate dehydrogenase complex, (f) succinyl CoA synthetase, (g) succinate dehydrogenase, (h) fumarase, (i) malate dehydrogenase, (j) nucleoside diphosphokinase. Abbreviations: TPP - thiamine pyrophosphate; CoASH - free coenzyme A; GDP - guanosine diphosphate; GTP - guanosine triphosphate; ADP - adenosine diphosphate; ATP - adenosine triphosphate.

7.2 Calculation of ATP yield

Π In the previous chapter we constructed an overall equation for the catabolism of glucose to two molecules of pyruvate. Construct a similar overall equation for one turn of the TCA cycle starting from acetyl CoA. It is actually quite difficult to balance the oxygen and hydrogen atoms on the two sides of the equation. Do not try to do this at this stage.

The answer is:

$$\text{acetyl CoA} + GDP + Pi + FAD + 3\,NAD^+ + 2H_2O$$

$$\rightarrow 2\,CO_2 + CoASH + GTP + FADH_2 + 3\,NADH + 3H^+$$

SAQ 7.2

Glycolysis yields two ATP (by substrate level phosphorylation) and $2NADH + 2H^+$ on degradation of glucose to pyruvate. Note that these two $NADH + H^+$ yield 2ATP each when reoxidised in the electron transport chain and oxidative phosphorylation apparatus, whereas $FADH_2$ and $NADH + H^+$ formed in the TCA cycle yield 2 and 3 ATP respectively.

Use the overall equation for the TCA cycle to help you fill in the following table and hence calculate the maximum number of ATP derived: a) by substrate level phosphorylation; b) by the electron transport chain and oxidative phosphorylation; when glucose is completely metabolised to $6CO_2$.

ATP production by:

pathway	substrate level phosphorylation	reoxidation of $FADH_2$ by ETC and Ox Phos	reoxidation of $NADH + H^+$ by ETC and Ox Phos
glycolysis	—	—	—
bridging reaction	—	—	—
TCA cycle	—	—	—
subtotal	—	—	—
		—	
final total			

key: ETC is the electron transport chain
Ox Phos is oxidative phosphorylation

Table to show the stage of ATP production by aerobic organisms using glycolysis and the TCA cycle when one glucose molecule is completely oxidised to CO_2.

7.3 The catabolic role of the TCA cycle in anaerobic organisms

reducing power is a hindrance in anaerobes

Let us now consider the catabolic role of the TCA cycle in anaerobic organisms. We must remember that these organisms do not have conventional electron transport chains linked to an oxidative phosphorylation system. Thus they cannot use these mechanisms to reoxidise the $3NADH + 3H^+$ and $1FADH_2$ produced during each turn of the cycle. In actual fact, this much reducing power, produced from a fully functional TCA cycle, would be impossible to deal with in an anaerobic environment (with some exceptions - see Chapter 9) and the cycle would grind to a halt.

There is something of a dilemma here because each turn of the cycle produces an ATP molecule (via GTP). This would be very useful but at a cost of far too much reducing power. However the strategy of the anaerobe is to avoid the use of the TCA cycle for energy production although they do use it in other ways, as we shall see. This potential ATP molecule has therefore been sacrificed due to the cell having to successfully balance its redox levels. This topic will be enlarged upon in a later chapter but an important principle arises which we must bear in mind that: 'it is vital that cells produce ATP but it is equally important that anaerobic organisms can precisely maintain their redox balance'.

Thus although molecular oxygen is not directly involved in the TCA cycle, the cycle as an oxygen-requiring and energy-yielding process and the catabolic function of the cycle is predominantly found in aerobes.

ATP yields in anaerobes

In the last SAQ we calculated the maximum amount of ATP derived from glucose to be 36 when glycolysis and the TCA cycle are used under aerobic conditions. The maximum in anaerobes not using a TCA cycle must be two, those derived from substrate level phosphorylation during glycolysis. Such cells have an immediate problem of reoxidising the two molecules of NADH + H^+ produced during glycolysis. This reoxidation generally involves the reduction of pyruvate or its reductive decarboxylation, the various possibilities being discussed in Chapter 9.

7.4 The anabolic role of the TCA cycle in aerobes

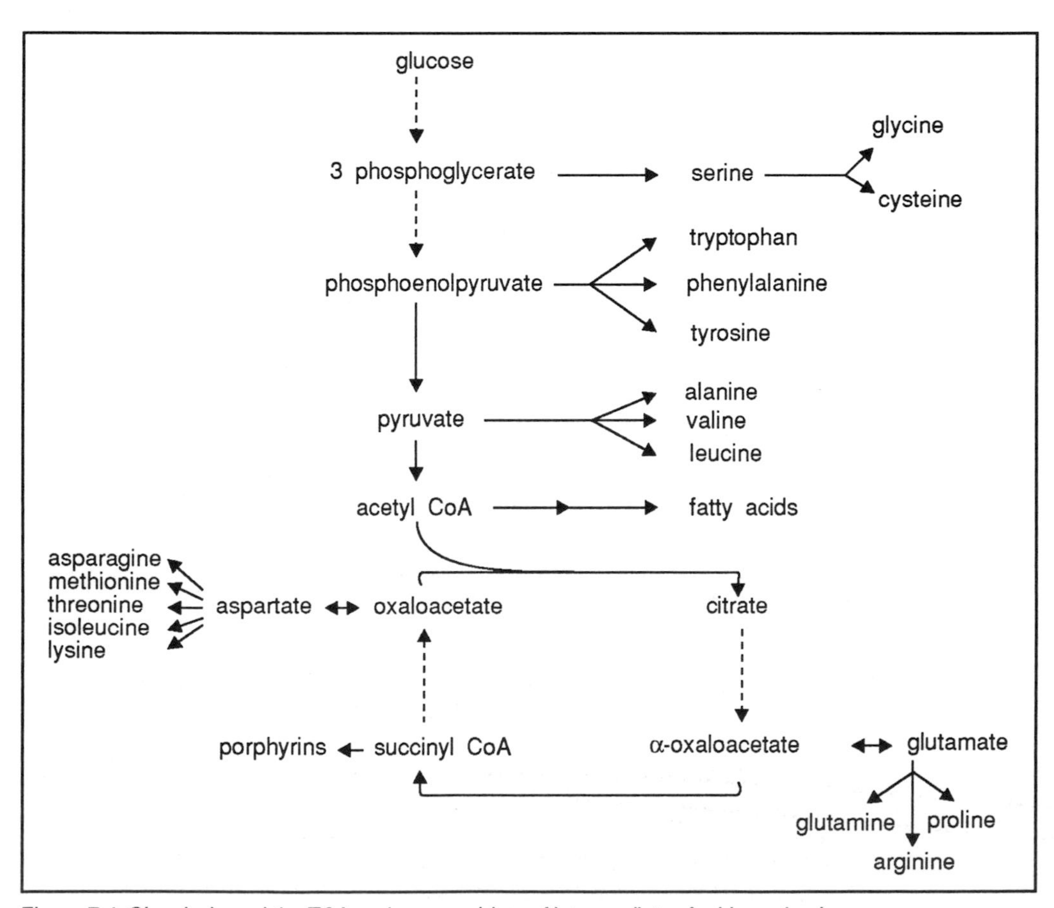

Figure 7.4 Glycolysis and the TCA cycle as providers of intermediates for biosynthesis.

Anabolically the TCA cycle is the provider of many intermediates for biosynthetic reactions, particularly amino acid precursors. The simplified diagram of glycolysis and the TCA cycle (Figure 7.4) shows the principal intermediates and their subsequent products.

Figure 7.4 helps to explain the claim that the TCA cycle is the central metabolic pathway. We have already shown that it is the major provider of energy in aerobic cells and here it is shown to be the major outgoing link between carbohydrates and the metabolism of fatty acids and amino acids. It is important to realise that the connecting arrows are reversible and it is possible to obtain input into the cycle from protein or fatty acid degradation. However when cells are growing and dividing with glucose as substrate there is a net outflow of intermediates from the system.

TCA cycle is at crossroads of carbohydrate, fatty acid and amino acid metabolism

This outflow in quantitative terms is enormous: acetyl CoA provides the carbon for fatty acids, three of the glycolytic intermediates are used to provide some or all of the carbon material in nine amino acids and two TCA cycle intermediates provide carbon for ten amino acids. Thus 19 of the 20 common amino acids, i.e virtually all of the protein of the cells, are synthesised from five metabolically-related carbohydrates.

Let us now consider how the cell copes with the demands on glycolysis and the TCA cycle to provide these intermediates. If a given cell had a demand for fatty acids then the way in which glycolysis would respond is obvious, there would be an increased catabolism of glucose to pyruvate and hence to acetyl CoA; i.e the flux of intermediates through glycolysis would speed up. Some acetyl CoA would proceed into the TCA cycle to produce energy and some would be diverted to fatty acid biosynthesis. The metabolic controls on the demand for acetyl CoA will be discussed later. Similar demands for any intermediate of the glycolytic pathway could be satisfied in the same way, that is by simply increasing glycolytic flux.

Now consider the situation where TCA cycle intermediates are required for biosynthetic precursors. Simply increasing flux through glycolysis and thus obtaining more acetyl CoA to increase the flux through the TCA cycle would not result in increased availability of TCA cycle intermediates.

SAQ 7.3

There is an important principle here. Can you identify why simply increasing the flux through the TCA cycle would not make intermediates available? (Answer this and check the response before reading on).

Consider again the initial reaction of the TCA cycle:

$$\text{oxaloacetate} + \text{acetyl CoA} \xrightarrow[\text{synthase}]{\text{citrate}} \text{citrate}$$

diversion of intermediates creates problems

For this reaction to proceed three things are required; the correct enzyme, acetyl CoA and oxaloacetate. We usually remember the first two essentials but sometimes overlook the requirement for oxaloacetate.

Π What is the solution to the problem of oxaloacetate regeneration when intermediates are diverted from the TCA cycle?

In broad terms we must put intermediates back to replenish the cycle. To maintain a metabolically functional TCA cycle replenishment and removal have to be maintained in a 1:1 ratio since for every molecule of intermediate removed, one must be put back. As we shall see each intermediate does not have to be replaced by an identical molecule but can be replaced by any other intermediate within the cycle.

anaplerotic reactions solve this problem

The replenishing reactions are known as anaplerotic reactions and are vital to the welfare of all living systems. There are many theoretical possibilities but the strategy in most living systems growing on hexoses can be summarised by the following generalised equation:

$$\text{phosphoenol pyruvate} + CO_2 \rightarrow \text{oxaloacetate}$$

Phosphoenol pyruvate, or occasionally pyruvate, compounds which we have already established as being readily available from glycolysis, are carboxylated to produce oxaloacetate, or occasionally malate. Thus the TCA cycle is replenished at the level of oxaloacetate but diversions may occur at any of its intermediates.

Five such anaplerotic reactions are known in living systems; not all occur in each organism although there is generally more than one of them present. These reactions together with the enzymes involved are shown in Figure 7.5.

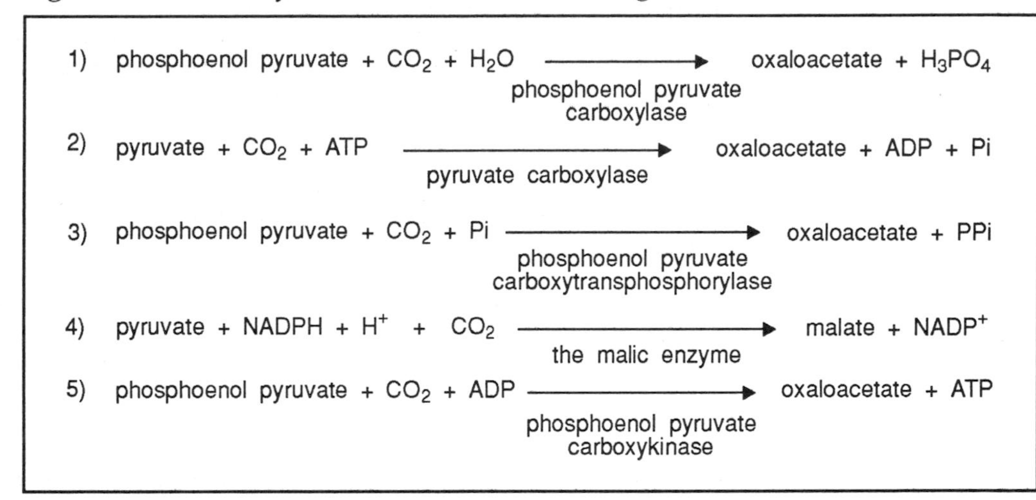

Figure 7.5 Anaplerotic reactions of the TCA cycle.

there are two main anaplerotic pathways

All five reactions are reversible and all, when progressing from left to right will serve to replenish the TCA cycle. However the last two enzymes often operate from right to left to provide ADP or NADPH + H^+. The reaction used in many bacteria and higher organisms is the first one, that catalysed by phosphoenol pyruvate carboxylase. However pyruvate carboxylase occurs in most living systems including bacteria. An NAD^+ -linked malic enzyme also occurs in addition to the $NADP^+$ -linked one shown in Figure 7.5.

As we have seen above there is generally a net outflow of intermediates from the TCA cycle. However there may be occasions when α-oxoglutarate or the four-carbon dicarboxylic acid intermediates may be temporarily in excess of requirements and are fed into the cycle. Aerobic cells would wish to remove them, preferably by oxidation to yield energy. However the intermediates themselves are not directly oxidised completely to CO_2. The assistance of the malic enzyme acting from right to left provides the answer.

SAQ 7.4

Draw out a scheme for the way in which aerobic cells can oxidise excess α-oxoglutarate completely to CO_2, using malic enzyme.

malic enzyme helps to degrade excess intermediates

The answer to SAQ 7.4 shows that the use of the TCA cycle enzymes plus malic enzyme plus pyruvate dehydrogenase enables C_5 or C_4 TCA cycle intermediates to be totally oxidised to CO_2 with concomitant production of a relatively large amount of energy.

7.5 The anabolic role of the TCA cycle in anaerobes

We established earlier that organisms growing anaerobically do not use the TCA cycle for energy production. They have the dilemma that on the one hand they wish to avoid the use of the TCA cycle, as it generates unacceptable levels of reducing power, but on the other hand they must provide intermediates for biosynthesis.

Anaerobes obtain biosynthetic precursors from glycolytic intermediates and acetyl CoA in exactly the same way as in aerobes. We have established that glycolysis is essentially an 'anaerobic pathway' and that its flux may be increased in anaerobes. The problem then to be resolved is to determine how anaerobes maintain their redox balance while producing sufficient TCA precursors for biosynthesis.

Π Given that the only missing enzyme of the TCA cycle in anaerobes is α-oxoglutarate dehydrogenase, how do anaerobes obtain sufficient α-oxoglutarate for anabolic reactions?

The answer is oxaloacetate is produced by one of the anaplerotic reactions and that α-oxoglutarate (and any intermediate between oxaloacetate and α-oxoglutarate via citrate) is produced by the TCA cycle operating in exactly the same manner as in aerobes.

Π Can you now show how anaerobes obtain succinyl CoA for anabolic reactions? Remember that α-oxoglutarate dehydrogenase is absent and there is no alternative direct route between α-oxoglutarate and succinyl CoA (check Figure 7.3 if you are stuck).

The answer relies on the fact that oxaloacetate is produced by an anaplerotic reaction as in aerobes but the TCA cycle now operates in an anticlockwise manner via malate, fumarate and succinate to succinyl CoA. This process is referred to as the reverse TCA cycle.

A summary of the forward and reverse phases of the cycle is shown in Figure 7.6.

Although the two halves of the TCA cycle are not in redox balance it is important to anaerobes that some of the excess reducing power produced in glycolysis, the bridging reaction and the production of α-oxoglutarate is taken up during the reduction of oxaloacetate to succinyl CoA.

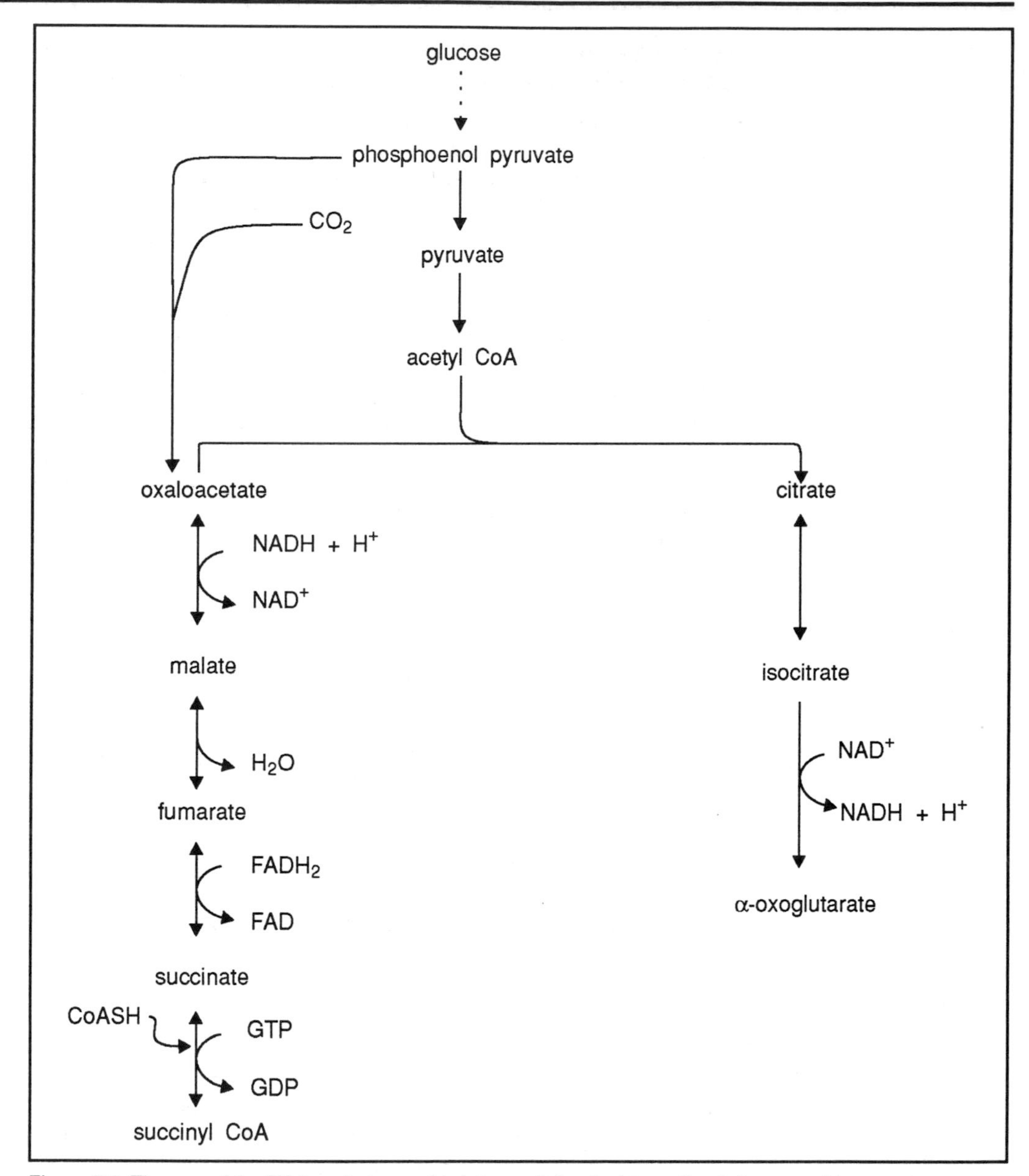

Figure 7.6 The use of the TCA cycle to provide intermediates for biosynthesis in anaerobic organisms.

7.6 Growth of micro-organisms on carbohydrates other than six-carbon compounds

One of the first considerations when deciding whether a compound is a suitable carbon and energy source is whether or not that compound can be absorbed through the cell membrane. Some compounds enter the cell by simple diffusion, for example glycerol, whereas the majority are taken up by some form of transport system. Others, which on the surface appear to be ideal substrates, cannot enter the cells of most organisms.

Dicarboxylic acids of the TCA cycle can only be metabolised by relatively few micro-organisms which have a specific energy dependent permease system.

All the examples of substrate utilisation so far have been based on glucose as the starting material. However we learnt earlier that living cells can interconvert hexoses. Thus glycolysis (or one of the alternatives) and the TCA cycle in aerobes would be suitable pathways for the degradation of a wide variety of hexoses.

utilisation of citrate

Six-carbon metabolic intermediates such as citrate will be utilised providing that the cell can take the compounds into its cytoplasm. Aerobically citrate enters the TCA cycle and is metabolised to oxaloacetate. Many organisms can decarboxylate oxaloacetate to form pyruvate and CO_2 by oxaloacetate decarboxylase. The pyruvate can then be either decarboxylated to acetyl CoA to re-enter the TCA cycle or used directly as a biosynthetic intermediate or converted to glucose via gluconeogenesis. Gluconeogenesis is discussed in the BIOTOL text, 'Biosynthesis and the Integration of Cell Metabolism'.

Citrate may also be used in organisms growing fermentatively (anaerobically). Remember that the TCA cycle is incomplete in these organisms and that the usual replenishing intermediate in anaerobic organisms is oxaloacetate. Such organisms have an enzyme citrate lyase which carries out the reaction:

$$\text{citrate} \rightarrow \text{acetic acid} + \text{oxaloacetate}$$

Acetic acid is an end-product.

Π Why is the enzyme citrate lyase useful in fermentative organisms utilising citrate as sole source of carbon and energy?

citrate lyase

Anaerobes must have a source of oxaloacetate and other TCA cycle intermediates but as we have discussed earlier they cannot use the TCA cycle in its full form due to the enormous amount of reducing power generated. The availability of citrate lyase means that citrate can be converted directly to oxaloacetate without the generation of reducing power. In actual fact citrate lyase is an essential enzyme in anaerobes growing on citrate.

7.6.1 Growth on five-carbon compounds

As indicated earlier many bacteria and yeasts are capable of growing on five-carbon carbohydrate compounds as their sole source of carbon and energy. Examples of such micro-organisms are diverse and include both aerobes, and anaerobes. We showed that in all cases the pentoses are converted to D-xylulose-5-phosphate. This compound could then be degraded by a variety of pathways eventually yielding pyruvate. Thus the TCA cycle will function as already described for organisms growing on six-carbon compounds.

novel anaplerotic pathway

The diversity of the group means inevitably that there are some novel anaplerotic mechanisms for replenishing the TCA cycle but the elaboration of these is beyond the scope of this text. However a relatively simple example is shown in Figure 7.7 in which a TCA cycle intermediate is formed from the substrate without the involvement of any other common metabolite.

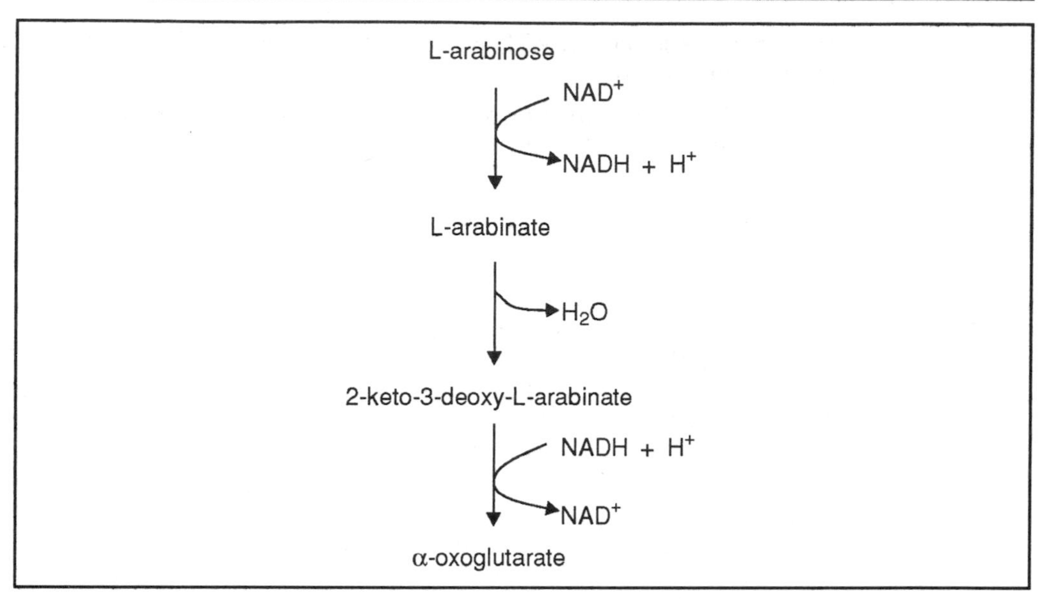

Figure 7.7 Anaplerotic pathway for replenishment of the TCA cycle by *Pseudomonas saccharophila* when grown on L-arabinose.

7.6.2 Growth on four-carbon compounds

Many organisms will incorporate four-carbon compounds into cellular metabolites and may possibly obtain a benefit in terms of energy. However, such compounds are mostly used in conjunction with additional substrates, such as six-carbon compounds, to satisfy their overall carbon and energy requirements.

7.6.3 Growth on three-carbon compounds

reduced 3C compounds used by aerobes

Several three-carbon compounds, particularly highly reduced examples such as glycerol or lactate, can be used by many bacteria as the sole source of carbon and energy. The fact that they are highly reduced makes them particularly attractive to aerobic organisms because their subsequent oxidation to CO_2 yields reducing power (as NADH + H^+) and thus yields useful cell energy by oxidative phosphorylation.

Remember that on entry into the cell, sugars have to be activated and this is achieved by phosphorylation. In the case of glycerol, following its entry into the cell by diffusion, glycerol phosphate is formed followed by oxidation to dihydroxyacetone phosphate. This compound you will recall is an intermediate of the glycolytic pathway and its further metabolism would be via the terminal steps of glycolysis and the TCA cycle as discussed earlier.

We should consider for a moment the use of lactate as sole source of carbon and energy by aerobically growing organisms.

Π If lactate is initially oxidised to pyruvate, how do such organisms obtain energy and how much ATP do they produce per mole of lactate?

lactate easily produces energy

Such organisms obtain energy from the TCA cycle and produce a total of 18 ATP per mole of lactate. 12 ATP are produced by one turn of the TCA cycle, one molecule of NADH + H^+ is produced, from the bridging reaction and one from the oxidation of lactate to pyruvate mediated by lactate dehydrogenase. These two NADH + H^+

molecules produce six ATP when oxidised through the electron transport pathway giving a total of 18.

The empirical formula of lactic acid is $C_3H_6O_3$, which is exactly half that of glucose; the yield of 18 ATP is half that produced from glucose.

Π Can you think of a potential difficulty in the way in which these organisms replenish the TCA cycle?

novel anaplerotic pathway needed when lactate is used

We have seen that the normal anaplerotic reaction to replenishing the TCA cycle involves phosphoenol pyruvate and not pyruvate. You would be excused for suggesting that lactate could be converted to pyruvate and this converted to phosphoenol pyruvate by reversal of the last reaction of glycolysis. However, this reaction is considered to be irreversible under physiological conditions; thus there is no obvious, ready supply of phosphoenol pyruvate in any of the reactions studied so far.

two new reactions to produce phosphoenol pyruvate

Two rather specialised enzymes have been identified which alleviate this problem. In *Enterobacteriaceae*, a phosphoenol pyruvate synthase can be detected when lactate is the sole source of carbon. This enzyme catalyses the conversion of pyruvate to phosphoenol pyruvate with the help of ATP. A second enzyme, found initially in *Acetobacter xylinium*, catalyses the reaction:

$$\text{pyruvate} + \text{ATP} + \text{Pi} \xrightarrow[\text{pyruvate phosphate dikinase}]{} \text{phosphoenol pyruvate} + \text{PPi} + \text{AMP}$$

The presence of these enzymes allows organisms to utilise lactate as the sole source of carbon and energy.

Other three-carbon compounds, for example propionate, may be used as sole carbon and energy source. However it is beyond the scope of this chapter to include all of these rather restricted pathways; sufficient for us to remember that all organic compounds can be degraded by one or more micro-organisms though not necessarily as sole source of carbon and energy.

7.6.4 Growth on two-carbon compounds

There are a variety of two-carbon compounds which can be utilised in some way by most living organisms, examples of which are shown in Figure 7.8.

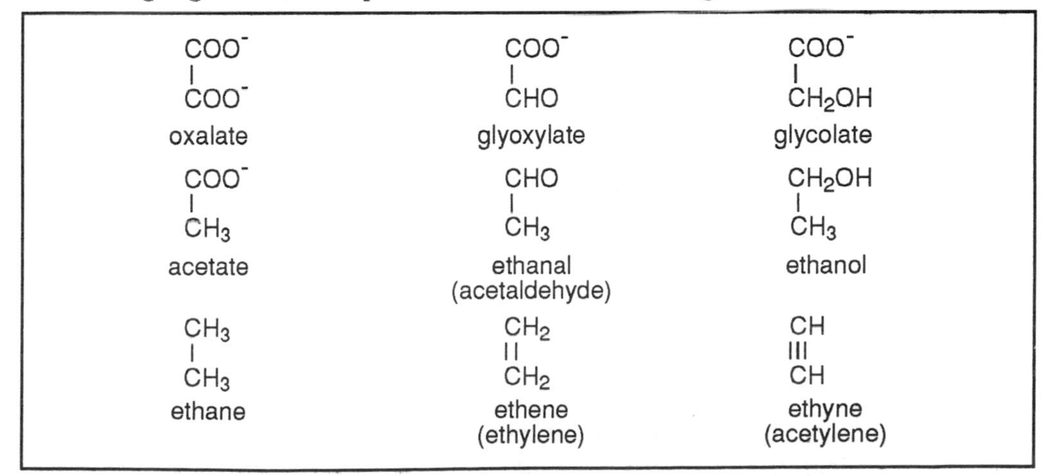

Figure 7.8 Two-carbon compounds which may be used as sole source of carbon and energy.

There are many more than the nine compounds shown in Figure 7.8, it is possible to substitute onto one or both of the carbons a variety of groups such as -NH_2; NO_2 etc. The most common of all the two-carbon compounds used by both micro-organisms and higher organisms is acetate. In the following discussion of acetate metabolism we will also consider ethanal (acetaldehyde) and ethanol as they are metabolised in a similar manner.

acetate can be used by micro-organisms and higher organisms

First we have to separate the use of a two-carbon compound in energy production from its use as an intermediate for biosynthesis. As we shall see, acetate can be used by many organisms as an energy source but its use as a carbon source is somewhat restricted.

Π Acetate, when used as a substrate, first has to be activated. Can you name the central metabolite to which acetate (and ethanol and ethanal) are first converted?

acetate is activated to acetyl CoA

The answer in all cases is acetyl CoA. Organisms growing on these compounds would carry out the following reactions:

$$\text{ethanol} \xrightarrow{NAD^+ \;\rightarrow\; NADH + H^+} \text{ethanol} \xrightarrow{NAD^+ \;\rightarrow\; NADH + H^+} \text{acetate} \xrightarrow[\text{CoASH}]{ATP \;\rightarrow\; AMP + PPi} \text{acetyl CoA}$$

Π How would the acetyl CoA be used as an energy source and by what types of micro-organisms?

acetyl CoA is fully oxidised in the TCA cycle

Acetyl CoA would enter the TCA cycle and be completely oxidised to $2CO_2$. For this process to occur a complete TCA cycle would be required together with the electron transport chain and oxidative phosphorylation apparatus. Therefore only aerobic organisms can use acetate, ethanol or ethanal as an energy source. In fact this ability to utilise acetate as an energy source is widespread amongst aerobes from bacterial systems through to higher plants and other higher organisms including ourselves.

The use of acetate as a carbon source is more restricted. Cells require some sort of process in which they can convert their only resource, acetate, into a TCA cycle and/or glycolytic pathway intermediate. The information given so far in this chapter has not revealed enzyme systems capable of doing this, thus extra enzymes are required. Such organisms growing on acetate have two specialised enzymes, isocitrate lyase and malate synthase.

$$\text{isocitrate} \xrightarrow[\text{isocitrate lyase}]{} \text{succinate + glyoxylate}$$

$$\text{glyoxylate + acetyl CoA} \xrightarrow[\text{malate synthase}]{} \text{malate + CoASH}$$

SAQ 7.5

Draw a simple diagram involving the two new enzymes shown above together with three TCA cycle enzymes to produce another cycle which we call the glyoxylate cycle. Write an overall reaction for your cycle indicating the TCA cycle intermediate produced.

glyoxylate cycle:acetate as carbon source

The succinate produced in the glyoxylate cycle functions as a TCA cycle intermediate either to be oxidised or converted into other intermediates directly or indirectly. The two key enzymes of the glyoxylate cycle are found only in certain aerobic organisms and many higher plants, some bacteria, some fungi, some algae and some protozoa. They are absent in higher animals.

In summary then, a restricted number of aerobes can use acetate as a sole carbon and energy source; most of the remaining aerobes can use acetate as an energy but not as a carbon source because they lack the glyoxylate cycle. Anaerobes can use acetate neither as an energy nor as a carbon source.

Of the other two-carbon compounds shown in Figure 7.8 the hydrocarbons are utilised by very few micro-organisms and will be discussed in the BIOTOL text, 'Energy Sources for Cells'. Glyoxylate and glycolate are used by relatively few bacteria, largely members of the family *Pseudomonadaceae* and have somewhat specialised metabolic pathways.

7.6.5 Growth on one-carbon organic compounds

It should not be surprising that growth on organic one-carbon compounds (such as methane and methanol) as the sole source of carbon and energy is even more specialised than the examples seen above. It is restricted to a few genera of bacteria and yeasts, the latter being able to utilise only methanol. Very specialised structures and metabolic pathways are found in these organisms. These are discussed in the BIOTOL text, 'Energy Sources for Cells'.

7.7 Control and subcellular location of the TCA cycle

7.7.1 Types of control mechanisms used

end-products of the TCA cycle?

The TCA cycle is the most important energy-producing pathway in the cell. Thus we would expect it to be controlled in some way by the energy status of the cell; for example by the adenylate charge or the ratio of reduced to oxidised pyridine nucleotides. The adenylate or energy charge of cells was discussed earlier and refers to the ratio of ATP to ADP plus AMP. However complications arise regarding the above fairly simple suggestion. Firstly the TCA cycle is an amphibolic pathway which provides both energy and intermediates. Thus control systems must operate to monitor these two factors. Secondly it is a cycle and not a linear pathway so the question of an end-product must be raised.

Π What would you consider to be the end-product(s) of the TCA cycle?

This complex question can be answered by several qualified statements. In aerobic cells we would have to say that ATP is an end-product together with any of the intermediates removed for biosynthesis. In the open ended anaerobic pathway then α-oxoglutarate and succinyl CoA can be considered to be end-products (Figure 7.6).

Owing to the pivotal role of the TCA cycle, the collective pressures and influences from metabolism in general are considerable and very complex. As we have seen in previous chapters many alternative control mechanisms are available to living cells. In this chapter we shall be dealing largely with those which affect enzyme activity rather than synthesis or degradation of enzyme molecules.

Π To get you thinking along the right lines, write down what you understand by:

- product inhibition;
- end product inhibition;
- regulation by covalent modification.

Product inhibition occurs when the product of an enzyme reaction, if allowed to build up, inhibits the enzyme.

End-product inhibition occurs when the activity of an enzyme in a pathway (usually the first enzyme) is inhibited by the end-product of that pathway.

Covalent modification regulation occurs by addition of a simple chemical group (often phosphate) to an amino acid of the enzyme. The enzyme complex is rendered inactive and is only reactivated by removal of the covalent group.

better to inhibit an enzyme than to degrade it

Remember that enzyme activation/inhibition is important as a short term response to changes in metabolic requirements whereas changes in the levels of key enzymes is important in the long term. Close examination of the enzyme inhibition/activation illustrates the advantages and disadvantages of various types of control. Substrate level regulation has a high response rate but requires large changes in substrate concentration to exert any effect. Allosteric modulation, the binding of a positive or negative effector molecule to increase or lower activity, has a short response time. Covalent modification is a very efficient but somewhat slower process. However, when considering the control of enzyme activity it is preferable to put an enzyme molecule on 'hold' rather than to degrade it. Enzyme degradation and, if necessary, resynthesis, is slow, expensive energetically and generally used only as a last resort.

The TCA cycle is a multi-enzyme system and one of the early questions posed by biochemists was, 'which of the enzymes are subject to control?'. Preliminary studies showed that all appeared to be very sensitive to one or more central metabolic cofactors and substances. When the reverse question was posed, 'what enzymes are not controlled?' - the answer was, 'none'.

Before launching into details of control mechanisms involved with the TCA cycle it is worth pausing to remind ourselves of its subcellular location and that of its closely related metabolic pathways.

SAQ 7.6

What is the subcellular location of the following pathways?

1) glycolysis; 2) gluconeogenesis; 3) other pathways which convert glucose to pyruvate; 4) the bridging reaction; 5) the TCA cycle; 6) β-oxidation, that is fatty acid degradation; 7) fatty acid synthesis.

7.7.2 Control of the bridging reaction

We are going to describe a rather complex series of control mechanisms. We suggest you read through Sections 7.7.2-7.7.4 and then do the in-text activity described at the end of Section 7.7.4. This will help you to remember and understand the content of these sections.

The formation of acetyl CoA from pyruvate is a key reaction in metabolism. It links glycolysis or its equivalent to the TCA cycle and, as we have seen is carried out irreversibly by a multi-enzyme complex. It should not be surprising that this complex is the subject of numerous control mechanisms. The multi-enzyme complex is controlled by:

- product inhibition. Increases in the acetyl CoA concentration inhibits transacetylase and increased NADH inhibits the dihydrolipoyl dehydrogenase. These effects are reversed by CoASH and NAD^+ respectively;
- feedback regulation. The cellular energy charge influences enzyme activity; high levels of GTP are inhibiting and AMP stimulates activity;
- regulation by covalent modification. A specific serine residue in the pyruvate dehydrogenase becomes phosphorylated, a process mediated by a specific kinase and requiring ATP. The phosphorylation step, which is slowed by high ADP and pyruvate concentrations, renders the pyruvate dehydrogenase inactive. A specific phosphatase is required for removal of phosphate and reactivation of the enzyme.

protein phosphorylation an important control method

Regulation of this enzyme in mammals and higher organisms by protein phosphorylation is thought to be the most important regulatory mechanism. It is less important in the simpler organisms, particularly bacteria.

SAQ 7.7

What general statement can you make about the status of the cell when the bridging enzyme complex is inhibited and when it is stimulated? How can you rationalise this situation?

7.7.3 Control of individual enzymes of the TCA cycle

Citrate synthase

A vast amount of research has been carried out into the control of this enzyme. It is present in virtually all living systems and is essentially the starting point of the TCA cycle. Krebs himself considered this reaction to be the rate-limiting step of the whole cycle.

ATP controls citrate synthesis

In higher organisms and many bacteria, citrate synthase is controlled by ATP concentration. ATP acts as an allosteric inhibitor increasing the K_M of the enzyme for acetyl CoA. Thus as the ATP concentration increases less of the enzyme is saturated with acetyl CoA and therefore less citrate will be formed. Some control in higher organisms is apparent by a high NAD/NADH ratio but this ratio may exhibit more of an effect controlling such enzymes as malate dehydrogenase and hence the availability of oxaloacetate for condensation to citrate.

There are apparently two citrate synthases in bacteria. In Gram-negative organisms the enzyme has a M.Wt of around 250 kDaltons and is inhibited by a high NADH + H^+ concentration. In Gram-positive organisms (and all eukaryotes) the citrate synthase has a M.Wt of around 100 kDaltons and is not subject to NADH + H^+ inhibition.

control of citrate synthesis in anaerobes and aerobes

The high M.Wt citrate synthases are further subdivided with regard to whether inhibition is relieved or not by increasing AMP concentrations. In facultative anaerobes during fermentation energy is produced solely by glycolysis, which is itself subject to strict catabolic regulation. Thus the TCA cycle is used only to produce intermediates, does not require catabolic regulation and thus will not exhibit relief of inhibition by AMP. Strict aerobes are largely dependent on the TCA cycle for energy and thus the cycle must be regulated directly by adenylate charge. It is therefore logical for AMP to relieve the NADH inhibition of citrate synthase.

Citrate synthase in facultative anaerobes generally is inhibited by α-oxoglutarate. This is because in the anaerobic model of the split TCA cycle (Figure 7.6), α-oxoglutarate is an end-product and overproduction is to be discouraged.

Aconitase

Aconitase is relatively free of control by metabolites. However, as we shall see later, it can be inhibited in certain microbes, particularly yeasts, by simple elements and compounds such as copper and hydrogen peroxide.

Isocitrate dehydrogenase

In higher organisms there are at least two isocitrate dehydrogenases, one is the cytosolic $NADP^+$ -linked enzyme and the other is NAD^+ -linked and found largely in the mitochondrion. The latter enzyme is mainly under the control of AMP and ADP, increased concentrations of which cause allosteric stimulation enhancing the affinity of the enzyme for its substrate. The binding of NAD^+, Mg^{2+}, ADP and isocitrate to the enzyme is mutually co-operative and high concentrations of NADH + H^+ are inhibitory since NADH + H^+ will displace NAD^+ on the enzyme. High concentrations of ATP can slow and possibly stop the enzyme. Overall then in eukaryotes the contribution to the control of the cycle by isocitrate dehydrogenase is through the energy charge.

isocitrate dehydrogenase controlled by energy charge

Very few bacteria contain both NAD^+ and $NADP^+$-linked enzymes, in fact the overwhelming majority contain the $NADP^+$-linked isocitrate dehydrogenase. This is probably because bacteria do not have the benefits of compartmentalisation afforded by mitochondria and this is a way of maintaining control independently of NAD^+.

One extra complication in certain organisms is the fact that isocitrate is the branch point of the TCA and glyoxylate cycles. The enzyme is known to be subject to synergistic inhibition by glyoxylate and oxaloacetate.

isocitrate lyase is starved of substrate unless the dehydrogenase is inhibited

Flux of isocitrate through isocitrate lyase, is largely dependent on the activity of isocitrate dehydrogenase. Under conditions of high energy charge isocitrate dehydrogenase would be inhibited by the high ATP concentration. Thus any isocitrate formed would be channelled into the glyoxylate cycle by isocitrate lyase.

α-Oxoglutarate dehydrogenase

Control of this enzyme is on similar lines to that exercised on pyruvate dehydrogenase. It is inhibited by succinyl CoA and high NADH + H^+ concentrations and stimulated by high AMP concentrations. There is therefore overall inhibition by high energy charge.

Succinate thiokinase

The enzyme found in animal cells utilises GDP whereas that in plant and bacterial cells utilises ADP. The latter enzyme is, however, thought to be more adaptable in its use of nucleotide cofactors.

Succinate dehydrogenase, fumarase and malate dehydrogenase

Two of these enzymes catalyse redox reactions; succinate dehydrogenase requiring FAD and malate dehydrogenase requiring NAD^+. All three reactions are reversible. Direction and control of flux appears to rest largely with the enzymes of the rest of the TCA cycle rather than these three.

7.7.4 Control of maintenance of the cycle

We have so far looked at control of the cycle largely from an energy point of view rather than from the point of view of the control of the correct amounts of intermediates.

Metabolic precursors will be withdrawn from the cycle, energy permitting, for biosynthesis as required. Therefore a balance has to be maintained between input and output to retain intermediate levels at their correct concentration.

oxaloacetate production is a major anaplerotic product

As we described earlier the major anaplerotic input into the cycle is oxaloacetate produced by phosphoenol pyruvate carboxylase. This enzyme is relatively slow acting in the presence of phosphoenol pyruvate alone but is strongly activated by increased acetyl CoA and/or fatty acids. We have seen that the normal glycolytic sequence is

phosphoenol pyruvate → pyruvate → acetyl CoA

Thus to avoid even more acetyl CoA production, the enzyme is stimulated to direct phosphoenol pyruvate to oxaloacetate formation which will help to mop up excess acetyl CoA.

Phosphoenol pyruvate carboxylase is also stimulated by fructose 1,6-bisphosphate (indicating a build up of glycolytic intermediates) but inhibited by malate and aspartate (to avoid the overproduction of four-carbon dicarboxylic acids).

The $NADP^+$ -linked malic enzyme can yield additional malate from pyruvate but is inhibited by $NADH + H^+$, acetyl CoA and oxaloacetate.

Π Section 7.7 has described a complex series of control mechanisms in different systems. It would be a helpful form of revision to draw out the TCA cycle, the bridging reactions, the glyoxylate cycle and the anaplerotic reactions onto a single sheet of paper. Then re-read Section 7.7 and write on the control mechanisms of each of the reactions.

7.8 Manipulation of micro-organisms to encourage production of high levels of TCA cycle intermediates

Several TCA cycle intermediates, for example citrate and to a lesser extent malate, have important commercial uses. It was shown as early as the last century that some micro-organisms have the power to produce and accumulate high levels of citrate and by the mid 1920s industrial scale production was in use. Let us have a look at certain aspects of these production systems.

7.8.1 Citrate production

Citrate is important as a food additive in soft drinks, jellies, sweets, jams etc., as a flavour enhancer, as a builder in detergents and disinfectants and as a sequestering agent in cosmetics. Citrate accumulation is largely a property of fungi and of yeasts in particular. The commercial organisms used include a mycelial fungus, *Aspergillus niger*, and certain yeasts. Briefly, there are several fungi which can utilise either carbohydrate sources, ethanol, acetate, fatty acids or n-paraffins to give high yields of citrate. Depending on the initial concentration of substrate there can be 85% conversion of carbon giving up to 200 g citrate l^{-1}.

Let us now briefly consider the intermediary metabolism of such organisms, so that we understand the process of citrate accumulation. All of the substrates mentioned would be converted eventually to acetyl CoA which would then enter the TCA cycle. The accumulation of citrate occurs only in stationary phase cultures when the nitrogen source has run out and the TCA cycle is, therefore, not being used to yield intermediates. The process is a strictly aerobic one.

Use the flow diagram shown in Figure 7.9 to help you understand the following discussion.

In these systems the substrates described enter intermediary metabolism at acetyl CoA and during growth will have maintained a fully functional glyoxylate cycle for TCA cycle intermediate replenishment. Aerobic organisms using carbohydrate sources will have used the phosphoenol pyruvate carboxylase for replenishment but even these organisms employ the glyoxylate cycle during citrate production. At the onset of citrate accumulation the cell is in a state of high energy charge but, as there is no nitrogen for biosynthesis, the ATP concentration stays high (since the synthesis of proteins and nucleic acids cannot proceed). The principal effect of this is to inhibit almost totally the NAD^+ -linked isocitrate dehydrogenase. However isocitrate lyase and malate synthase are still active so that extra succinate is produced by the glyoxylate cycle. This bypasses the need for oxoglutarate dehydrogenase which in any case is inhibited by the increased $NADH + H^+$ levels. The inhibition of citrate synthase by high ATP concentration is apparently overruled as citrate continues to be produced.

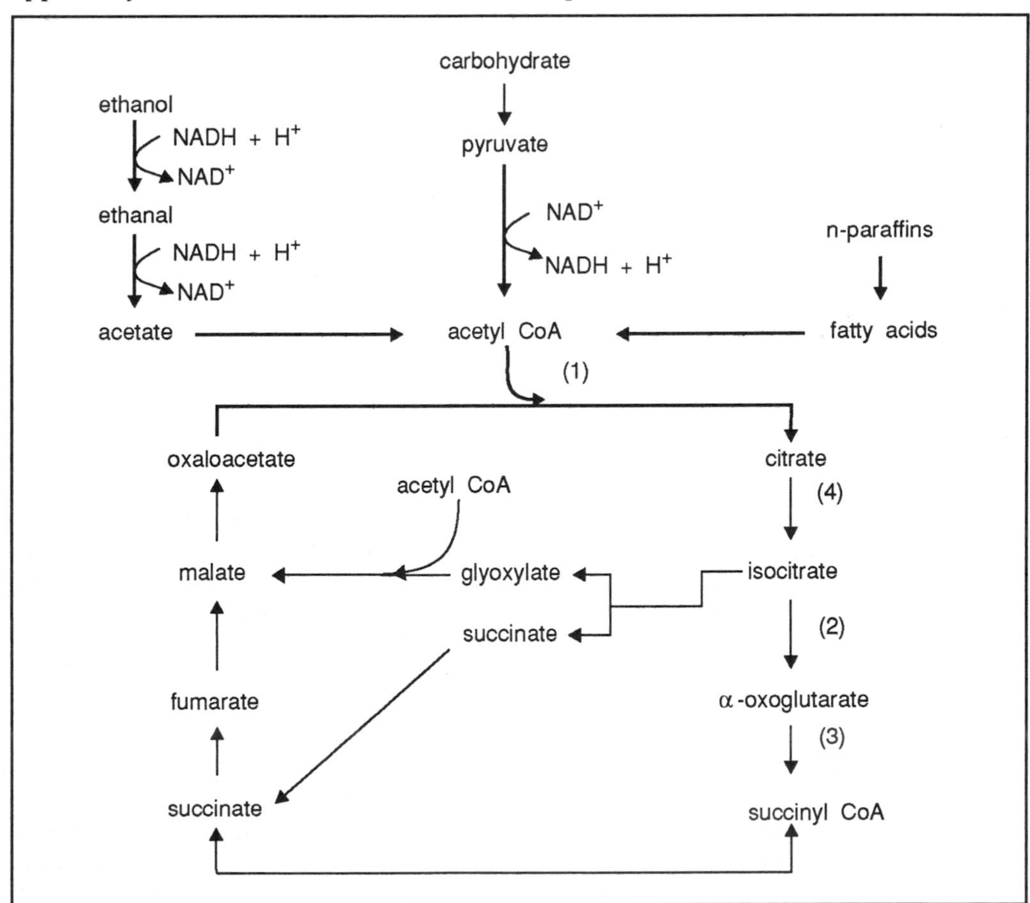

Figure 7.9 Substrates utilised and pathways involved in citrate production.
Key: 1) Citrate synthase; 2) Isocitrate dehydrogenase; 3) α-oxoglutarate dehydrogenase; 4) Aconitase.

Π Draw a diagram as in Figure 7.9 but use bold arrows to highlight the parts of the pathways which will be used in citrate production. (Do this before looking at our following response.)

Really three grades of 'arrows' should be used. The thick arrows show the conversion to citrate of whichever substrate is being used. The thin arrows show the recycling of the isocitrate to be converted back to citrate. Lines without arrows show areas where there is little or no flux.

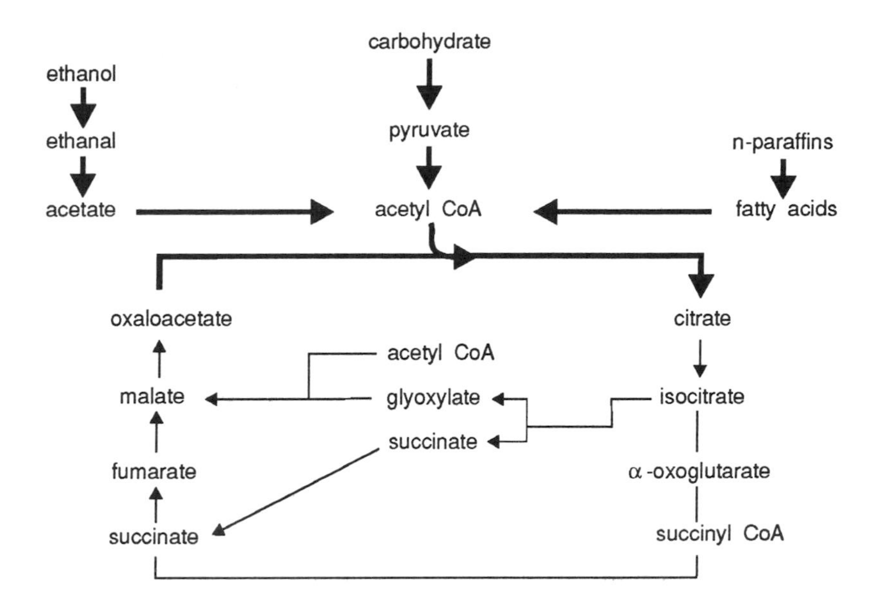

Π From your diagram there are two complications which need to be resolved; one connected with the actual product and one connected with a cofactor problem. Can you identify either or both of them?

citrate production requires low aconitase activity

The first problem is that from your diagram isocitrate and not citrate will be the end-product. The former is not nearly as useful or in demand industrially. This problem has been resolved in part in several ways; firstly by obtaining yeast species or mutant strains with very low aconitase activity, secondly by adding copper ions which inhibit aconitase activity and thirdly by trying to virtually eliminate ferrous ions which are essential for aconitase activity. In practice the last is expensive and difficult to accomplish satisfactorily.

The second, cofactor, problem is that whichever substrate is used, varying but significant amounts of NADH + H^+ will be produced. As the cell already has a very high energy charge there is little chance of reoxidising the NADH + H^+ by conventional electron transport and oxidative phosphorylation. Although not completely clear it is thought that the electron transport chain and oxidative phosphorylation are somewhat dissociated (that is uncoupled) allowing reoxidation of NADH + H^+ without ATP production.

7.8.2 Malate

This compound is of industrial interest as an alternative to citrate. A species of yeast, *Candida brumptii*, will grow on n-paraffins and, after the growth phase has finished, will produce largely malate though significant amounts of succinate are also produced. In this organism the NAD^+ -linked isocitrate dehydrogenase is absent and during malate production vastly elevated levels of isocitrate lyase and malate synthase are noted. Effectively these cells must be channelling one molecule of isocitrate and one of acetyl (from acetyl CoA) to yield two malates, the isocitrate itself originating from three acetyl CoA molecules.

Summary and objectives

The TCA cycle is the principal energy-yielding pathway in living systems growing aerobically. In anaerobic systems it is incomplete and never used to produce energy.

The cycle is used in all living systems to supply intermediates for biosynthesis. The diverting of intermediates acts as a drain on the cycle and intermediates must be replaced by anaplerotic reactions. Organisms growing on two-carbon compounds have specialised anaplerotic mechanisms, for example the glyoxylate cycle.

Control of the direction and flux of the TCA cycle is complex but depends largely on the energy charge of the cell.

Under certain conditions some cells produce very large amounts of certain TCA cycle intermediates, for example citrate and malate. These quantities are sufficient for commercial exploitation.

Now that you have completed this chapter you should be able to:

- distinguish between the terms catabolic, anabolic and amphibolic;
- list the reactions of the TCA cycle and discuss its use as a producer of ATP in aerobically-growing organisms;
- describe the meaning of anaplerotic reactions and discuss their use, with the TCA cycle, to provide intermediates for biosynthesis;
- state the subcellular location of TCA cycle enzymes in eukaryotes and prokaryotes and discuss mechanisms to control direction and flux within the cycle;
- give industrially-useful examples of ways in which micro-organisms can be used to produce large quantities of individual TCA cycle intermediates.

The oxidation of pyridine nucleotides in an aerobic environment

The oxidation of pyridine nucleotides in an aerobic environment

In previous chapters we saw that ATP can be produced in a cell without the presence of oxygen. For example, in glycolysis 2 molecules of ATP are produced by substrate level phosphorylation for every molecule of glucose oxidised. This represents the conservation of only a small fraction of the total free energy potentially available from the oxidation of glucose. In the presence of oxygen, most cells produce around 36 molecules of ATP for each molecule of glucose oxidised to CO_2 and H_2O; this is termed complete aerobic respiration.

In this chapter we shall examine the processes that enable cells to be more efficient at producing ATP in the presence of oxygen. You will recall that NADH + H^+ and $FADH_2$ are formed by primary catabolic pathways and by the TCA cycle. These are energy-rich molecules because each contains a pair of electrons that have a high transfer potential. When these electrons are transferred to molecular oxygen, a large amount of energy is liberated. The released energy can be conserved in the form of the phosphate-bond energy of ATP, in the process called oxidative phosphorylation.

8.1 The cellular location of the respiratory machinery

The respiratory machinery consists of proteins embedded in a membrane, the actual membrane used in respiration depending on the type of cell.

In eukaryotic cells the respiratory machinery is located in mitochondria, which are present in all eukaryotic cells. These organelles are the 'power-houses' of the cell since they produce large amounts of ATP. They are characterised by having extensive internal membrane systems which reflects their role as compartments of high metabolic activity. Figure 8.1 will remind you of the structural organisation of mitochondria.

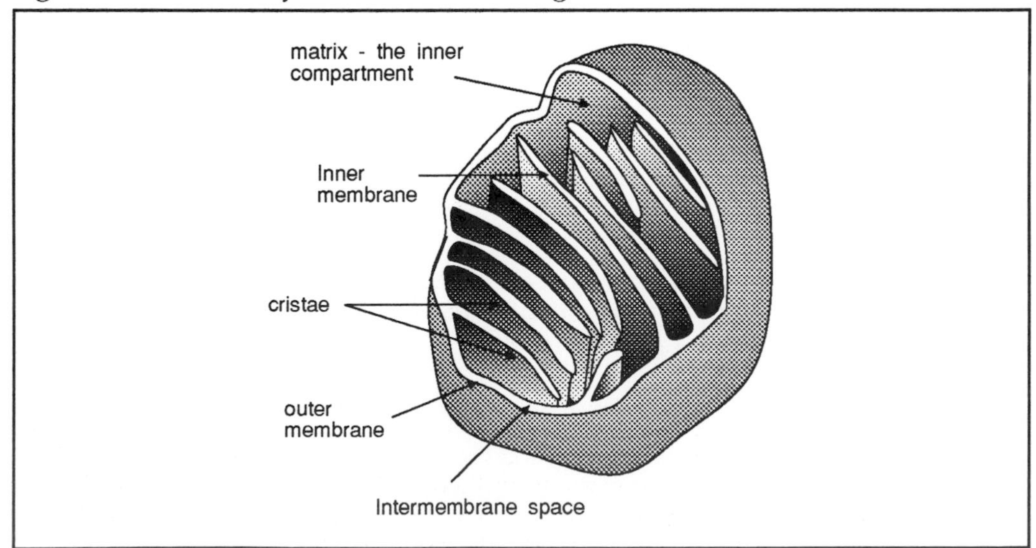

Figure 8.1 Structural organisation of mitochondria.

We can see that the inner membrane of mitochondria is folded, forming what are called cristae, which greatly increases its total surface area. This membrane contains the respiratory machinery and ATP is produced on its inward facing side.

ATP formed inside mitochondria in eukaryotes

As we have seen in previous chapters the respiratory apparatus is fuelled by the products of both the primary catabolic pathways, such as glycolysis, and the TCA cycle, the glycolytic pathway being located in the cytoplasm whereas the TCA cycle is located in the matrix of mitochondria.

ATP formed in cytoplasm in prokaryotes

By comparison, prokaryotic cells do not have membrane-bound organelles and the respiratory machinery is located on the plasma membrane. The primary catabolic pathways and the TCA cycle are located, almost exclusively, in the cytoplasm.

Π Now attempt to complete the following statements:

ATP formed by a membrane bound enzyme complex is produced in the ________ of mitochondria and in the ________ of prokaryotic cells.

The missing words are matrix and cytoplasm respectively.

Primary catabolic pathways, such as glycolysis, are located mainly in the soluble ________, whereas the TCA is located in the ________ of mitochondria in eukaryotic cells.

The missing words are cytoplasm and matrix respectively.

The NADH + H^+ formed by glycolysis and ________ and ________ formed by the TCA cycle can transfer their ________ to the respiratory apparatus located on the ________ membrane of mitochondria.

The missing words are NADH + H^+ , $FADH_2$, electrons and inner respectively.

8.2 How the respiratory machinery works

In this section we will consider in general terms the mechanism of aerobic respiration. Details of the individual components involved will be considered in later sections.

The mechanism of aerobic respiration in mitochondria and bacteria is the same. It involves the generation of an electrochemical gradient across a membrane. This is formed by pumping protons from one side of a membrane to the other. The energy stored in the gradient is then used to drive reactions catalysed by enzymes embedded in the membrane.

generation of electrochemical gradient

In eukaryotic cells, most of the energy is used to convert ADP and Pi to ATP, although some is used to transfer specific metabolites into and out of organelles. In bacteria the electrochemical gradient is again used to generate ATP but is also an important store of directly usable energy, it drives many transport processes in the plasma membrane and, in motile bacteria, drives the rotation of flagella. These points are summarised in Figure 8.2. (Note that sunlight too can generate an electrochemical gradient in photosynthetic organisms and the gradient can be used to convert ADP and Pi to ATP - we will not deal with photosynthetic ATP generation any further here except to remark that it is analogous to respiratory ATP generation in so far as both are driven by an electrochemical proton gradient. Photosynthetic ATP synthesis is described in greater detail in the BIOTOL text, 'Energy Sources for Cells').

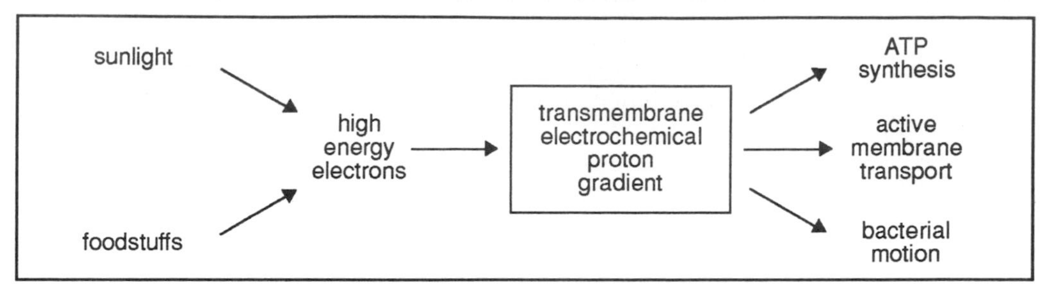

Figure 8.2 Energy harnessing by a transmembrane electrochemical gradient.

The formation of a transmembrane electrochemical gradient involves the movement of both electrons and protons. We will first consider the movement of electrons.

8.2.1 The movement of electrons along the respiratory chain

NADH + $FADH_2$ carry high energy electrons

Nearly all the energy available from the oxidation of carbohydrates, fats and other foodstuffs in the early stages of their metabolism is initially held in the form of high-energy electrons carried by NADH and $FADH_2$. These electrons are then combined with molecular oxygen by means of the respiratory chain. This consists of a series of electron carriers embedded in a membrane as shown diagrammatically in Figure 8.3.

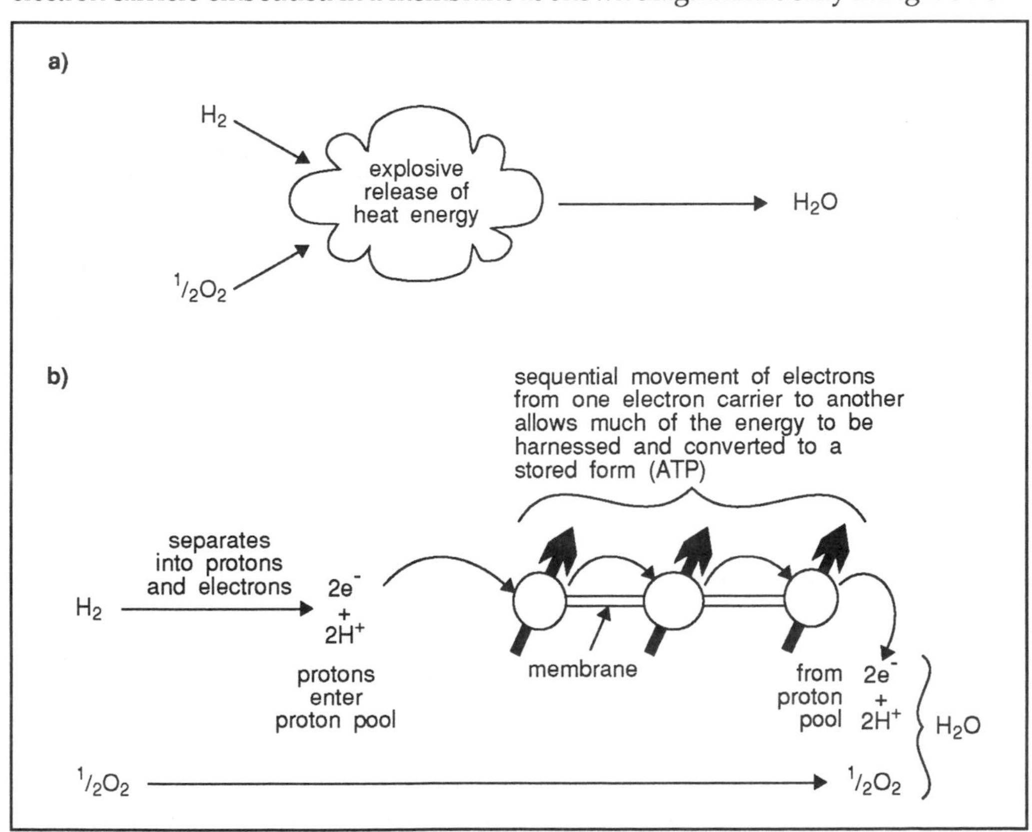

Figure 8.3 The reduction of oxygen. a) In a single step in the absence of an electron-transport chain. b) In multiple steps by movement of electrons along an electron-transport chain.

The reaction $H_2 + ½O_2 \rightarrow H_2O$ is made to occur in many small steps, so that most of the energy released can be converted into a storage form instead of being lost to the environment as heat.

electrons and protons separate

electrons and protons merge again

The respiratory chain is unique in that the hydrogen atoms are first separated into protons and electrons. For example, the hydride ion (H^-, which consists of $H^+ + 2e^-$) is removed from NADH to generate NAD^+ and then is separated to H^+ and $2e^-$. The two electrons are passed to the first of the electron carriers of the respiratory chain. Each component in the chain has a greater affinity for electrons than its predecessor, and electrons pass sequentially from one complex to another until they are finally transferred to oxygen, which has a high affinity for electrons. Protein molecules are associated with most of the electron carriers and these guide the electrons along the respiratory chain so that the electrons move sequentially from one component to another - with no short cuts. When the electrons reach the end of the electron-transport chain, the protons are added back to neutralise the negative charges of the electrons.

Before examining the details of the transfer of electrons along the respiratory chain let us have a look at the movement of protons into and out of the mitochondrial matrix.

8.2.2 The movement of protons across the mitochondrial membrane

proton pumping

As the high-energy electrons are being transported down the respiratory chain, the energy released as they pass from one carrier molecule to the next is used to pump protons (H^+) across the membrane from the inside to the outside. The movement of protons has two major consequences:

- it generates a pH gradient across the membrane, with the pH lower on the outside than on the inside (ie more protons on the outside);
- it generates a voltage gradient (membrane potential) across the membrane, with the inside negative and the outside positive. This is caused by the net flow of positive ions.

protonmotive force

Together, these two forces form an electrochemical proton gradient which exerts a protonmotive force (pmf). The two components of the pmf are illustrated in Figure 8.4.

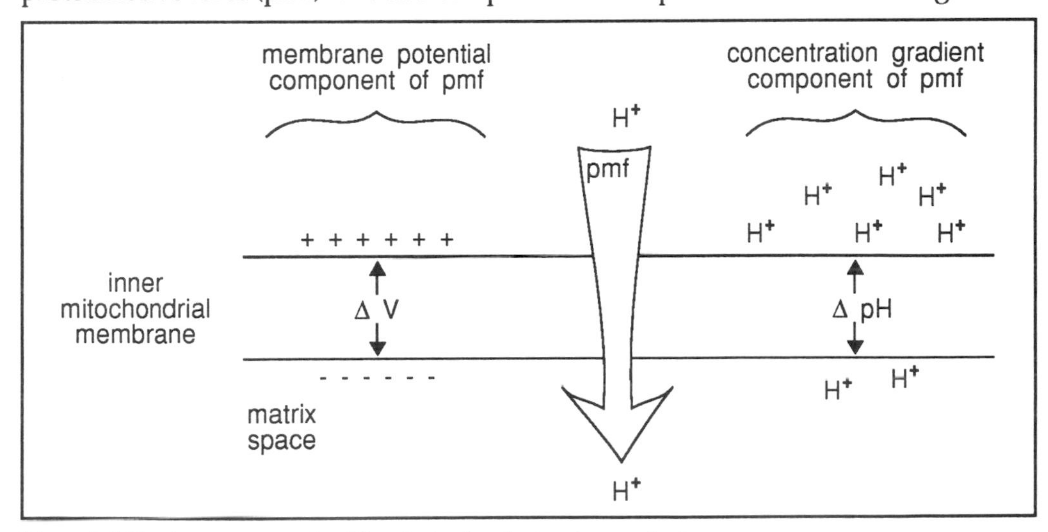

Figure 8.4 The two components of the electrochemical proton gradient.

The pmf can be measured in millivolts (mV). Each pH difference of 1.0 generates an effect equivalent to a membrane potential of about 60mV.

Π In a typical cell the pmf across the inner membrane of respiring mitochondria is made up of a membrane potential of 160 mV and a pH difference of -1 pH unit. What is the magnitude of the pmf in suitable units?

-1 pH unit = 60mV

pmf of respiring mitochondria = 160 + 60 = 220mV

SAQ 8.1

How large is the pmf (in volts) of protons across a membrane in the following cases:

1)	membrane		2)	membrane	
outside	:	inside	outside	:	inside
pH = 6	:	pH = 8	pH = 7	:	pH = 9
potential = 0V	:	potential = -0.12V	potential = -0.06V	:	potential = +0.06V

The pH gradient acts to drive H^+ back across the membrane to the inside of the mitochondria leaving OH^- on the outside. Similarly, the membrane potential attracts positive ions to the inside and tends to push negative ions out.

The energy stored in the pmf is used for two main purposes:

- to drive ATP synthesis;
- to transport metabolites and inorganic ions across the membrane.

The synthesis of ATP is catalysed by the enzyme ATP synthetase which is a major protein component of the membrane. It is a large protein complex through which protons flow down their electrochemical gradient to the inside compartment. The action of ATP synthetase is illustrated in Figure 8.5.

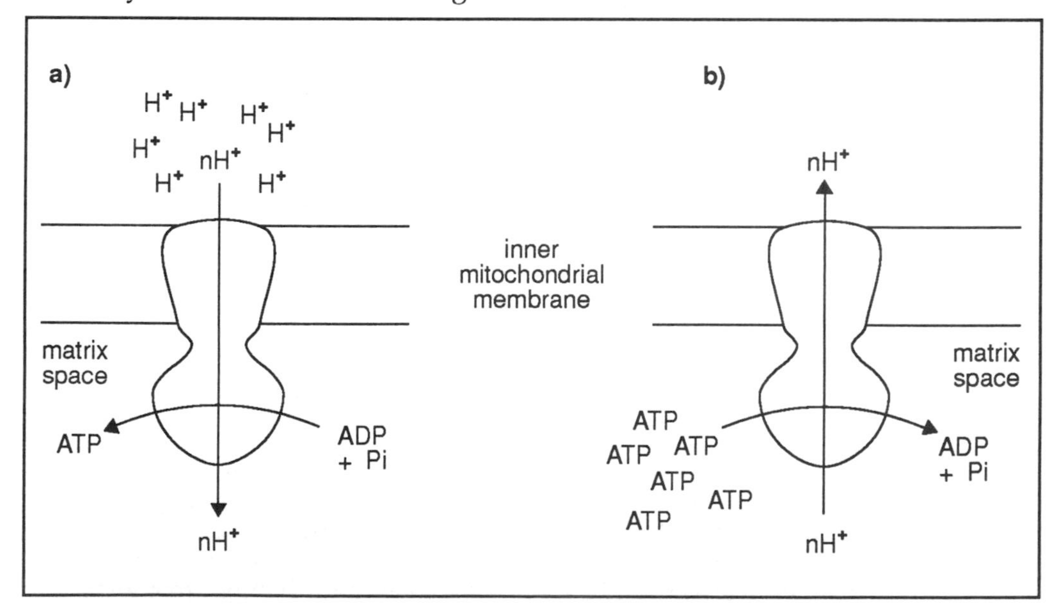

Figure 8.5 The reversible action of ATP synthetase. a) ATP synthesis driven by hydrogen ions passing from the outside of the inner mitochondrial membrane into the mitochondrial matrix; b) ATP hydrolysis coupled to the pumping of H^+ ions across the mitochondrial membrane.

ATP synthetase is reversible

We can see that ATP synthetase can work in both directions: it can use the energy of ATP hydrolysis to pump protons across the inner membrane, or it can harness the flow of protons down an electrochemical proton gradient to make ATP. It can thus be regarded as a turbine, interconverting electrochemical proton gradient and chemical bond energies.

membrane transport

The action of the pmf in membrane transport is also essential for the generation of ATP. In mitochondria, for example, the TCA cycle and other metabolic reactions take place in the mitochondrial matrix. These must be supplied with high concentrations of substrates and ATP synthetase must be supplied with ADP and phosphate. These requirements are fulfilled by various membrane carrier-proteins, located on the inner mitochondrial membrane, which transport specific molecules into the matrix from the soluble cytoplasm. The molecules are transported against a concentration gradient by a process termed active transport. The concentration in the matrix becomes greater than that in the soluble cytoplasm and this requires the input of energy which is supplied by the pmf. The energy of the pmf is coupled to transport because the carrier-proteins involved actually transport two molecules at the same time; one molecule moves up its electrochemical gradient while another moves down its electrochemical gradient. This process is termed co-transport and is illustrated in Figure 8.6.

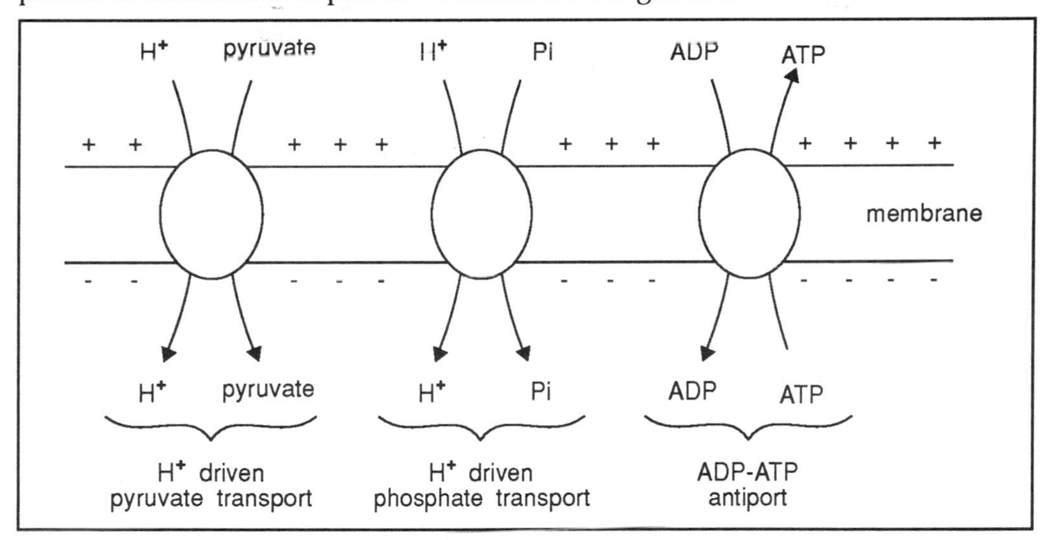

Figure 8.6 Some of the active transport processes driven by the pmf.

Π Consider Figure 8.6 and decide whether the mitochondrial matrix is located above or below the membrane.

The mitochondrial matrix is located below the membrane. The carrier-proteins transport ADP, pyruvate and Pi into the matrix where they are needed for the operation of the TCA cycle and the synthesis of ATP.

ATP/ADP exchange

We can see from Figure 8.6 that ADP molecules produced by ATP hydrolysis in the soluble cytoplasm enter the mitochondria for recharging, while the ATP molecules formed in the mitochondrial matrix by oxidative phosphorylation are rapidly pumped into the soluble cytoplasm, where they are needed. This is achieved by an ADP-ATP transport protein embedded in the inner mitochondria membrane which maintains a concentration of ATP in a cell about 10 times higher than that of ADP.

Π Why do cells maintain a high concentration of ATP relative to ADP?

You should have reasoned that this is necessary to drive energetically unfavourable reactions by coupling them to the energetically favourable hydrolysis of ATP. If the activity of the mitochondria is halted, ATP levels fall and eventually energetically unfavourable reactions can no longer be driven by ATP hydrolysis and as a consequence many biosynthetic reactions would stop and even run backwards.

The pmf is also used to transport Ca^{2+} into a mitochondrial matrix where it is thought to be important in regulating the activity of certain enzymes. The import of Ca^{2+} into mitochondria may also be important for removing Ca^{2+} from the soluble cytoplasm when its concentration becomes too high.

Π Can you think of a reason why isolated mitochondria might cease ATP production completely if they are incubated in a high concentration of Ca^{2+}?

Under these conditions large quantities of Ca^{2+} are pumped into the mitochondria suggesting that all the energy in the pmf is diverted to pumping Ca^{2+}.

ATP can be used to generate a proton gradient

So far we have considered respiration only in mitochondria whereas bacteria (prokaryotes) do not possess mitochondria. The plasma membrane of the vast majority of bacteria contains an ATP synthetase that is very similar to that in mitochondria. Certain anaerobic bacteria lack an electron-transport chain and can therefore generate ATP only by substrate level phosphorylation. They must, however, still generate a pmf in order to drive the transport of nutrients into the cell.

Π Can you suggest a mechanism whereby anaerobic bacteria could generate a pmf in the absence of a respiratory chain?

They do this by using the ATP synthetase in the reverse direction, utilising ATP produced by substrate level phosphorylation to pump protons from the cytoplasm to the outside.

SAQ 8.2

Identify each of the following statements as True or False. If False justify your response.

1) The reduction of $\frac{1}{2}O_2$ to H_2O via the respiratory chain results in a smaller change in Gibbs function (ΔG) compared to a single step reduction of $\frac{1}{2}O_2$.
2) A PMF is generated by pumping protons out of the mitochondrial matrix.
3) The respiratory apparatus creates a voltage gradient by directly moving electrons across the inner mitochondrial membrane.
4) The mitochondrial matrix is negatively charged in comparison with the outside of the inner membrane.
5) The inner-mitochondrial membrane is freely permeable to H^+.
6) The energy required to drive pyruvate into the mitochondrial matrix against a concentration gradient is supplied directly by ATP hydrolysis.
7) ATP synthetase simultaneously pumps protons out of the mitochondrial matrix and synthesises ATP.

Having considered in general terms how ATP synthetase functions, we will now examine the functioning of its individual components.

8.3 ATP synthetase

We will have seen that the respiratory apparatus includes an ATP-driven proton pump. Since the primary function of this pump is ATP synthesis, in which the gradient of protons supplies the energy required, it is called ATP synthetase. We find quite similar ATP synthetases in the inner membrane of mitochondria and in the plasma membrane of bacteria.

ATP synthetase, in all cases, is an enzyme complex consisting of two components:

- the component F_o which is an integral membrane protein;
- the component F_1 (F_1 ATPase) which protrudes from the membrane on the side on which ATP hydrolysis takes place.

The structure of ATP synthetase has been elucidated to a large extent by means of studies on the isolated complex. During purification of ATP synthetase the whole complex is made soluble by using a detergent. The purified preparation can be reconstituted in an artificial phospholipid membrane and will regain its properties. A procedure for the preparation of sub-mitochondrial vesicles, F_o and F_1 components and reconstituted vesicles is shown in Figure 8.7.

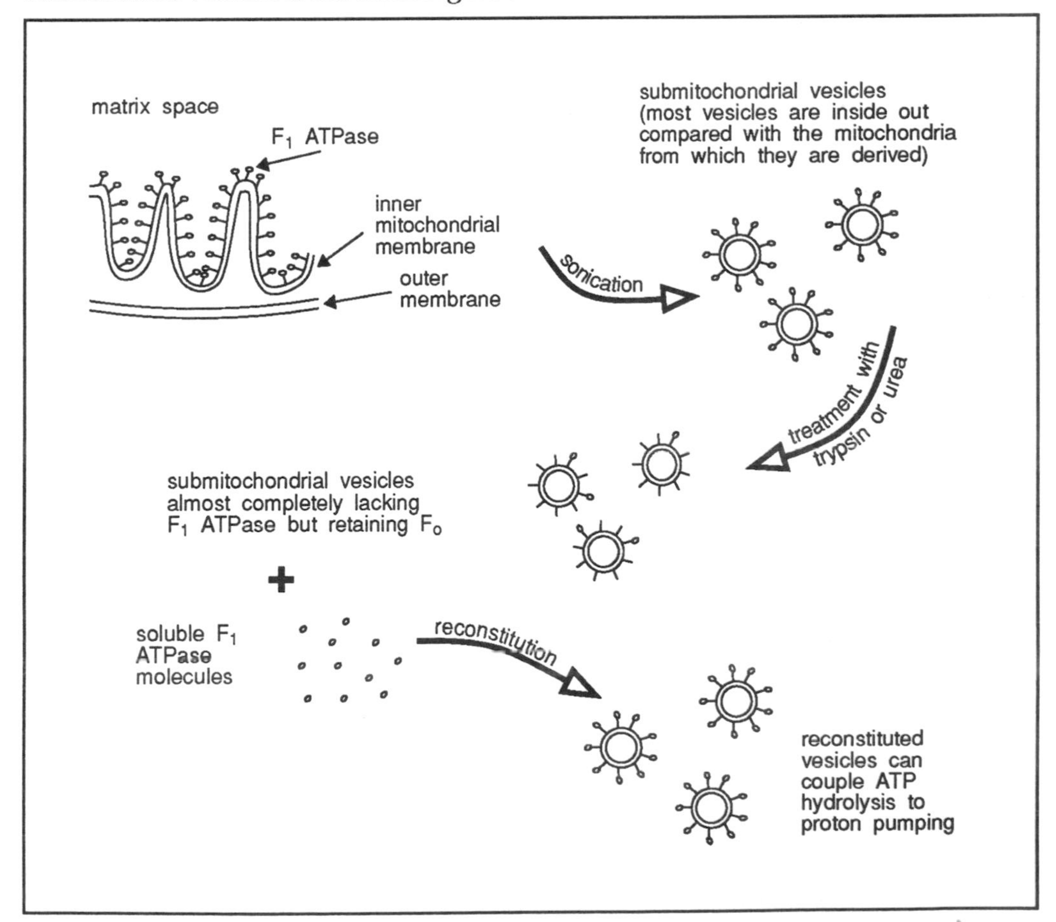

Figure 8.7 Schematic representation of the preparation of phosphorylating sub-mitochondrial particles.

The two parts of the ATP synthetase, F_o and F_1, can be further purified. F_1 appears to dissolve well in water, whereas F_o can only be kept in solution in detergent micelles. F_0, present in biological or artificial membranes, is capable of binding F_1 under the right experimental conditions. The two parts of the complex have functions which are clearly distinguishable, and not surprisingly, their structure differs too (Figure 8.8). We will now examine the structure in more detail.

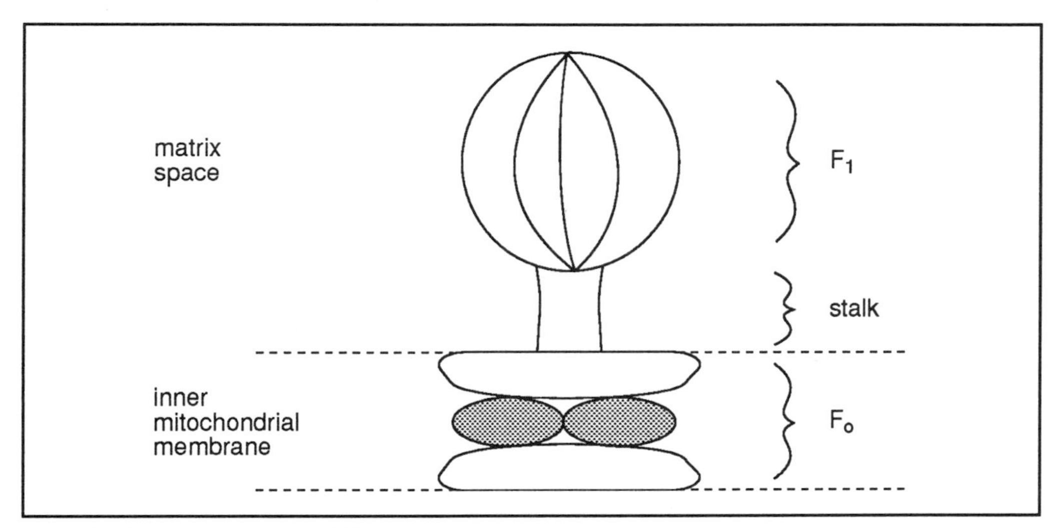

Figure 8.8 The components of ATP synthetase.

8.3.1 Soluble ATPase: F_1

The isolated soluble part of ATP synthetase (F_1) is capable of catalysing ATP hydrolysis. Since no proton gradient can be created in this situation (as there is no membrane), hydrolysis will take place until all the ATP has been converted to ADP and Pi and then stop. F_1 consists of five different polypeptides, two of which are much larger than the others and are referred to as the α and β polypeptides. The ATP binding sites and thus the enzymatic active centre involves both α and β subunits.

ATP synthesis involves both α and β subunits

There are six ATP binding sites but interaction must occur between them because specific inhibitors are known for which only one molecule of inhibitor is required per molecule of F_1 for total inhibition of ATP hydrolysis. Binding of a molecule of the inhibitor to one site renders the other sites inactive for ATP hydrolysis.

F_1 inhibitor

A small protein can be isolated from mitochondria, which has an inhibitory influence on the ATPase activity of F_1. This protein, called the F_1 inhibitor, is thought to play a role in the regulation of the activity of the ATP synthetase. The binding of the inhibitory protein to F_1 is influenced by a number of factors. ATP, for example, reinforces the binding, whereas ADP weakens it. The result is that the kinetic properties of the ATP synthetase are such that ATP synthesis is relatively better catalysed than ATP hydrolysis.

8.3.2 The membrane-bound part: F_o

Like F_1, the membrane-bound part of ATP synthetase consists of several subunits (see Figure 8.8). These subunits are strongly hydrophobic, as would be expected of polypeptides which are integral to the membrane.

oligomycin blocks F_o

F_o acts as a channel for protons in the membrane. This can be demonstrated by using specific inhibitors, for example oligomycin, which block proton movement through F_o. F_1 can be separated from biological membranes by means of several treatments (see Figure 8.7). The remainder of the membrane still contains F_o and is permeable to protons but permeability is decreased greatly on the addition of oligomycin.

OSCP is the stalk linking F_o and F_1

The binding of F_1 to F_o in the membrane requires an additional protein factor, called oligomycin sensitivity conferring protein (OSCP), which forms the stalk anchoring F_o to F_1. Its name already indicates its function: the hydrolysis of ATP by F_1 is inhibited by oligomycin only when it is bound to F_o by means of OSCP, thus oligomycin inhibits ATPase activity of the complete ATP synthetase complex.

SAQ 8.3

Identify each of the following statements as true or false. Justify your response in each case.

1) ATP will be hydrolysed when added to a suspension of F_1.

2) Addition of ATP to a suspension of sub-mitochondrial vesicles will lead to a decrease in pH on the inside of the vesicles.

3) ATP synthetase can hydrolyse six ATP molecules simultaneously.

4) The F_1 inhibitor protein binds strongly to F_1 in the presence of a high concentration of ADP.

5) Oligomycin will inhibit both ATP synthesis and ATP hydrolysis by ATP synthetase in intact mitochondria.

8.4 The respiratory chain (electron-transport chain)

many carriers have a specific absorption spectrum

The respiratory chain constitutes a series of membrane-embedded electron carriers which transfer electrons from one to another and eventually to oxygen. Concurrently with electron transport along the chain protons are transported out of the matrix thereby generating the pmf. The driving force for this process is the redox potential. Many of the electron carriers absorb light and change colour when they are oxidised or reduced. In general, each has an absorption spectrum that is distinct enough to allow assay of the electron carrier in cell extracts. There are two major families of electron carriers:

- cytochromes. These are coloured proteins that are related by the presence of a bound haem group. The haem group consists of a porphyrin ring with a tightly bound iron held by four nitrogen atoms, this is shown in Figure 8.9. The iron atom changes from the ferric (Fe III) to the ferrous (Fe II) state whenever it accepts an electron;

- iron-sulphur proteins. In these proteins either two or four iron atoms are bound to an equal number of sulphur atoms and to cysteine side chains, forming an iron-sulphur centre on the protein (see Figure 8.9).

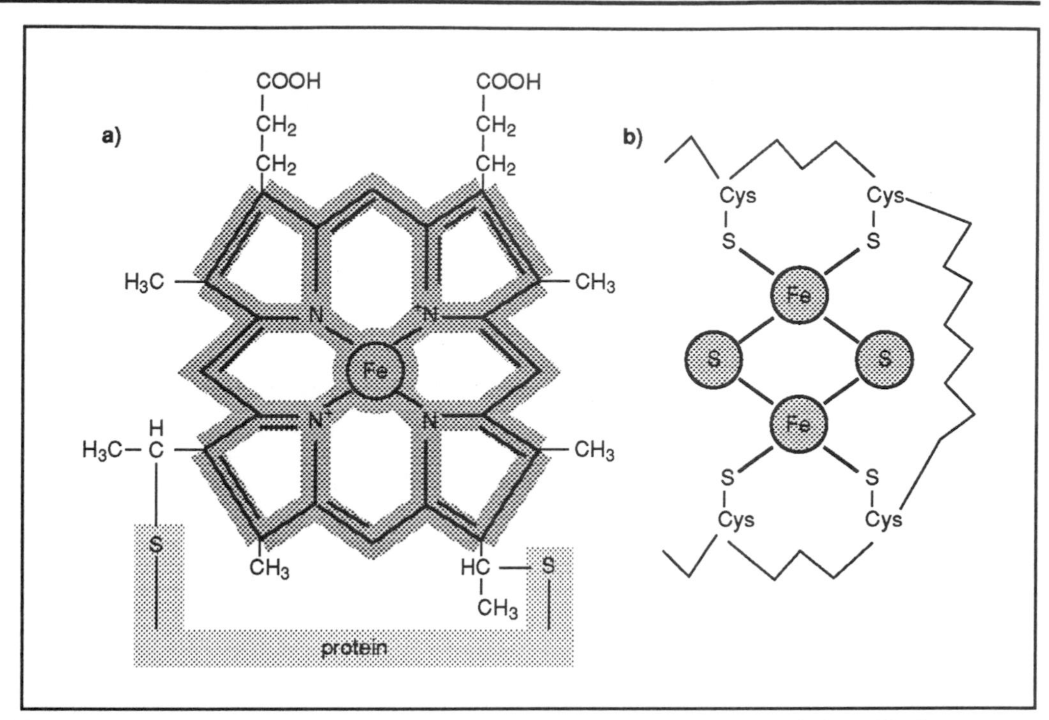

Figure 8.9 The structure of some of the electron carriers in the electron transport chain. a) The haem group attached covalently to cytochrome C and b) a 2Fe2S type of iron-sulphur centre.

There are also other types of electron carriers involved in the electron transport chain. The simplest is a small hydrophobic molecule dissolved in the lipid bilayer known as ubiquinone, or coenzyme Q. You will recall from an earlier chapter that ubiquinone can pick up or donate either one or two electrons, and it temporarily picks up a proton along with each electron that it carries. There are also two copper atoms and a flavin serving as electron carriers tightly bound to respiratory-chain proteins in the pathway from NADH to oxygen.

three large enzyme complexes

The pathway of electron transport involves about 40 different proteins in all. These are organised into three large enzyme complexes:

- the NADH dehydrogenase complex (flavoprotein 1; FP_1) is the largest of the respiratory enzyme complexes, with a molecular weight of 800 kDaltons and more than 22 polypeptide chains. It accepts electrons from NADH and passes them through a flavin and at least five iron-sulphur centres to ubiquinone (Q). This then transfers its electrons to a second respiratory enzyme complex, the B-C_1 complex;

- the B-C_1 complex contains at least eight different polypeptide chains and is thought to function as a dimer with a molecular weight of 500 kDaltons. Each monomer contains an iron-sulphur protein and three haems each bound to a cytochrome. The complex accepts electrons from ubiquinone and passes them through cytochromes B and C_1 on to cytochrome C, a small protein that carries its electrons to the cytochrome oxidase complex;

- the cytochrome oxidase complex (cytochrome A, A_3) is composed of at least eight different polypeptide chains and is isolated as a dimer with a molecular weight of about 300 kDaltons. Each monomer contains two cytochromes and two copper atoms. The complex accepts electrons from cytochrome C and passes them through cytochromes A and A_3 to oxygen.

Π Using the information provided above write the sequence of major electron carriers involved in the transfer of electrons from NADH to oxygen.

NADH $\frac{1}{2}O_2$

The correct sequence is:

$$NADH \rightarrow FP_1 \rightarrow Q \rightarrow B - C_1 \rightarrow C \rightarrow A - A_3 \rightarrow \frac{1}{2}O_2$$

Note we have not included all of the iron-sulphur centres.

The sequence of electron carriers in the respiratory chain is again shown in Figure 8.10 which also shows the relative sizes and shapes of the three respiratory enzymes complexes.

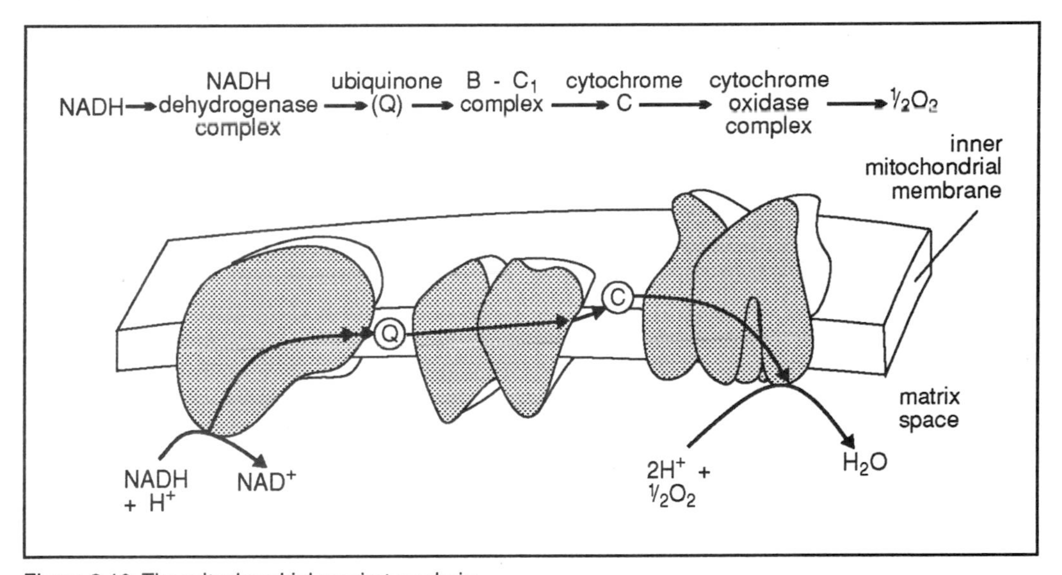

Figure 8.10 The mitochondrial respiratory chain.

The path of electrons along an electron-transport chain can be determined by observing spectral changes in the carriers which differ in their oxidised and reduced states. Under normal conditions all carriers are in a partially oxidised state. A number of inhibitors are available which block the movement of electrons along the respiratory chain at a specific site. These act on specific carriers in the chain. Addition of one of these inhibitors causes the carriers downstream of it to become more oxidised and the carriers upstream of it to become more reduced.

Π Antimycin is an inhibitor that prevents the passage of electrons from the B-C_1 complex. Would you expect cytochrome C, ubiquinone and cytochrome A to be more oxidised or more reduced in the presence of antimycin?

Ubiquinone is upstream of the site of action of the inhibitor and would become more reduced. Both cytochrome C and cytochrome A are downstream of the inhibitor and would become more oxidised.

8.4.1 The mechanism of electron transfer

functional interactions among electron transfer chain components

It is easy to imagine a structurally ordered chain of electron-transport proteins in the lipid bilayer to account for the pathway of electron-transport. However, the three enzyme complexes exist as independent entities and can migrate in the plane of the inner membrane of the mitochondrion. The orderly transfer of electrons is due entirely to the specificity of the functional interactions among the components of the chain, rather than to any structural arrangements.

Π Would you expect the various components of the respiratory chain to be present in the same or in quite different amounts?

They are, in fact, present in quite different amounts. This is consistent with the orderly transfer of electrons being due entirely to specificity of functional interactions. It is not consistent with the idea of a structurally ordered chain of electron-transfer proteins.

Two components carry electrons between the three major enzyme complexes of the respiratory chain, ubiquinone and cytochrome C. These diffuse in the plane of the membrane and collisions between these mobile carries and the enzyme complexes can account for the observed rates of electron transfer - each complex donates and receives an electron about once every 5 to 20 milliseconds.

specificity of electron transfer

The specificity of the functional interactions among the components of the chain depends upon the specificity of the major complexes and their relative affinity for electrons. Pairs of compounds such as H_2O and 0.5 O_2, or NADH and NAD^+, are called conjugate redox pairs, since one compound is converted to the other by adding one or more electrons plus one or more protons. You will recall from Chapter 3 that by placing electrodes in contact with solutions that contain the appropriate conjugate redox pairs, the redox potential (E) can be determined, and is a measure of the electron carrier's affinity for electrons. Under standard conditions (E_o) those pairs of compounds that have the most negative redox potential have the weakest affinity for electrons and therefore have the greatest tendency to donate electrons and the least tendency to accept them.

Since E_o gives a relative measure of electron donating or accepting 'power' a spontaneous reaction will occur between two coupled half reactions if there is a positive difference in E_o values (ie ΔE_o is positive). If ΔE_o has a negative value, the reaction will not proceed spontaneously unless there is an input of energy.

relation between $\Delta G^\circ + \Delta E_o$

We also know that the change in Gibbs function, ΔG°, (often not quite correctly called standard free energy change) indicates the tendency of a reaction to proceed spontaneously under standard conditions. As you might recall ΔG° and ΔE_o are related by:

$$\Delta G^\circ = -nF\Delta E_o$$

where n is the number of electrons transferred per molecule; F is the Faraday constant.

So E_o emerges as an alternative notation to G° when describing oxidation reduction reactions.

An outline of the redox potentials measured along the respiratory chain and the free-energy released is shown in Figure 8.11. Some E_o values are recorded in Table 8.1.

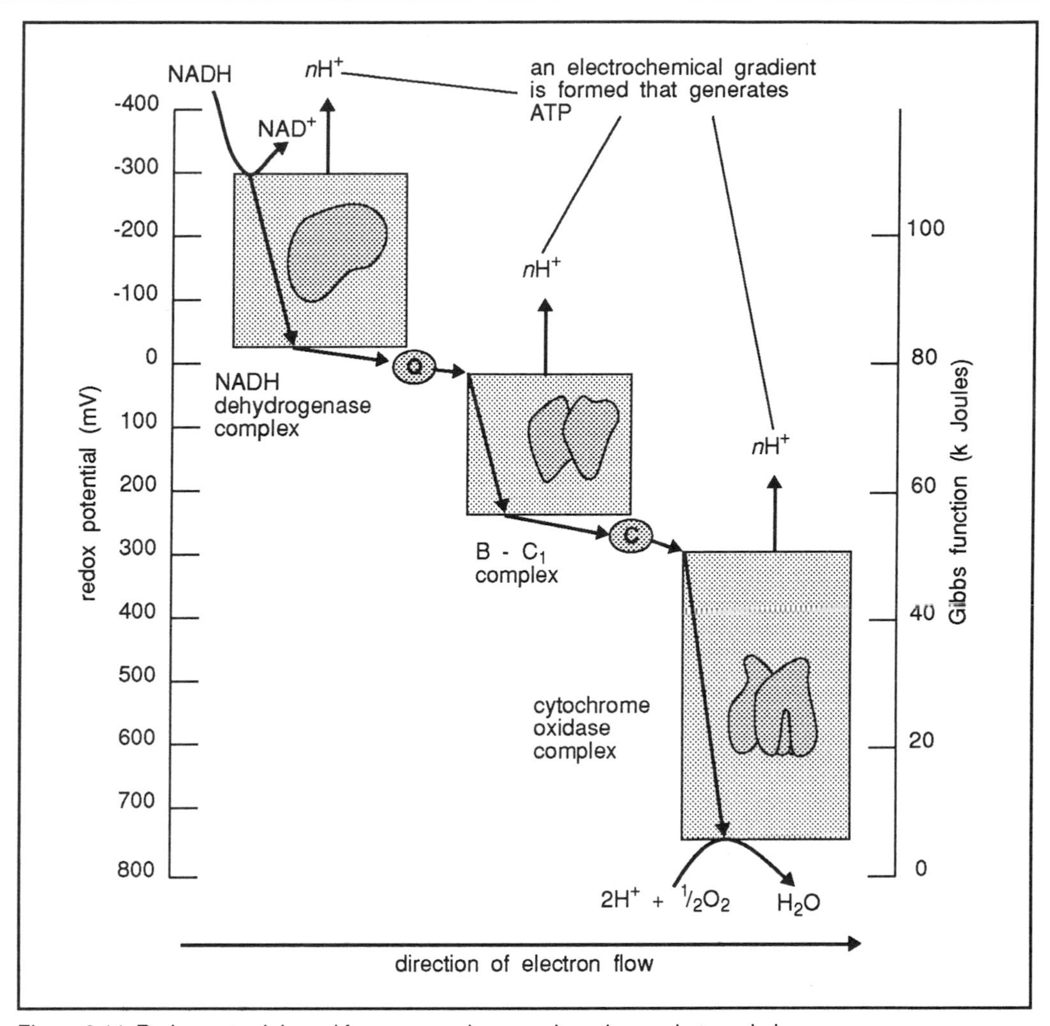

Figure 8.11 Redox potentials and free-energy changes along the respiratory chain.

SAQ 8.4

Examine Table 8.1 and answer the following questions:

1) Which redox couple is the strongest oxidant?

2) Which redox couple has the greatest tendency to donate electrons?

3) Will the reaction shown below proceed spontaneously under standard conditions (pH 7.0, 25°C)?

cytochrome C (Fe II) + cytochrome A (Fe III) →

cytochrome C (Fe III) + cytochrome A (Fe II)

4) Under standard conditions (pH 7.0, 25°C) what reaction will occur between the couples $FAD^+/FADH_2$ and fumarate/succinate?

5) What will be the ΔE_o for the reaction in part 4)?

			E_o(V)
$NAD^+ + H^+ + 2e^-$	→	NADH	-0.32
$0.5O_2 + 2H^+ + 2e^-$	→	H_2O	+ 0.82
cytochrome C (FeIII) + e^-	→	cytochrome C (FeII)	+ 0.22
cytochrome A (FeIII) + e^-	→	cytochrome A (FeII)	+ 0.29
fumarate + $2H^+ + 2e^-$	→	succinate	+0.03
$FAD^+ + 2H^+ + 2e^-$	→	$FADH_2$	-0.06

Table 8.1 Some standard redox potentials of major electron transport components.

8.4.2 The coupling of electron transport to proton pumping

H^+ pumping conserves energy

We can see from Figure 8.11 that as electrons move along the respiratory chain there are three positions where there is a large change in Gibbs function (ie free energy is released) - one across each major enzyme complex. Each complex acts as an energy - conserving device to harness this free-energy, pumping H^+ across the inner membrane to create and electrochemical proton gradient as electrons pass along the respiratory chain.

This linking of electron transport to H^+ pumping and the generation of ATP is referred to as the chemiosmotic hypothesis. When it was first proposed by Mitchell in 1961, this mechanism explained a long-standing puzzle in cell biology. It was originally believed that the energy of oxidation was thought to generate a high-energy bond between a phosphate group and an intermediate compound. The conversion of ADP to ATP was thought to be driven by the energy released when this bond was broken. Despite intensive efforts, however, the expected intermediates could not be detected. According to the chemiosmotic hypothesis, the high-energy chemical intermediates are replaced by a link between chemical processes ('chemi') and transport processes ('osmotic') - hence chemiosmotic coupling. This hypothesis is now widely accepted and Mitchell was awarded the Nobel prize for his work. The molecular mechanism linking proton pumping with electron flow is not understood. However, it is considered significant that the three major enzyme complexes stretch right across the inner mitochondrial membrane. Thus they are exposed both to the matrix and to the inter membrane space, a fact which presumably facilitates the binding of a proton on one side and the release of a proton on the other. This could conceivably be achieved by a change in conformation of a subunit of the major enzyme complexes but little is known about this.

Π During electron transport, which of the following sequences can involve the synthesis of ATP?

1) cytochrome C → cytochrome A
2) cytochrome A, A_3 → O_2
3) NADH → FP
4) FP_1 → Q
5) Q → B-C_1
6) B-C_1 → C
7) malate → NADH

To be fully confident of the answer to this you probably needed more information than we have given. You probably spotted 2) and 6) as sequences involving ATP synthesis as we suggested that ATP synthesis associated with each of the major electron transport components. You might however have found it difficult to decide between reactions 3) and 4). The answer is 4). Look back at Figure 8.11, you will see that reactions 2), 4) and 6) are associated with one of the mitochondrial complexes and these are involved in proton pumping. It is the pumping of the protons which drives ATP synthesis.

SAQ 8.5

Complete the following statements which summarise the main postulates of the chemiosmotic coupling hypothesis.

1) The mitochondrial inner membrane is ———————— to H^+, OH^- and generally to anions and cations.

2) The respiratory chain pumps ———————— out of the mitochondrial ———————— when ———————— are transported along the chain.

3) ATP synthetase is a reversible coupling device - it can use the energy of ———————— to pump H^+ out of the mitochondrial matrix, but if a large pmf is present protons flow into the matrix and drive ————.

8.5 Energetic aspects of the movement of protons

In this section we will consider the factors that govern the direction of action of ATP synthetase. This depends on:

- the balance between the steepness of the electrochemical proton gradient;
- the local free-energy change (ΔG) for ATP hydrolysis.

free energy changes and ATP production

You will recall from an earlier chapter that we saw that the ΔG for proton translocation (ΔG_{H+}) across a membrane is the sum of 1) the ΔG for moving a mole of ion through a difference in membrane potential ΔV and 2) the ΔG for moving a mole of molecule between two compartments in which its concentration differs. We already know that the pmf (Section 8.2.2) is essentially composed of two components. Furthermore, the two components of G_{H+} and those of the pmf are essentially the same - replacing the concentration difference by an equivalent 'electrochemical potential' for the proton. Thus the ΔG for proton translocation (ΔG_{H+}) and the pmf measure the same potential, one in kilojoules and the other in millivolts. The conversion factor between them is:

$$\Delta G_{H+} = -0.096 \text{ pmf}$$

where ΔG_{H+}. is in kilojoules per mole and the pmf is in millivolts.

Π What is the value for ΔG_{H+}., in suitable units, for a pmf of 220mV?

$$\Delta G_{H+} = 220 \times -0.096 = -21.12 \text{ kJ mol}^{-1}.$$

Π What is the value for the pmf, in suitable units, for a ΔG_{H+} of -10 kJ $mole^{-1}$?

$$\text{pmf} = -10/-0.096 = 104 \text{ mV}$$

Having established that ΔG for proton translocation (ΔG_{H+}) and the pmf are closely related, we will now consider how the action of ATP synthetase is dependant on (ΔG_{H+}).

First we will work out how many protons can be pumped across a membrane which has a particular pmf value. Obviously pumping protons against a concentration gradient requires energy. Let us assume that we have an oxidation:reduction reaction that results in a change in Gibbs function (or free energy) of ΔG.

Thus a free energy change can be used to drive protons across the membrane. The energy required to pump protons across the membrane is, of course, (ΔG_{H+}).

Thus the oxidation:reduction reaction could pump $\frac{\Delta G_{ox.red}}{\Delta G_{H+}}$ protons across the membrane

or, since ΔG_{H+} is related to pmf, $\frac{\Delta G_{ox.red}}{-0.096\ \text{pmf}}$.

This will of course, give the maximum number of protons which can be pumped as it assumes that all of the energy derived from the oxidation:reduction reactions has been used to drive protons up a proton gradient.

Use this approach to calculate how many protons can be driven across a mitochondrial membrane using NADH + H^+ oxidation (SAQ 8.6).

SAQ 8.6

The free-energy change in oxidation of NADH + H^+ by oxygen is -220 kJ mol^{-1}. If the pmf across the mitochondrial inner membrane is +0.21V, how many protons can be pumped at the most by the respiratory-chain complexes during oxidation of 1 mol NADH + H^+?

In SAQ 8.6 you should have calculated that NADH + H^+ oxidation could drive up to about 11 protons for each NADH + H^+ oxidised by oxygen. In practice the number is always lower than this. Usually it is about 9.

We can ask the question, how many protons have to be driven through ATP synthetase in order to get a molecule of ATP produced? We can tackle this from an energetic standpoint.

Let us write the change in Gibbs function for ATP synthetase as ΔG_{ATP}.The energy produced when a proton crosses a membrane depends upon the membrane potential and the difference between proton concentrations on either side of the membrane.

Thus the energy release is proportional to the membrane potential and ΔpH.

The membrane potential and ΔpH are of course the same as pmf which in turn is equivalent to ΔG_{H+}.

Thus if a proton is allowed down the pH gradient, it will release ΔG_{H+} energy. The maximal amount of ATP that can be synthesised follows from $\left(\frac{\Delta G_{ATP}}{\Delta G_{H+}}\right)$.

In practice, this value is close to 3.

Now attempt SAQ 8.7. It uses the same sort of information but in a slightly different way.

SAQ 8.7

The equilibrium of ATP synthesis with the proton gradient is measured in isolated mitochondria. It was found to be at $\Delta pH = -0.5$, membrane potential = 0.18 V and ΔG_{ATP} synthesis = -60 kJ mol^{-1}. How many protons are pumped per ATP hydrolysed?

(ΔG_{H+} =-0.096 pmf; -1pH = 60mV)

8.5.1 The stoichiometry of oxidative phosphorylation

In oxidative phosphorylation a maximum of about three ATP molecules can be synthesised for each NADH molecule oxidised, that is, for every two electrons that pass through all three of the enzyme complexes in the respiratory chain. Since two electrons are required to reduce each oxygen atom, the above relationship is usually referred to as the P/O ratio - the number of ATP molecules synthesised by electron transport per oxygen atom reduced.

Π We previously made the assumption that three protons flow through ATP synthetase for every ATP molecule synthesised. If the P/O ratio is about three, calculate the number of protons that an average enzyme complex would pump across the membrane for each electron it transports.

Since the P/O ratio is about three, and three protons are required for each ATP synthesised, then about 9 protons are pumped for each O consumed. But there are three enzyme complexes along the transport chain. Thus for each pair of electrons transferred each enzyme complex pumps three protons. It follows that about 1.5 protons would be pumped by the average enzyme complex for each electron it transports. This is an average number of protons because some enzyme complexes pump one proton per electron, while others pump two.

The P/O ratio can be predicted by considering the number of protons moving across the membrane, as follows:

$P/O = n_o/n_p$

where n_o is the number of protons pumped out of the matrix per redox reaction.

n_p is the number of protons required to produce an ATP via ATP synthetase.

For NADH, $n_o/n_p = 9/3 = 3$

some protons leak out

If the mitochondrial membrane was completely impermeable to protons, we could precisely predict the P/O ratio. However, this is not the case in practice because there is as a certain leakage of protons. The result is that the actual P/O ratio is always lower than the P/O ratio calculated from the stoichiometries of the proton pump.

Π We established in SAQ 8.6 that the oxidation of NADH by oxygen would pump a maximum of 10.9 protons across membrane with a pmf of 220mV. What P/O ratio would this support?

The answer is 10.9/3 = 3.63, since 3 protons are required for each ATP produced. Because of the natural leakage of protons through the inner mitochondrial membrane the P/O ratio is always below that predicted from the number of protons pumped and is usually nearer 3.

SAQ 8.8

During oxidative phosphorylation the formation of ATP is coupled to the oxidation of NADH by oxygen. Calculate, using the information provided below, the standard free-energy change (ΔG°) for the reaction:

$$0.5\ O_2 + NADH + H^+ + 3ADP + 3Pi \rightarrow NAD^+ + 3ATP + 4H_2O$$

$$(NADH + H^+ + 0.5\ O_2 \rightarrow NAD^+ + H_2O;\ \Delta E_o = +1.14\ V)$$

$$(ATP + H_2O \rightarrow ADP + Pi;\ \Delta G^\circ = -30.5\ kJ\ mol^{-1})$$

$$(\Delta G^\circ = -nF\Delta E_o;\ \ F = 96.5)$$

It follows that the P/O ratio will decrease if the number of protons that can leak through the membrane increases. This can be induced by the addition of a protonophore, which is a H^+ carrier (H^+ ionophore); several lipophilic weak acids can fulfil this role. The protonophores are also called uncoupling agents because ATP synthesis stops but O_2 uptake continues. ATP synthesis is said to be uncoupled from electron transport. The explanation of this is that protonphores provide a pathway, in addition to that of ATP synthetase, for the flow of H^+ across the membrane. As a result of this the pmf is completely dissipated, and ATP can no longer be synthesised. Figure 8.12 illustrates the action of a widely used uncoupler, 2,4-dinitrophenol.

uncouplers dissipate the pmf

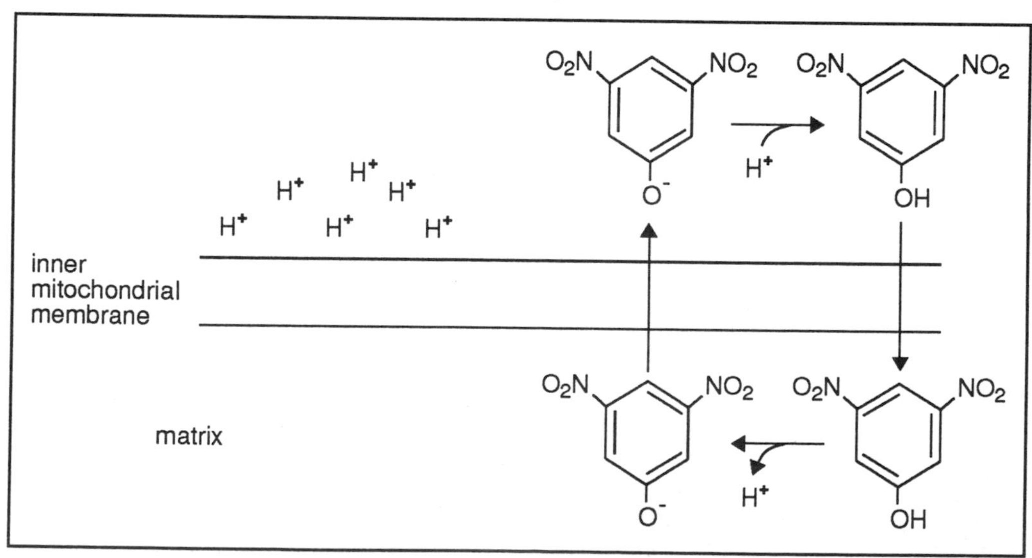

Figure 8.12 The action of an uncoupling agent.

8.5.2 Respiratory control

When an uncoupler such as dinitrophenol is added to cells, mitochondria increase their oxygen uptake substantially because of an increased rate of electron transport. This increased rate reflects the existence of respiratory control. The control acts by a direct inhibitory effect of the electrochemical-proton gradient on the rate of electron transport. When the gradient is collapsed by an uncoupler, electron transport is free to run unchecked at the maximum rate. As the proton gradient increases as in the absence of an uncoupler, electron transport becomes more difficult and the process slows.

respiratory control

Respiratory control is just one part of an elaborate interlocking system of feedback controls that co-ordinate the rates of glycolysis, fatty acid breakdown, the TCA cycle and electron transport. The rates of all these processes are adjusted to the ATP/ADP ratio, which will increase whenever increased utilisation of ATP causes the ratio to fall. The ATP synthetase in the inner mitochondrial membrane, for example, works faster as the concentration of its substrates ADP and Pi increases. As it speeds up, the enzyme lets more protons flow into the matrix and thereby dissipates the electrochemical proton gradient more rapidly. The falling gradient, in turn, enhances the rate of electron transport.

Similar controls act to adjust the rates of NADH production to the rate of NADH utilisation by the respiratory chain.

SAQ 8.9

Suggest physiological explanations for the following observations:

1) Uncouplers of oxidative phosphorylation have been used as slimming agents.

2) Injection of an uncoupler into a rat causes an immediate rise in body temperature.

3) A patient was admitted to hospital with the following symptoms: oxygen intake twice normal for age and weight, profuse sweating, and extreme thinness despite a high food intake. After removing a small amount of muscle and isolating mitochondria, a P:O ratio of between 1.5 and 2.6 was found.

4) In the absence of added ADP and Pi, isolated mitochondria maintain a steady low rate of oxygen consumption.

5) A sudden drop in the pmf followed by a recovery to a constant 220 mV.

8.6 The supply of reducing equivalents

use of NAD-linked dehydrogenases generates 3 ATP

The respiratory chain is 'nourished' by reactions with reduced substrates. This is illustrated in Figure 8.13 which also shows the sites of action of specific inhibitors of electron transport. Many of these 'nourishing reactions' react via dehydrogenases to create NADH. Examples of the substrates used are pyruvate, 2-oxoglutarate, malate and isocitrate. For each of these substrates the reducing equivalent (NADH) goes through the whole respiratory chain and thus three molecules of ATP are synthesised per NADH oxidised. However, some substrates eg succinate, do not have sufficient 'reducing power' to be able to reduce NAD^+. These will react further down the respiratory chain. For such substrates we are dealing with dehydrogenases which have flavin as a prosthetic group. The redox potential of these flavins is such that they react well with ubiquinone (Q). The result is that only two ATP per oxidised substrate can be synthesised for each pair of electrons.

use of flavin linked dehydrogenases generates 2 ATP

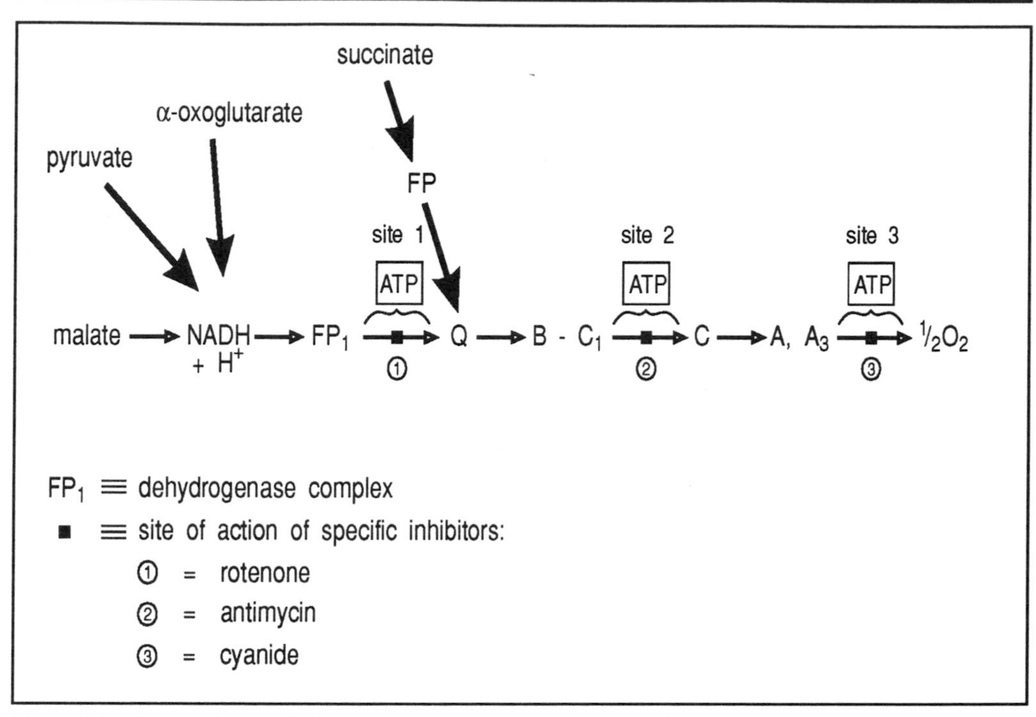

Figure 8.13 The supply of reducing equivalents to the respiratory chain.

SAQ 8.10

What will be the influence of rotenone, antimycin or cyanide on the oxidation of isocitrate, succinate or cytochrome C by mitochondria (+ = inhibition; - = no inhibition)?

SAQ 8.11

Isolated mitochondria are incubated with an oxidisable substrate and the oxygen consumption rate is measured. For each of the following additions predict whether the oxygen consumption rate increases or decreases. Justify your decision in each case.

1) Oligomycin.

2) Dinitrophenol.

3) Oligomycin and dinitrophenol.

4) ADP.

The reducing equivalents derived from the substrates shown in Figure 8.13 are generated in the matrix of mitochondria. This relies on the transport of pyruvate from the soluble cytoplasm into the matrix. We have already seen (Figure 8.6) that pyruvate can be transported across the inner mitochondrial membrane. Many NAD^+-linked dehydrogenases in the cytosol generate NADH and this must also pass into the matrix if it is to be reoxidised to NAD^+ via the electron transport chain.

However, the inner mitochondrial membrane is impermeable to NADH. So, how is extramitochondrial NADH oxidised by the electron transport chain?

glycerol phosphate shuttle

While extramitochondrial NADH cannot itself penetrate the mitochondrial membrane, electrons derived from it can enter by indirect routes called shuttles. For example, the glycerol phosphate shuttle, shown in Figure 8.14.

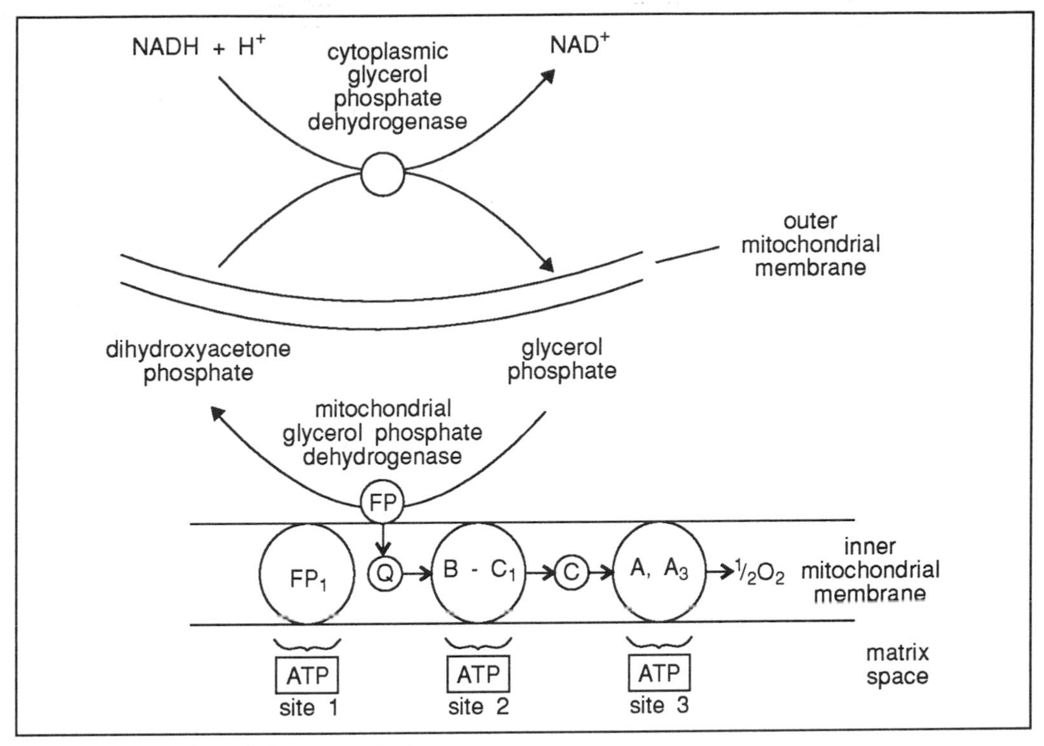

Figure 8.14 The glycerol phosphate shuttle

Two enzymes are involved in this shuttle:

- Cytosolic glycerol phosphate dehydrogenase. This enzyme reduces dihydroxyacetone phosphate (one of the intermediates of glycolysis) using NADH and forms glycerol phosphate. The product so formed passes rapidly through the outer membrane.
- Membrane bound glycerol phosphate dehydrogenase. This enzyme is located on the outer surface of the inner mitochondrial membrane and oxidises glycerol phosphate. The mitochondrial enzyme is not NAD^+ linked but is a flavin containing enzyme. The reducing equivalents of the reduced flavin are then transferred to ubiquinone, from which they pass down the electron transport chain to oxygen.

malate/aspartate shuttle

The glycerol phosphate shuttle is unidirectional and only two molecules of ATP are produced from electron transport since the electrons enter the chain after the first energy-conserving site. In some tissues another shuttle operates that is bidirectional the malate/aspartate shuttle. This can transport electrons from extramitochondrial NADH into the mitochondrion and from intramitochondrial NADH into the cytosol. This shuttle, unlike the glycerol phosphate shuttle, yields three ATP per molecule of cytosolic NADH oxidised.

8.7 Energy balances in aerobic catabolism

In this section we will consider the amount of ATP produced by the complete oxidation of a glucose molecule. This allows an estimation of the efficiency of aerobic respiration to be made.

The respiratory chain is the final stage of aerobic carbohydrate catabolism, so we can now depict the whole process as shown in Figure 8.15, where only the net ATP synthesis is shown.

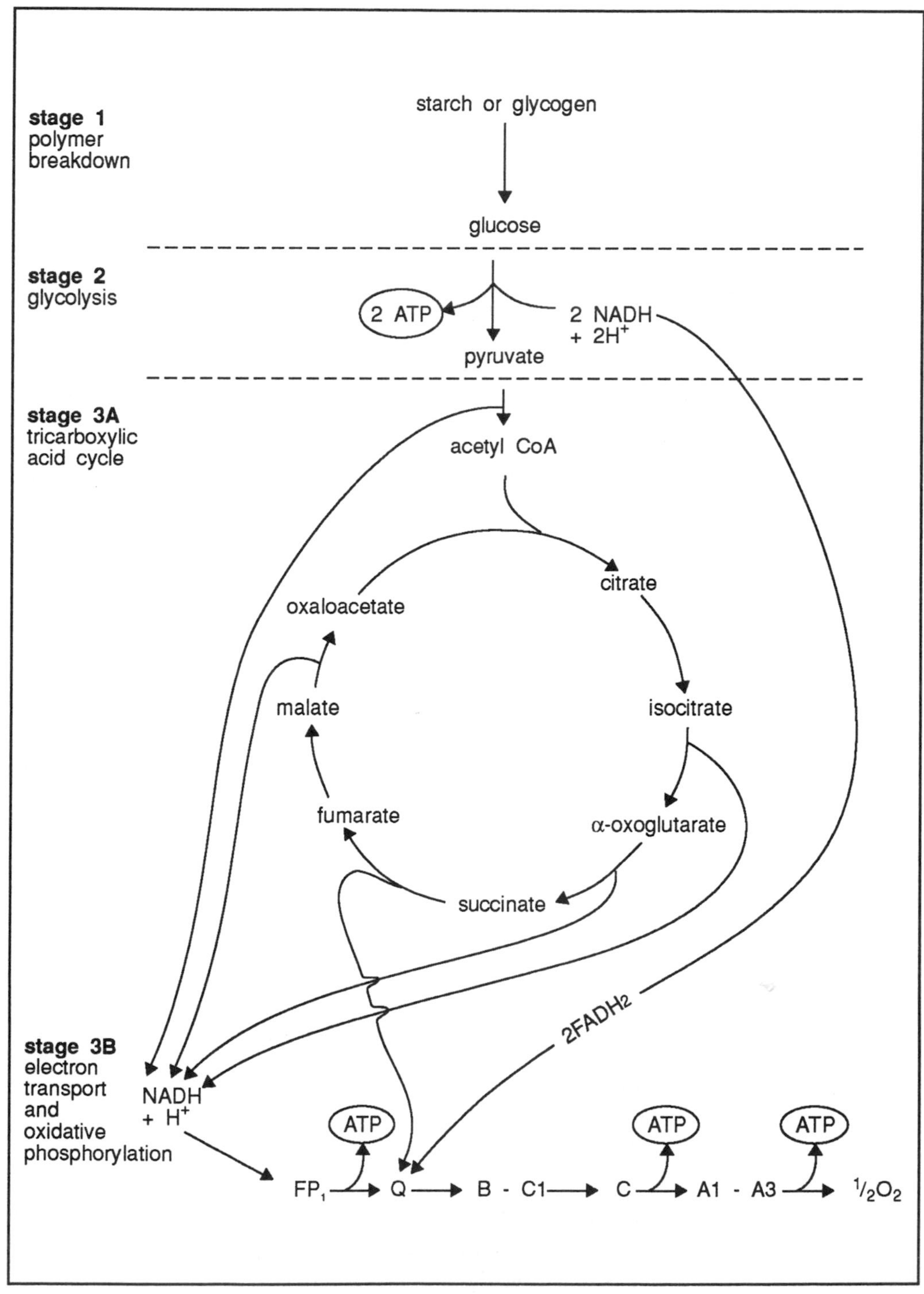

Figure 8.15 Complete aerobic respiration of glucose via glycolysis.

SAQ 8.12

Fill in the missing numbers in the following paragraph:

The net result of the TCA cycle is the complete oxidation of pyruvate to CO_2 with the production of —— molecules of NADH. Each of the NADH molecules can be oxidised to NAD^+ through the electron transport chain, producing —— ATP molecules per NADH oxidised. In addition, —— molecules of ATP are formed during the flavin-linked oxidation of succinate and —— molecule of ATP is formed by the substrate level phosphorylation at the expense of succinyl CoA. Thus a total of —— ATP molecules are synthesised for each turn of the cycle. If we suppose that the glucose is being metabolised via glycolysis then —— molecules of pyruvate are formed per molecule of glucose. This means that there are —— turns of the tricarboxylic acid cycle which produces —— molecules of ATP. In addition, glycolysis generates —— ATP's by substrate-level phosphorylation and a further —— NADH molecules. Oxidation of each of these extramitochondrial NADH molecules, via the electron transport chain, may generate —— or —— molecules of ATP depending on the shuttle used to transfer reducing equivalent into the mitochondrial matrix. If we assume that the glycerol phosphate shuttle is operating then —— ATPs are generated from the extramitochondrial NADH. This means that aerobic respiration produces —— ATP molecules per molecule of glucose oxidised.

SAQ 8.13

Metabolism of stearic acid to acetyl -CoA is described by the following reaction.

$$\text{Stearic acid} + ATP + 8FAD + 8NAD^+ + 8H_2O$$

$$\rightarrow 9 \text{ acetyl CoA} + 8FADH_2 + 8NADH + 8H^+ + AMP + 2Pi$$

The reaction with ATP, in which stearate is activated occurs in the cytoplasm and consumes two high energy bonds. Subsequent breakdown occurs in the mitochondrial matrix. The acetyl CoA enters the TCA cycle and a conventional electron transport chain is in operation. What is the net yield of ATP per mole of stearic acid oxidised by complete aerobic respiration?

energy conservation in glucose oxidation is 38% efficient

Since we can predict the amount of ATP generated per mole of glucose we can estimate how efficient cells are at harnessing energy during complete aerobic respiration. The high-energy phosphate bond of ATP releases 30.5 kJ mol^{-1} when hydrolysed. This means that 1100 kJ $mole^{-1}$ is converted to high-energy phosphate bond energy in the complete oxidation of glucose (30.5 x 36). The total amount of energy available from the complete oxidation of glucose by oxygen is about 2879 kJ $mole^{-1}$. It follows that the overall efficiency of energy recovery in the complete oxidation of glucose is 1100/2879 x 100 = 38 percent.

Π In what other way could you estimate the energy capturing efficiency of cells?

By measuring the heat released since the energy not trapped in ATP is lost as heat. Thus the heat loss would consist of approximately 1779 kJ mol^{-1}.

SAQ 8.14

The complete aerobic respiration of palmitic acid is described by the following equation:

$$\text{Palmitic acid} + 23\,O_2 + 130\,ADP + 130Pi \rightarrow 16\,CO_2 + 146\,H_2O + 130\,ATP$$

ΔG° for the oxidation of palmitic acid is -9790 kJ mol^{-1}. What is the energy capturing efficiency of palmitic acid oxidation in this system:

(ΔG° ATP hydrolysis = -30.5 kJ mol^{-1}).

Summary and objectives

The common pathway by which mitochondria, chloroplasts and bacteria harness energy for biological purposes operates by a process known as chemiosmotic coupling. It begins when strong electron donors, such as NADH and $FADH_2$, pass their 'high-energy' electrons along a series of electron carriers embedded in an ion-impermeable membrane. In the course of travelling along this electron-transport chain, the electrons, which have been either excited by sunlight or derived from the oxidation of electron-rich foodstuffs, fall to a successively lower energy level. In mitochondria and bacteria part of the energy released is harnessed by three major enzyme complexes that pump protons from one side of the membrane to the other. This generates a proton-motive force across the membrane and the energy stored in this gradient is then used to drive reactions catalysed by enzymes embedded in the membrane. One such enzyme is ATP synthetase and is a reversible coupling device that normally coverts a backflow of protons into phosphate-bond energy in ATP, but it can also hydrolyse ATP to pump protons in the opposite direction.

Now that you have completed this chapter you should be able to:

- demonstrate an understanding of the hypothesis of the chemiosmotic coupling of electron transport and ATP synthesis;
- calculate pmf from ΔpH and membrane potential measurements;
- describe the structure of ATP synthetase and its involvement in oxidative phosphorylation;
- name components of the respiratory chain and list them in order of electron transfer;
- explain what is meant by the P:O ratio and how it is calculated;
- show an understanding of the concept of respiratory control and predict the physiological effect of named inhibitors of the respiratory apparatus;
- apply a knowledge of the mechanism of operation of the respiratory apparatus to the interpretation of experimental data;
- determine energy balances and energy capturing efficiencies for the complete aerobic respiration of substrates.

Oxidation of reduced pyridine nucleotides in an anaerobic environment

Oxidation of reduced pyridine nucleotides in an anaerobic environment

Introduction

Some micro-organisms can grow and reproduce under anaerobic conditions (in the absence of oxygen). There are numerous anaerobic environments throughout the biosphere and an abundant variety of micro-organism thrive in these niches. Deep wells and the bottom of ponds are examples of anaerobic environments. The rumen (stomach) of cows and the large intestine of humans are also anaerobic environments where micro-organisms grow.

The micro-organisms found in anaerobic environments are either obligate anaerobes or facultative anaerobes. Obligate anaerobes are intolerant of oxygen, they grow only in oxygen-free environments. The facultative anaerobes can grow in the absence or presence of oxygen. When oxygen is present facultative organisms use oxygen as a terminal electron acceptor in aerobic respiration (this was considered in the previous chapter) but they are able to alter their metabolism to grow when oxygen is absent.

In this chapter we will examine the mechanisms whereby organisms are able to grow and reproduce under anaerobic conditions. You will recall from the preceding chapter that during aerobic respiration ATP is generated and reduced nucleotides (NADH and $FADH_2$) are reoxidised by passing electrons to oxygen. The generation of ATP and maintenance of redox balance must still occur even when an organism is growing in the absence of oxygen. Anaerobic micro-organisms achieve this by the mechanisms of anaerobic respiration and fermentation and we shall see that many biotechnological processes exploit this ability of micro-organisms. In addition, the chemical transformation of certain elements, which is carried out on a global scale in soil and in water, is a consequence of the growth of micro-organisms in the absence of oxygen. Thus this is a very important topic.

9.1 Anaerobic respiration

Anaerobic respiration is an ATP generating process in which an inorganic compound other than oxygen serves as the terminal electron acceptor. It involves essentially the same biochemical pathways as in the aerobic metabolism of heterotrophs, differing principally in the compound that serves as the final (or terminal) electron acceptor of the electron-transport chain.

anaerobic respiration uses inorganic acceptors

Terminal electron acceptors for anaerobic respiration include nitrate, sulphate and carbon dioxide. We will now examine some of the mechanisms and discuss the importance of anaerobic respiration using different electron acceptors.

9.1.1 Nitrate as electron acceptor

denitrification

Nitrate (NO_3^-) is one of the most common alternative electron acceptors to oxygen. All bacteria that reduce nitrate are facultative anaerobes since they will transfer electrons to

O_2 if it is present and to NO_3^- only when O_2 is absent. Thus they possess a full respiratory chain.

use of nitrate releases 2 ATP

Some bacteria reduce nitrate only to nitrite while others can reduce nitrite to nitrous oxide (N_2O) and to nitrogen gas (N_2), a process called denitrification. Nitrate is reduced to nitrite by nitrate reductase and this enzyme utilises electrons which come from cytochrome. Thus the transfer of electrons along the respiratory chain is diverted at cytochrome b and two rather than three ATP molecules are generated by oxidative phosphorylation (see Figure 9.1). Nitrate reductase is strongly inhibited by O_2.

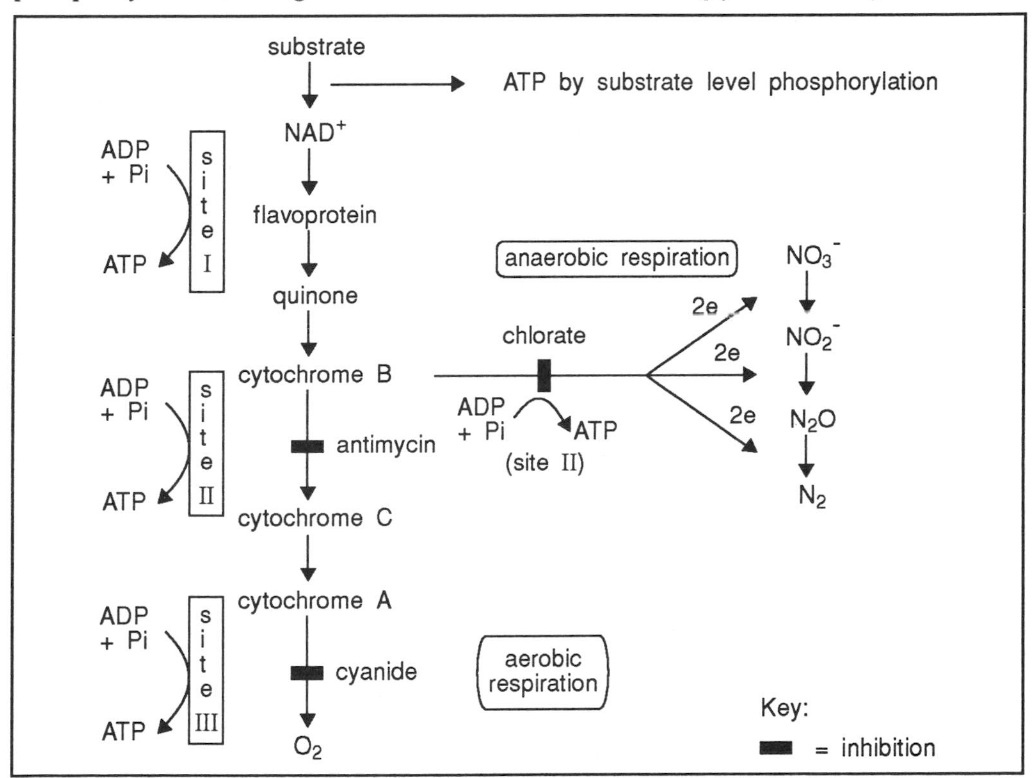

Figure 9.1 The electron-transport chain in *Paracoccus denitrificans*. This bacterium carries out complete aerobic respiration via a conventional electron-transport chain when oxygen is present and can use nitrate, nitrite or nitrous oxide in the absence of oxygen. The sites of action of certain inhibitors of respiration are shown.

SAQ 9.1

1) Would you expect the redox potential for nitrate reduction to be lower or higher than that for reduction of oxygen? What is the significance of your answer?

2) Explain why the inhibition of nitrate reduction by O_2 is advantageous to a denitrifying bacterium as far as energy generation is concerned.

3) How many ATP molecules are generated by the reduction of nitrate to nitrous oxide? Explain why this reduction should be regarded as being less energy efficient that the reduction of one atom of oxygen.

4) Would *Paracoccus denitrificans* be able to grow in nitrate containing medium under a) anaerobic conditions in the presence of chlorate and b) aerobic conditions in the presence of antimycin? Explain your reasoning.

The process of denitrification is detrimental to soil fertility. Whenever organic matter is decomposed in soil and oxygen is exhausted as a result of aerobic respiration, perhaps coupled with waterlogging of the soil, some organisms will carry out denitrification. Denitrification is a process of major ecological importance because it depletes the soil of nitrate, which is essential for plants, thereby decreasing agricultural productivity. Such losses are particularly large from fertilised soils. Although precise values are not known, under certain conditions, the amount of nitrogen-based fertiliser lost through denitrification may approach 50% of that applied to land.

denitrification involves loss of nitrogen as N_2.

Not all the consequences of denitrification are detrimental. Denitrification is vital to the continued availability of combined nitrogen on the land masses on the Earth. The highly soluble nitrate ion is constantly washed from the soil by rainwater, and it is eventually carried to the oceans. Without denitrification, the Earth's supply of nitrogen, including N_2 of the atmosphere, would eventually accumulate in the oceans. The involvement of denitrification in the geological cycling of nitrogen compounds in nature is illustrated in Figure 9.2.

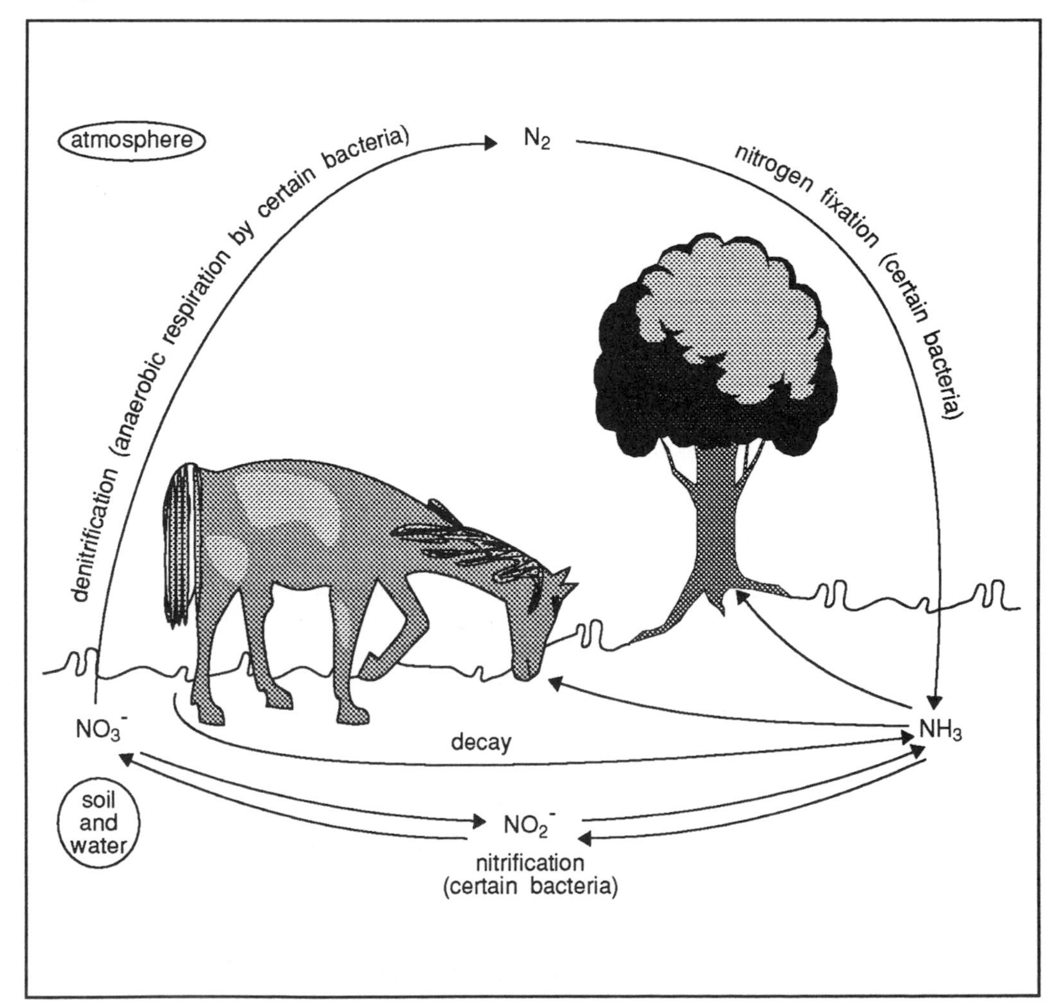

Figure 9.2 The involvement of denitrification in the geological cycling of nitrogen. Nitrification is the conversion of NH_3 to NO_3^-. It is carried out by certain aerobic autotrophs and is discussed in the BIOTOL text, 'Energy Sources for Cells"

9.1.2 Sulphate as electron acceptor

sulphate-reducing bacteria are obligate anaerobes

The use of sulphate as an electron acceptor in anaerobic respiration is called dissimilatory sulphate reduction. In this process, sulphate is reduced to sulphide (H_2S) and it is carried out by the sulphate-reducing bacteria. These are obligate anaerobes, unable to grow in the presence of O_2. Indeed, many species are actually killed by O_2. There are only seven genera of bacteria that can use sulphate as an electron acceptor in respiration. They include *Desulphovibrio* and *Desulphomonas.*

The reduction of sulphate to sulphide involves the transfer of four pairs of electrons. The initial two-electron transfer involves the utilisation of ATP and the formation of a key intermediate, adenosylphosphosulphate (APS). It is believed that individual reductase enzymes are responsible for the transfer of electron pairs which is accompanied by the recycling of sulphite. The sequence is shown below:

$$SO_4^{2-} \xrightarrow{ATP \;\; 2\,Pi} APS \xrightarrow[2e]{AMP} SO_3^{2-} \xrightarrow[2e\, +\, 6H^+]{2SO_3^{2-} \;\; 3H_2O} S_3O_6^{2-} \xrightarrow[2e]{SO_3^{2-}} S_2O_3^{2-} \xrightarrow[2e]{SO_3^{2-}} S^{2-}$$

The sulphate-reducing bacteria have a complex and specialised electron transport chain that couples organic substrate oxidation with the various steps of sulphate reduction. Some ATP synthesis has been shown to be associated with electron transfer through the chain but many details of this system are still poorly understood.

□ The redox potential for the reduction of sulphate is even lower that for nitrate reduction. What does this suggest concerning the energetics of the process?

A lower redox potential indicates that less Gibbs free-energy will be liberated in the passage of electrons to sulphate than to nitrate. This, in turn, suggests that ATP produced per pair of electrons transferred will be lower using sulphate than with nitrate as electron acceptor.

sulphate reduction releases less ATP than nitrate reduction

Typical habitats of the sulphate-reducing bacteria are anaerobic sediments which contain organic matter and sulphate. The activities of these organisms result in massive generation of H_2S which can form insoluble metal sulphide (eg FeS). These bacteria are largely responsible for the formation of metal sulphide ores and, indirectly, for natural deposits of elemental sulphur (formed by secondary oxidation of sulphide).

In waterlogged (and hence largely anaerobic) soils the production of H_2S by sulphate-reducing bacteria may cause damage to plants. This is sometimes a serious economic problem in the cultivation of rice, which is sown in flooded fields (rice paddies). H_2S is also produced when leaves fall into ponds and this can be detrimental to pond life. H_2S has a pungent smell (like rotten eggs), often associated with stagnant water.

9.1.3 Other important electron acceptors

The organic compound fumarate is used by a wide variety of bacteria (eg *Escherichia coli*) as an electron acceptor. Fumarate is reduced to succinate, which is the reverse reaction to that found in the TCA cycle. The reduction of fumarate can be coupled with NADH oxidation and a variety of other electron donors. The reaction with NADH is shown below:

$$\begin{array}{c} COO^- \\ | \\ CH \\ || \\ HC \\ | \\ COO^- \end{array} \quad \xrightarrow{NADH + H^+ \quad\quad NAD^+} \quad \begin{array}{c} COO^- \\ | \\ CH_2 \\ | \\ CH_2 \\ | \\ COO^- \end{array}$$

fumurate succinate

fumarate reduction can oxidise NADH and release ATP

The energy yield from the reduction of fumarate is sufficient for the synthesis of one ATP. However, a small number of bacteria do not couple fumarate reduction with ATP synthesis via oxidative phosphorylation but merely use it to reoxidise NADH formed during the oxidation of organic substrates.

CO_2 reduction produces methane

Carbon dioxide, CO_2 can be used as an electron acceptor and is reduced to methane, CH_4. The organisms that carry out this process, called the methanogenic bacteria, are a diverse group of extremely oxygen-sensitive anaerobes (O_2 kills then). Most of the methanogenic bacteria can grow on H_2 as sole electron donor and CO_2 as sole electron acceptor. The reduction of CO_2 is linked to oxidative phosphorylation via an unusual electron transport chain containing a novel type of electron carrier called F_{420}.

These organisms are widespread in anaerobic muds, animal intestinal tracts, and in anaerobic sewage treatment installations and they are responsible for the formation of vast amounts of methane in nature, (so called natural gas is composed of methane). The metabolism of methanogenic bacteria is described in more detail in the BIOTOL text, 'Energy Sources for Cells'.

Iron III (Fe^{3+}) can be used as an electron acceptor, being reduced to iron II (Fe^{2+}). Reduction of iron III is carried out by many organisms that reduce nitrate, although it has not yet been shown to be coupled to the generation of ATP by oxidative phosphorylation. The process contributes to the formation of iron ore deposits in nature. Iron III is present in soils and rocks, often as the insoluble $Fe(OH)_3$, and, when conditions become anaerobic, reduction to the iron II state can occur. Iron II is much more soluble than iron III, and reduction by micro-organisms thus leads to the solubilisation of iron. This is an important first step in the formation of the type of ore deposit called bog iron.

SAQ 9.2

Identify each of the following statements as True or False. Justify your response in each case.

1) *Desulphovibrio* can use sulphite (SO_3^{2-}) as an electron acceptor.

2) Sulphate reduction to sulphite would cost the cell energy if only one ATP is produced by oxidative phosphorylation per pair of electrons transferred.

3) Some bacteria are able to generate one ATP per pair of electrons transferred by oxidative phosphorylation during anaerobic growth using fumarate as terminal electron acceptor.

4) Methanogenic bacteria carry out complete aerobic respiration in the presence of oxygen because it is more energy efficient than the reduction of CO_2.

5) Iron II is an end-product of anaerobic respiration in some organisms.

9.2 Fermentation

organic compounds donate and accept electrons in fermentation

Fermentation is defined as an ATP-generating process in which organic compounds serve as both electron donors (becoming oxidised) and electron acceptors (becoming reduced). The compounds that perform these two functions are usually different metabolites derived from a single fermentable substrate (such as a sugar). You should note that the term fermentation is used by the non-scientific community to describe any process involving the controlled growth micro-organisms. The fates of electrons and carbon skeletons in fermentation, aerobic respiration and anaerobic respiration processes is shown below.

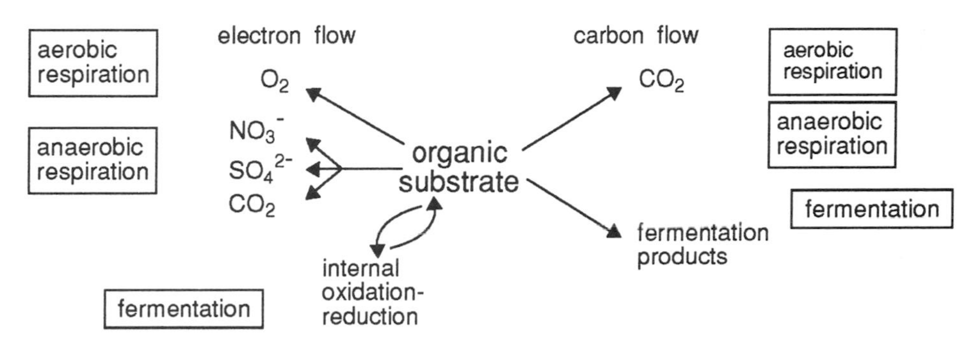

SAQ 9.3

Select which modes of metabolism listed below are applicable to each of the statements labelled 1 to 6:

1) The terminal electron acceptor is always a metabolically derived molecule.

2) The terminal electron acceptor may be organic.

3) The carbon growth substrate provides electrons.

4) ATP is generated only by substrate level phosphorylation.

5) It can be performed by facultative anaerobes.

6) CO_2 may be an end-product.

Modes of motabolism:

a) aerobic respiration; b) anaerobic respiration; c) fermentation.

Fermentation processes have two essential features:

- they always maintain a strict oxidation-reduction balance;
- because the respiratory chain is not working, substrate-level phosphorylation is the only mode of ATP synthesis in fermentation.

oxidation states of products must equal that of reactants

In order to maintain the oxidation reduction balance the average oxidation state of the end-products must be identical to that of the substrate. Similarly, pyridine nucleotides reduced in one step of the process must be oxidised in another.

The substrate-level phosphorylation characteristics of the fermentation of organic compounds are the same as those which operate during the breakdown of glucose to pyruvate.

Many of the organisms that generate ATP by fermentation are obligate anaerobes. Others are facultative anaerobes, being able to grow either in the presence or absence of air. Virtually all facultative anaerobes change their mode of ATP generation on exposure to air, the presence of molecular oxygen induces a metabolic shift from fermentation to aerobic respiration. Lactic acid bacteria are an exception to this because ATP generation by fermentation continues even in the presence of oxygen.

end products are released into the medium

pyruvate is formed in all fermentations

Carbohydrates are the principle substrates of fermentation. In all cases the end-products are organic molecules, formed as a result of the fermentation and released into the surrounding medium. With glucose as oxidisable substrate all fermentation types oxidise glucose first to pyruvate and you will recall from an earlier chapter that a number of different pathways can be used for this.

In fermentative metabolism, glycolysis (Embden Meyerhof pathway) is used most with the pentose phosphate pathway used by some bacteria. The resulting pyruvate is processed further in a number of ways via fermentation reactions but in all cases it becomes reduced and NADH becomes oxidised, with the formation of an organic end-product. The main types of carbohydrate fermentation are shown in Table 9.1.

fermentation type	principle end-products	main groups of micro-organism	importance
ethanolic	ethanol, CO_2	some yeasts and some moulds	manufacture of beer, wine and bread
homolactic	lactate	streptococci some lactobacilli	manufacture of butter, cheese and yogurt
heterolactic	lactate, CO_2 ethanol	other lactobacilli *Leuconostoc*	
mixed acid	lactate, acetate formate, H_2, CO_2, ethanol, succinate	many enteric bacteria eg *Escherichia, Salmolella*	fermentative patterns can be used to distinguish genera within the enteric group
butanediol	as mixed acid plus 2,3-butanediol	some enteric bacteria eg *Serratia*	
butyrate	butyrate, acetate CO_2, H_2	some clostridia *Butyribacterium*	used in the retting process for the manufacture of linen, jute and rope - the plant pectin, freeing the fibre
solvent	as butyrate plus butanol, acetone, isopropanol	some clostridia	have been used for manufacture of industrial solvents acetone and butanol
propionate	acetate, propionate, CO_2, succinate	several genera of anaerobic bacteria eg *Propionibacterium*	

Table 9.1 The main types of carbohydrate fermentation.

We will now consider these pathways in turn and you may find it useful to return to the table from time to time as we proceed.

You should remember that, for an organism to grow, the fermentation process must:

- generate some ATP;
- maintain oxidation-reduction balance.

In the descriptions that follow, glycolysis will be represented by a single reaction even though many enzymes are involved. Other pathways are also summarised in this way, although for these the number of enzymatic steps involved in the conversion will be given in the text.

9.2.1 Ethanolic fermentation

Facultative anaerobic yeasts are by far the most important group of organisms that have the property of producing ethanol as the sole organic end-product of fermentation. These organisms ferment sugars under anaerobic conditions and carry out complete aerobic respiration in the presence of O_2. The pathway for the complete fermentation of glucose by yeasts is summarised in Figure 9.3.

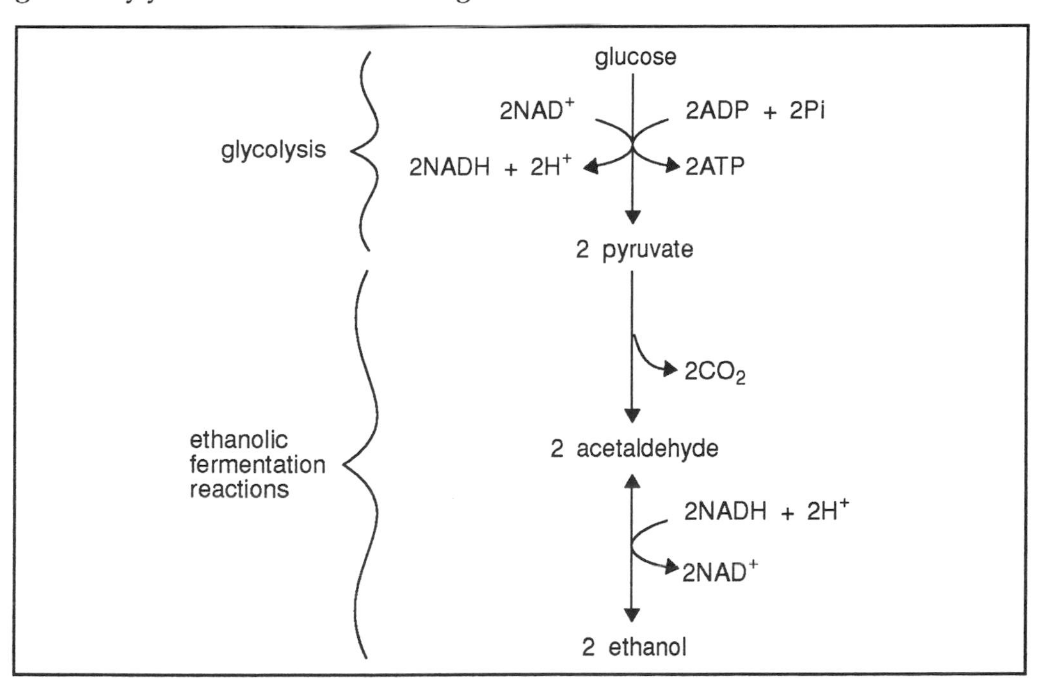

Figure 9.3 Pathway of ethanolic fermentation by facultative anaerobic yeasts.

2ATP + 2NADH + 2H$^+$ for every ethanol produced

Oxidation of glucose to pyruvate is via glycolysis, this generates two ATP and two NADH + H$^+$ molecules. You will recall from earlier chapters that although two ATP molecules are used initially to phosphorylate the sugar, two ATP molecules are synthesised from each of the two C3 fragments. The net gain for ATP is thus two molecules per molecule of glucose fermented. Similarly, in the oxidation step of the glycolytic pathway, NAD$^+$ is reduced to NADH + H$^+$ for each of the C3 fragments. The net gain for NADH + H$^+$ is thus two molecules per molecule of glucose fermented.

The cell has only a limited supply of NAD^+, and oxidation of glucose would stop, since the oxidation of each molecule of glyceraldehyde-3-phosphate can proceed only if there is a molecule of NAD^+ to accept the released electrons. This 'roadblock' is overcome in the complete fermentation by the oxidation of NADH + H^+ to NAD^+ through reactions involving the conversion of pyruvate to ethanol and CO_2. The first step is the decarboxylation of pyruvate to acetaldehyde and CO_2. Electrons are then transferred to acetaldehyde (the terminal electron acceptor) from NADH + H^+ forming ethanol and NAD^+. The NADH produced in glycolysis is thus oxidised back to NAD^+.

acetaldehyde - the terminal acceptor

The final products, ethanol and CO_2, must also be in oxidation-reduction balance with the carbohydrate substrate. The degree of oxidation or reduction of a carbon compound is expressed by the oxidation state of carbon in the molecule. In simple organic molecules the total oxidation state of the carbon must balance. The total oxidation states of hydrogen and oxygen atoms, are respectively H = +1 and O = -2 but for carbon it is variable. In glucose, for example, the sum of oxidation numbers for oxygen atoms is 6 x -2 = -12 and for H atoms is 12 x +1 = +12. The oxidation state of carbon in glucose is therefore zero.

Π Write the overall reaction for the ethanolic fermentation of glucose by yeasts.

We can see from Figure 9.3 that the overall reaction is:

$$\text{glucose} + 2\text{ADP} + 2\text{Pi} \rightarrow 2\ \text{ethanol} + 2\text{CO}_2 + 2\text{ATP}$$

oxidation reduction balance

You will note that NADH does not feature in the overall reaction because there is reoxidation of NADH formed and thus no net gain or loss.

Π What are the oxidation states of carbon in ethanol and CO_2? Explain how the ethanolic fermentation described above is in oxidation-reduction balance.

For ethanol: (6 x +1, for hydrogens) + (-2, for oxygen) = +4, there are two carbon atoms in ethanol and their oxidation state must balance +4. The oxidation state of each carbon in ethanol is thus -2, (total -4).

For CO_2: 2 x -2 (for oxygen) = -4. The oxidation state of the single carbon atom is thus +4.

The overall oxidation state of carbon in the products is therefore zero. We have previously established that the oxidation state of carbon in glucose is also zero and that all NADH produced is reoxidised. Thus, the fermentation is in oxidation-reduction balance.

Π Just to show that the oxidation state of carbon is extremely variable complete the table below:

compound	**carbon oxidation state**
methane CH_4	??
ethanol	??
carbon monoxide CO	??
carbon dioxide	??

The figures are -4, -2, +2 and +4 down the column. Notice that carbon is progressively more oxidised as we go down the column.

Calculation of the oxidation-reduction balance provides a check to ascertain that the reaction is correctly written.

Pasteur effect

For the yeast cell the crucial product is ATP. Ethanol and CO_2 are merely waste products. You will have noted that the yield of ATP per molecule of glucose is far less by ethanolic fermentation (2) than by complete aerobic respiration (35-38). This difference is one of the explanations for the so-called 'Pasteur effect'. Pasteur observed that a yeast cell consumes much more glucose under anaerobic conditions than under aerobic conditions. We are now able to understand this because the cell utilises different amounts of glucose to produce the same amount of ATP.

We will now consider some thermodynamic aspects of the ethanolic fermentation. For this it is best to think of the overall reaction as consisting of two sub-reactions:

$$\text{glucose} \rightarrow 2\ \text{ethanol} + 2\ CO_2 \quad : \Delta G^{o'} = -244\ \text{kJ mol}^{-1}$$

$$\text{ADP} + \text{Pi} \rightarrow \text{ATP} \quad : \Delta G^{o'} = +49\ \text{kJ mol}^{-1}$$

Π Consider the two sub-reactions shown above and determine 1) the $\Delta G^{o'}$ of the overall reaction and 2) the energy capturing efficiency of the fermentation.

1) $\Delta G^{o'}$ of the net reaction = (2 x (+49)) -244 = -146 kJ mol^{-1} glucose.

2) The energy capturing efficiency = (98/244) x 100 = 40% (approximately).

ethanol fermentation is 40% efficient

We saw in an earlier chapter that aerobic respiration has an energy capturing efficiency of about 38%. We can see, therefore, that fermentation is as efficient at harvesting energy as aerobic respiration. Much less energy, however, is released overall because much of it is still locked up in the ethanol produced.

9.2.2 Lactic acid fermentations

lactic acid bacteria are obligate fermenters

Lactic acid bacteria produce lactic acid as a major end-product of glucose fermentation. Unlike the yeasts they are obligate fermenters and can not utilise oxygen in respiration. Lactic acid fermenters can be divided into two biochemical subgroups, according to the end-products formed during fermentation of glucose:

- homolactic fermenters convert sugars almost completely to lactic acid. This means that they produce two moles of lactate per mole of glucose and lactate is thus the end-product of the fermentation of glucose. These reactions are also carried out by mammalian muscle under the anaerobic conditions generated by strenuous exercise. In muscle the accumulation of lactate leads to muscle fatigue;
- heterolactic fermenters convert sugars to an equimolar mixture of lactic acid, ethanol and CO_2.

The pathways of lactic acid fermentations are shown in Figure 9.4.

heterolactic fermenters use the pentose phosphate pathway

Homolactic fermenters metabolise glucose via glycolysis. The NADH formed in this pathway is oxidised by converting pyruvate to lactate. In heterolactic fermenters the glycolytic pathway is incomplete because they lack one of the enzymes, fructose-bisphosphate aldolase, which cleaves the six carbon sugar. These organisms oxidise glucose via the pentose phosphate pathway which initially oxidises glucose-6-phosphate to ribulose-5-phosphate and CO_2 followed by a C3/C2 cleavage giving glyceraldehyde-3-phosphate and acetyl phosphate.

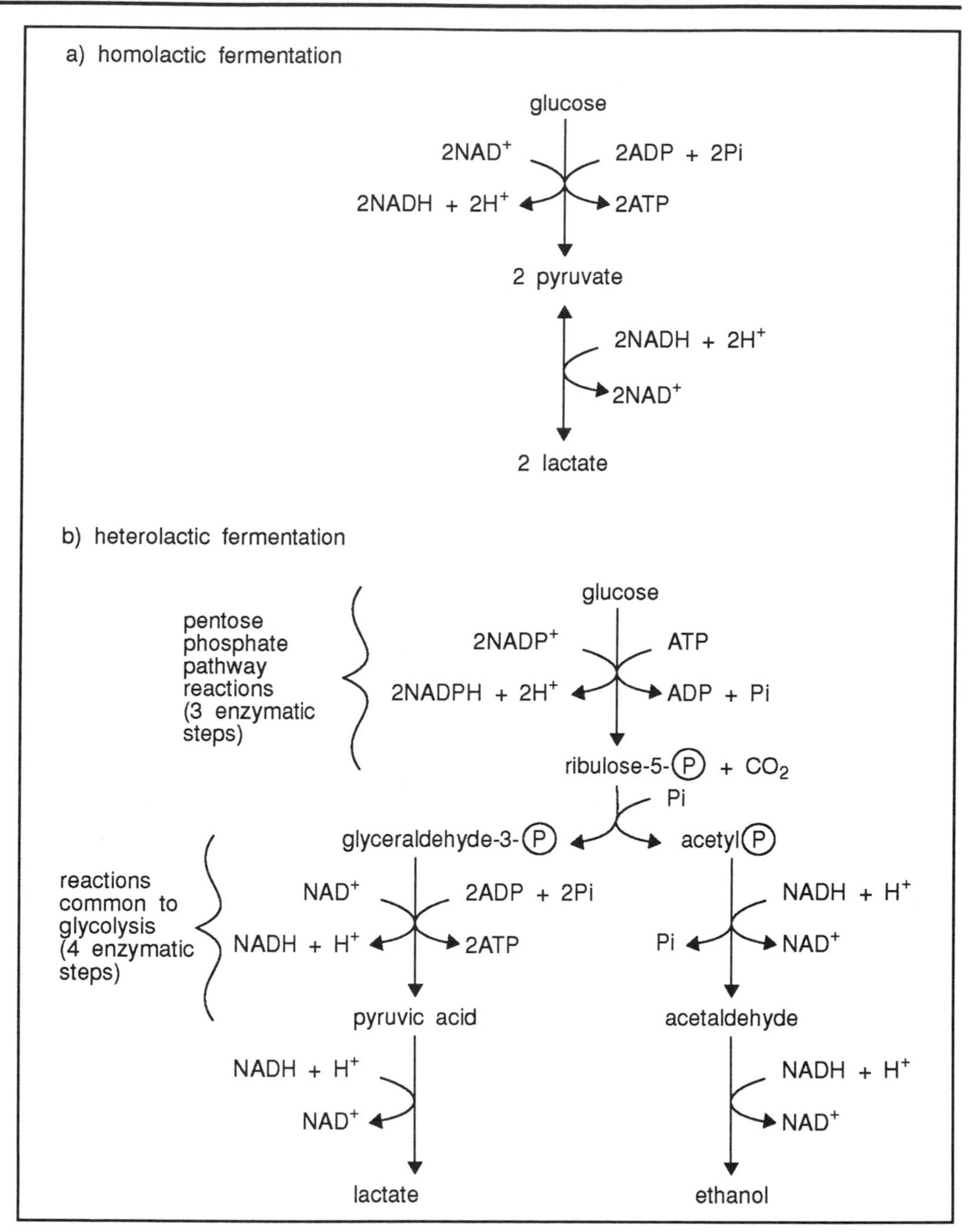

Figure 9.4 Pathways of homolactic and heterotactic fermentations.

Glyceraldehyde-3-phosphate is converted to lactate through a sequence of reactions identical with those of homolactic fermenters. The acetyl-phosphate is reduced to ethanol and serves as a means of oxidising 2NADH. This is necessary because 2NADPH are formed in the conversion of glucose to ribulose-5-phosphate and CO_2. You may wonder how the organism gets around the potential problem of generating NADPH but oxidising NADH. In the cell this does not present a problem since all cells contain a transhydrogenase enzyme that interconverts the two types of pyridine nucleotide:

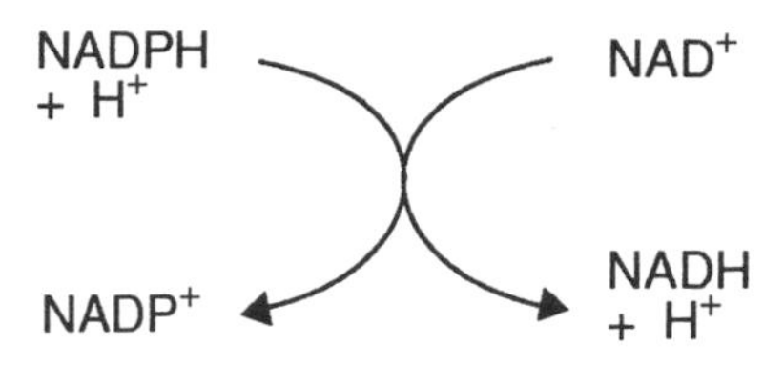

SAQ 9.4

1) Write out the overall reactions for homolactic and heterolactic fermentation of glucose.

2) Calculate $\Delta G^{o'}$ and the energy capturing efficiency of the homolactic fermentation.

(glucose $\rightarrow$ 2 lactate, $\Delta G^{o'}$ = -137 kJ mol^{-1})

(ATP $\rightarrow$ ADP + Pi, $\Delta G^{o'}$ = -49 kJ mol^{-1})

3) Establish whether or not the heterolactic fermentation is in oxidation-reduction balance.

9.2.3 Mixed acid and butanediol fermentations

Mixed acid fermentations have lactate, succinate, formate (or CO_2 and H_2) and ethanol as end-products. They are carried out by nine genera of facultatively anaerobic bacteria, which include *Escherichia* and *Salmonella*.

During mixed acid fermentations pyruvate is formed from glycolytic breakdown of sugars and the end-products arise through three independent pathways. These are summarised in Figure 9.5. (It might be helpful for you to draw out this figure but include the structures of the named intermediates).

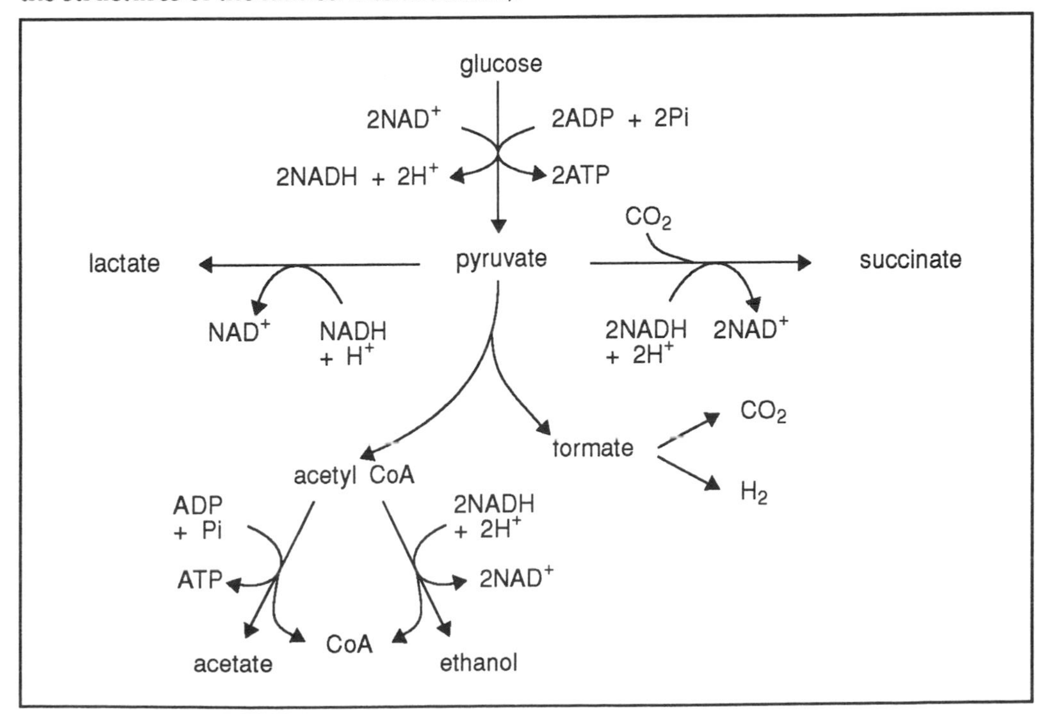

Figure 9.5 Pathways of mixed acid fermenters.

These fermentations have one characteristic biochemical feature, which is not encountered in any other bacterial fermentation. This is the cleavage of pyruvate to give formate and acetyl CoA. Formate is, therefore, frequently a major fermentative end-product. However, it does not always accumulate because some mixed acid fermenters possess the enzyme formic hydrogenlyase, which splits formate to CO_2 and H_2. In such organisms, formate may be largely replaced as a fermentative product by equimolar quantities of H_2 and CO_2.

acids not produced in equal amounts

The ratio of the end-product varies considerably, not only between organisms but also in a single organism grown under different environmental conditions. For example, environmental pH may have an influence. We already know that formate may be an end-product or it may be converted to CO_2 and H_2. Formic hydrogenlyase is inactive at alkaline pH and formate accumulates under these conditions. However, if the pH of the medium drops due to the accumulation of the organic acids, succinate, lactate and formate to around pH6, formic hydrogenlyase is reactivated and starts to convert formate to CO_2 and H_2 which slows the drop in pH. This mechanism is considered to prevent extremes of pH.

In butanediol bacteria, in addition to the products shown in Figure 9.5, sugar fermentation gives rise to an additional major end-product, 2,3-butanediol, which is formed from pyruvic acid.

$$2 \times CH_3{-}CO{-}COOH \xrightarrow[\text{NADH + H}^+ \rightarrow \text{NAD}^+]{} CH_3{-}HCOH{-}HCOH{-}CH_3 + 2CO_2$$

pyruvate → 2,3-butanediol

butanediol formation stops the reduction in pH

Butanediol formation in these bacteria is pH dependent and is again thought to be a mechanism of resisting extremes of pH. As the pH of the medium drops due to the accumulation of the acids, at around pH 6.3 some bacteria switch to butanediol formation. Butanediol is a neutral end-product formed from two molecules of pyruvate with the liberation of two CO_2 molecules. So, much less acid is formed in butanediol fermentation than in simple mixed acid fermentation.

SAQ 9.5

A bacterium grows anaerobically with glucose as substrate. Lactate, acetate and formate are found as fermentation products. Other products must have been formed as well: why? Use Figure 9.5 as a guide.

SAQ 9.6

The conversion: glucose $\rightarrow$ acetate + ethanol + 2 formate has a $\Delta G^{o'}$ of -306 kJ mol^{-1}. If each of the reactants is present at a concentration of 1mmol l^{-1}. What is the energetic efficiency of the fermentation?

($\Delta G'$ for ATP hydrolysis is -49 kJ mol^{-1}) (RT ln = 5.7 log).

$$\Delta G' = \Delta G^{o'} + RT\ln(\text{reactants})/(\text{products})$$

SAQ 9.7

You have an absent minded friend who unfortunately is not a very good experimenter. He was asked to run a fermentation but neglected to weigh the glucose he added and also forgot to analyse for ethanol. He has found that the fermentation produces the compounds shown below. He knows that all the glucose is oxidised to pyruvate via glycolysis and that ethanol is the only additional compound formed. He comes to you for help in salvaging something out of this mess. How many moles of ethanol should you tell him have been formed and how much glucose was added to the fermentation medium? Figure 9.5 will help you.

lactate = 10 moles; acetate = 5 moles; carbon dioxide = 10 moles; hydrogen = 10 moles;

SAQ 9.8

Organisms a), b), c) and d) were grown in a medium (pH 7.5) containing glucose as sole source of carbon and energy. The cultures were analysed for fermentation products and the results are shown below:

	products, moles per 100 moles of glucose fermented			
	a)	b)	c)	d)
ethanol	50	70	64	46
2,3-butanediol	—	66	—	64
acetic acid	36	0.5	62	4
lactic acid	79	3	43	10
succinic acid	—	—	22	8
formic acid	2.5	17	105	48
H_2	80	35	—	—
CO_2	88	172	—	116

1) Compare results for organisms a and b and give two reasons why organism b) should be regarded as a butanediol fermenter.

2) Which of the organisms have attempted to resist a drop in pH of the medium? By what mechanism(s) did each of these organisms attempt this?

9.2.4 Butyrate and solvent fermentations

These fermentations are mainly carried out by obligate anaerobic bacteria from the genus *Clostridium*.

H_2, CO_2, acetate and butyrate = basic butyrate fermentation

The butyric acid bacteria ferment soluble carbohydrates to form acetate, butyrate, CO_2 and H_2. This is known as basic butyrate fermentation. The fermentation is initiated by the conversion of sugars to pyruvate via glycolysis. The pathway for the formation of end-products from pyruvate is shown in Figure 9.6.

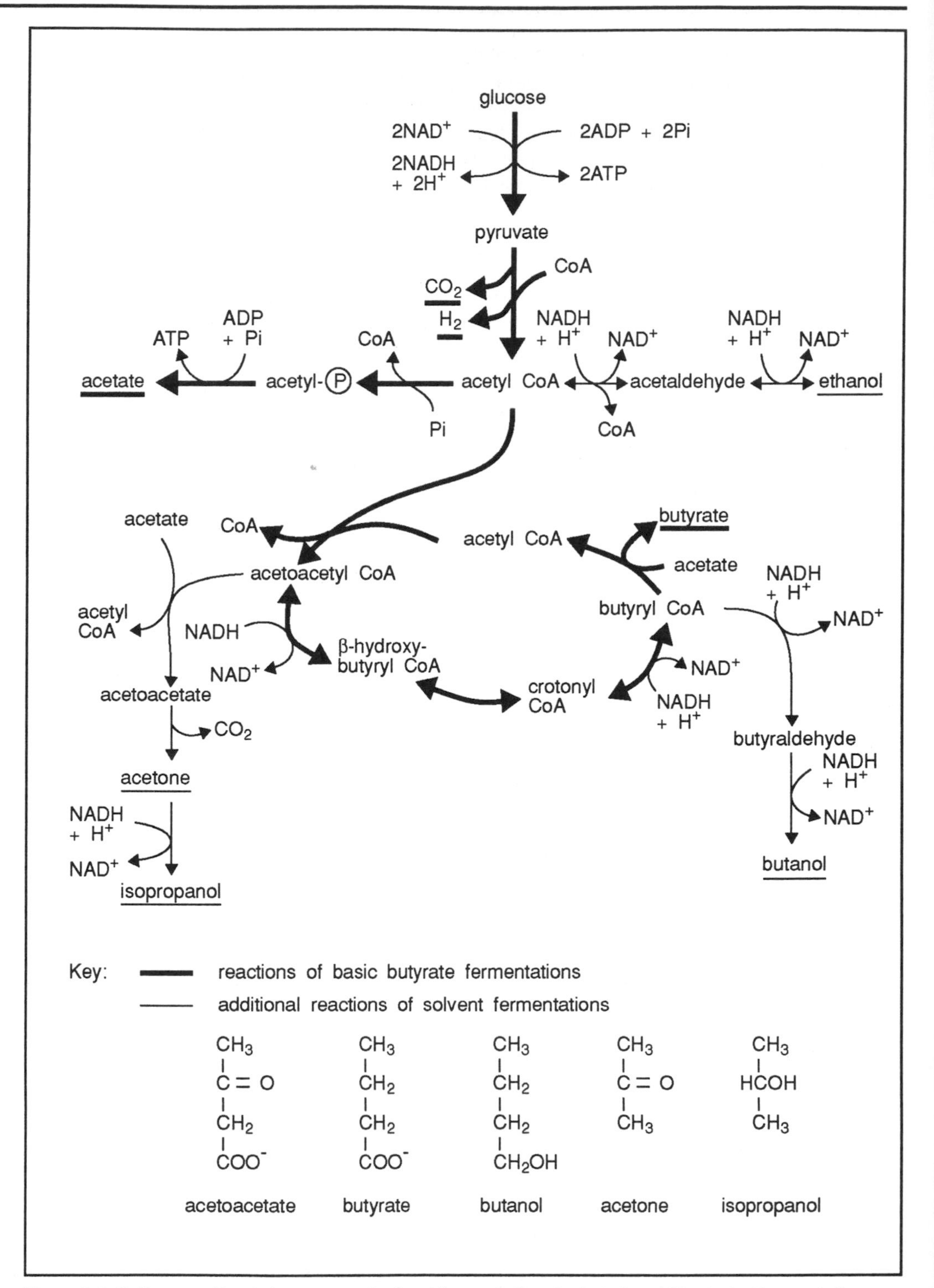

Figure 9.6 Pathways of basic butyrate and solvent fermentations by butyric acid bacteria. Fermentation products are underlined.

There are two notable stages:

CO_2 and H_2 formed but no formate

- a phosphoroclastic reaction in which pyruvate is converted to acetyl CoA, CO_2 and H_2. We can see that, unlike the mixed acid fermenters, formate is not an intermediate in this. Inorganic phosphate is then used to form acetylphosphate. The phosphate linkage in acetylphosphate is a high-energy bond and is used to synthesise ATP in a substrate-level phosphorylation, which also forms acetate;
- butyrate synthesis occurs through condensation of two moles of acetyl CoA, to form acetoacetyl CoA, which is then converted in a three-stage sequence to butyryl CoA. Butyrate is then produced by the transfer of CoA to acetate. Figure 9.6 shows that butyrate formation is cyclic, involving a re-synthesis of acetyl CoA from free acetate. The net effect of this cycle is to convert one mole of acetyl CoA and one mole of free acetate into 1 mole of butyrate.

Π Now use Figure 9.6 to complete the overall reaction for basic butyrate fermentation of glucose:

2 glucose + acetyl CoA + (—) ADP + (—) Pi →

(—) CO_2 + (—) butyrate + (——) + CoA + (—) H_2 + (—) ATP

The completed reaction is:

2 glucose + acetyl CoA + 7ADP + 7Pi →

$4CO_2$ + 2 butyrate + acetate + CoA + $4H_2$ + 7ATP

high H_2 favour solvent production

Some of the butyric acid bacteria form additional neutral compounds from sugar fermentation. These are butanol, acetone, isopropanol and small amounts of ethanol. With the exception of ethanol produced by the reduction of acetyl CoA formed from pyruvate neutral compounds arise by divergence from the normal pathway of butyrate formation. These modified butyric acid fermentations are known as solvent or acetone-butanol fermentations (see Figure 9.6). The neutral compounds typically arise in the later stages of growth. Their accumulation is accompanied by a reduction in the amount of butyrate and H_2 formed. Some of the H_2 initially produced is reutilised and serves as a reductant for NAD^+, this is necessary to maintain oxidation-reduction balance since as you can see in Figure 9.6 a great deal of NADH is oxidised in the formation of the neutral end-products. The accumulation of neutral end-products is favoured by maintaining high levels of H_2 in the culture, and it can be largely prevented by removal of the H_2 as it is formed.

SAQ 9.9

1) Identify each of the following statements as true or false. Justify your response in each case.

a) The pathway of ethanol formation used by solvent fermenters is the same as that used for ethanolic fermentation in yeasts.

b) Basic butyrate fermenters produce butyrate, acetone, CO_2 and H_2 as end-products.

c) Acetate is a precursor in the formation of acetone and butyrate.

d) High H_2 levels favour the formation of neutral end-products.

SAQ 9.10

1) Select the correct name for the structures shown below from the list provided.

Pyruvic acid, butyric acid, lactic acid, ethanol, butanol, acetone, formic acid, isoproponol, glucose, acetic acid and succinic acid.

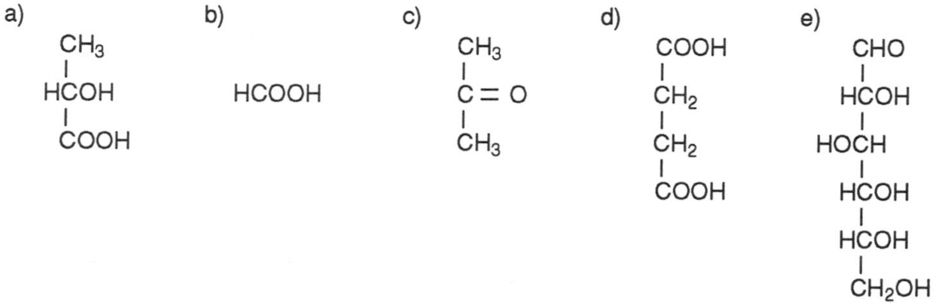

a)
CH_3
|
HCOH
|
COOH

b)
HCOOH

c)
CH_3
|
C = O
|
CH_3

d)
COOH
|
CH_2
|
CH_2
|
COOH

e)
CHO
|
HCOH
|
HOCH
|
HCOH
|
HCOH
|
CH_2OH

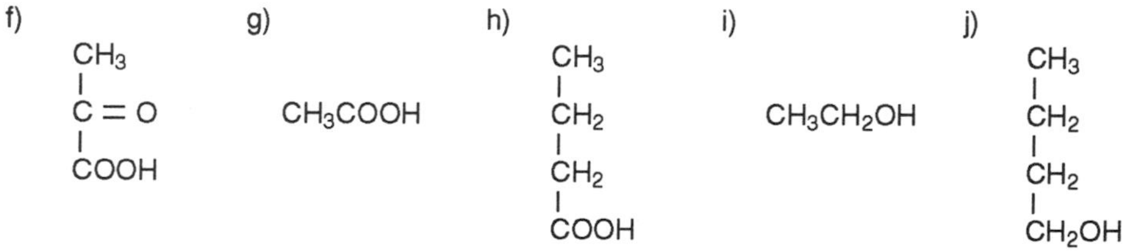

f)
CH_3
|
C = O
|
COOH

g)
CH_3COOH

h)
CH_3
|
CH_2
|
CH_2
|
COOH

i)
CH_3CH_2OH

j)
CH_3
|
CH_2
|
CH_2
|
CH_2OH

9.2.5 Propionate fermentations

Propionate fermentations of sugars are performed by several genera of obligate anaerobic bacteria (eg the genus *Propionibacterium*). Sugars are utilised via glycolysis and the end-products of the fermentation, derived from pyruvate, are propionate, acetate and CO_2. Some succinate may also be formed.

several pathways lead to propionate

Several different fermentative pathways yield propionate as a major end product. In all of these pathways some of the pyruvate is oxidised to acetyl CoA and CO_2, with ATP being produced in the conversion of acetyl CoA to acetate. In some cases this is the only ATP producing reaction which occurs.

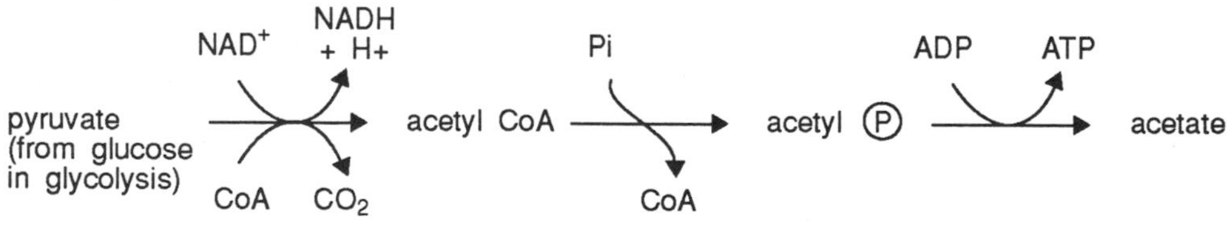

pyruvate (from glucose in glycolysis) → (NAD⁺, CoA in; NADH + H+, CO_2 out) → acetyl CoA → (Pi in; CoA out) → acetyl Ⓟ → (ADP → ATP) → acetate

The formation of acetate and CO_2, which are the oxidised products of the fermentation, is balanced by the reductive formation of propionate. Some of the pathways to propionate include such intermediates as fumarate and succinate. These are formed by initial carboxylation of C3 pyruvate to C4 compounds, oxaloacetate or malate. These are then converted by reversal of TCA cycle reactions to fumarate and succinate.

Propionate, a C3 compound, is then formed by further conversion of succinate and involves decarboxylation of a C4 intermediate. The pathway is summarised below:

CH_3–C=O–COO^- (pyruvate) → [+ CO_2, 2NADH + H^+ → 2NAD^+, – H_2O; 3 or 4 enzymes involved] → COO^-–CH_2–CH_2–COO^- (succinate) → [– CO_2; 3 enzymes involved] → CH_3–CH_2–COO^- (propionate)

randomisation of labelled carbons

We can see that succinate has an axis of symmetry - carbon atoms C1 and C4 are indistinguishable as are carbon atoms C2 and C3. Hence pyruvate radioactively labelled in carbon two is converted to fermentation products that are labelled in two carbons, a process termed randomising. The pathway is thus a randomising pathway.

(3) CH_3 – (2) *C=O – (1) COO^- → [+ CO_2] → (4) COO^- – (3) CH_2 – (2) *CH_2 – (1) COO^-

→ [carbon 1 CO_2] → COO^-–CH_2–*CH_3

→ [carbon 4 CO_2] → CH_3–*CH_2–COO^-

propionic acid labelled in both C_2 and C_3

* indicates ^{14}C carbon

Another pathway to propionate exists which does not pass through symmetrical intermediates, and hence is termed non-randomising. In this pathway propionate is formed from lactate via pyruvate so four carbon intermediates are not involved.

Π Write the overall reaction for propionate fermentation using lactate as the sole growth substrate. You will need to refer to the first two paragraphs of Section 9.2.5 to answer this. (Try this before checking with our equation below).

$$3 \text{ lactate} + ADP + Pi \rightarrow 2 \text{ propionate} + \text{acetate} + CO_2 + ATP$$

The production of acetate must be included because it is the only ATP-generating reaction when these bacteria grow on lactate.

The formation of succinate is strongly influenced by the content of CO_2 in the growth medium.

SAQ 9.11

A skeleton flow-diagram of the derivation of some of the major end-products of the bacterial fermentations of sugars from pyruvate is shown below. The end-products are represented by numbers and certain reactions are coded with letters. The following questions are not easy. See how many you can get right.

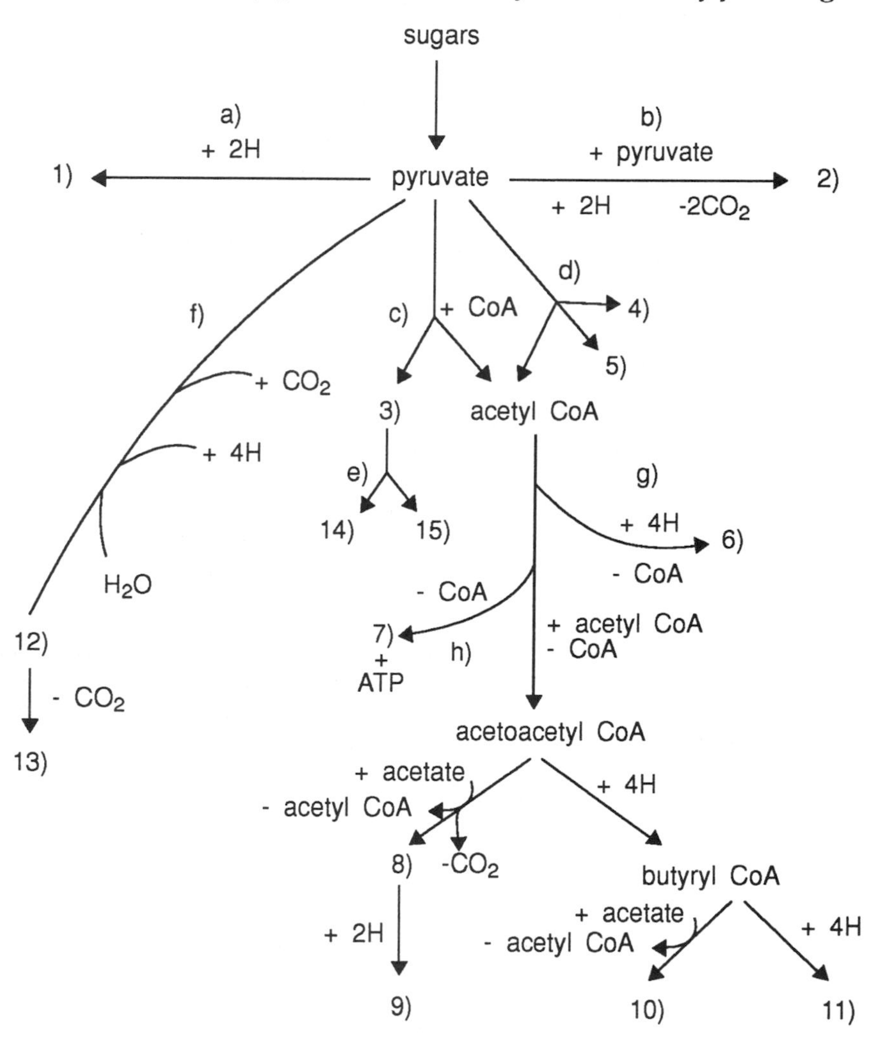

1) Identify each of the numbered (1)-15)) end-products shown on the diagram.

2) For each of the reactions labelled with letters (a)-h)) identify which of the fermentation classes listed below use the reaction:

ethanolic; homolactic; heterolactic; mixed acid; butanediol; solvent; propionate.

9.3 Other fermentations

Many compounds other than glucose can be fermented by bacteria. These include most sugars, many amino acids, certain organic acids, purines and pyrimidines. For a compound to be fermentable it must be neither too oxidised or too reduced. If it is too oxidised, it will be a poor electron donor and therefore be unable to complete very favourable oxidation-reduction reactions. Similarly, if it is too reduced, it will be a poor electron acceptor and only slightly favourable redox couples will be possible.

Another requirement is that the substrate must be converted to an intermediate that can undergo substrate-level phosphorylation, since this is the only form of ATP generation possible by fermentation. Non-fermentable compounds include hydrocarbons and fatty acids. In some cases, an organic compound cannot be fermented unless another organic compound is present as the electron acceptor. The requirement for mannitol by some heterolactic fermenters, is an example of mixed substrate fermentation.

9.4 Growth yields and fermentation

We have seen that net ATP per mole of substrate fermented depends on the type of fermentation. In this section we will examine the relationship between net ATP produced by fermentation and the amount of biomass formed.

biomass measured as grams dry weight

The growth yield is the amount of biomass produced per unit amount of substrate consumed. Biomass can be measured in a variety of ways, the most common is to measure dry weight in grams. We can relate biomass produced to ATP consumed because the energy required to polymerise carbon building blocks into macromolecules does not vary among micro-organisms. Approximately 10 g of cell material are produced per mole of ATP. The amount of biomass produced per mole of ATP is known as Y_{ATP}.

biomass mol^{-1} ATP = Y_{ATP}

When certain fermentative micro-organisms are grown in rich media, radioactive tracer experiments show that little or no carbon from the fermentable substrate is converted into cellular material. Under these conditions, cellular material is derived from the other medium components (amino acids, purines, pyrimidines, etc), and the fermentable substrate is used solely as a source of energy. So, we should be able to predict the growth yield from a knowledge of the fermentative pathway. This is normally expressed as the molar growth yield which is defined as grams of biomass produced per mole of substrate. Let us consider ethanolic fermentation using glucose as substrate:

- If g biomass produced per mole of ATP consumed = 10.5 (ie Y_{ATP}= 10.5) and net ATP yield = 2 mol mol^{-1} glucose then the growth yield for glucose = 2 x 10.5 = 21 g biomass mol^{-1} glucose. Thus the molar growth yield on glucose is 21.

SAQ 9.12

1) Three kinds of bacteria were grown on glucose and they yielded the following fermentation products respectively:

   a) lactate; b) ethanol + CO_2; c) acetate + ethanol + formate.

   Which is likely to have the highest molar biomass growth yield?

   Hint: what class of fermentation do these organisms belong to and how much ATP do they produce per mole of glucose?

2) Two kinds of lactic acid bacteria were grown on glucose. The molar growth yield for organism A) was 22 and for organism B) was 33. What fermentation pathways were being used by these organisms?

Summary and objectives

We have seen that in the absence of oxygen certain micro-organisms are able to grow by anaerobic respiration or by fermentation. These processes generate some ATP whilst maintaining oxidation-reduction balance. Anaerobic respiration uses alternatives to oxygen as terminal electron acceptor. The acceptor used depends on the organism and may be inorganic or organic. Nitrate, sulphate, fumarate, carbon dioxide and iron III are commonly used. In nature, anaerobic respiration is essential for the cycling of elements (N and S) and also forms methane and deposits of iron ore. The use of nitrate as an electron acceptor reduces soil fertility by the process of denitrification.

Fermentation uses organic, metabolically derived, compounds as electron acceptors. During fermentation ATP can only be produced by substrate level phosphorylation. The ATP yield per mole of glucose depends on the pathway used but is always far less than that produced by complete aerobic respiration. Glucose is fermented via glycolysis with the exception of heterotactic fermenters which use the pentose phosphate pathway. Pathways from pyruvate leading to end-products of the fermentations are diverse and only some generate ATP. However, they all reoxidise the NADH produced in the pathways leading to pyruvate and the end-products are in oxidation-reduction balance with the carbon substrate.

On completion of this chapter you should be able to:

- distinguish between aerobic respiration, anaerobic respiration and fermentation processes;
- describe the use of alternatives to oxygen as electron acceptors in metabolism and their importance in nature;
- write overall reactions, determine ATP yields and the energy-capturing efficiencies for carbohydrate fermentations;
- determine the oxidation-reduction balance for carbohydrate fermentations;
- interpret end-product data derived from carbohydrate fermentations and predict the fermentation pathway used from the molar growth yield.

Responses to SAQs

Responses to Chapter 1 SAQs

1.1

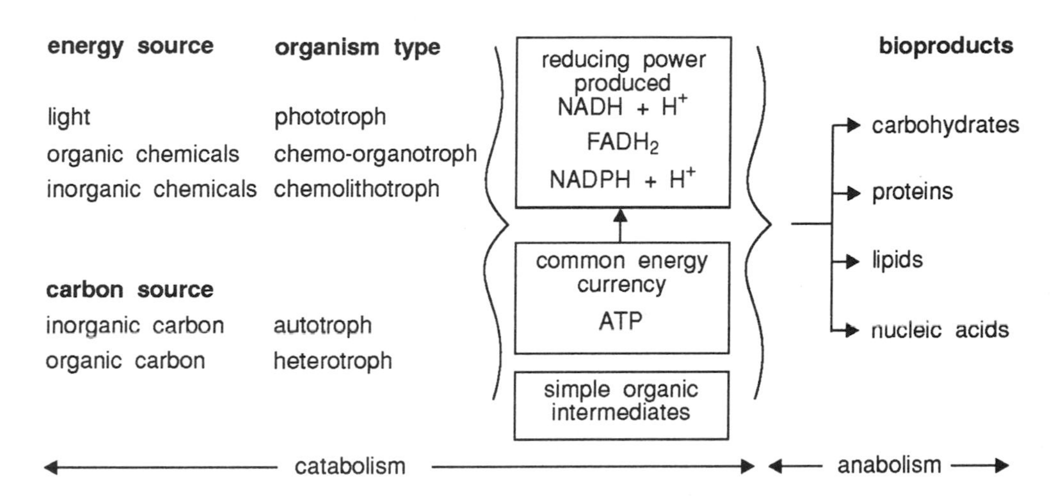

Your answer could have included several other compounds on the right hand side of the figure and still have been correct. The four listed are the major but by no means the only macromolecules produced by living cells.

Responses to Chapter 2 SAQs

2.1 1) c and d; 2) b; 3) a.

A living organism is an open system. It must take up nutrients (matter) from its environment and expel waste products.

2.2 1)

type of energy	intensive factor	extensive factor
mechanical	pressure	volume
electrical	electrical potential	electrical charge
thermal	temperature	entropy
chemical	chemical potential	mass

2) The statement is false.

At equilibrium the intensive factors are the same and there is consequently no tendency to change. If we take thermal energy as an example; as the total entropy of a system is a function of the size of the particular system, a large system at low temperature may have the same thermal energy as a smaller system at a higher temperature. They would not however be in equilibrium. If thermal contact is made between them, heat would flow from the high temperature to the low temperature system. They would be at equilibrium when their intensive factor (temperature) is the same.

2.3

$$\Delta U = -154.90 \text{ kJ}$$

Here is how we calculated this:

ΔU, the change in energy of the system, is given by:

$$\Delta U = \Delta Q + W$$

where ΔQ is the heat absorbed by the system and W is the work done on the system by its surroundings.

$$\Delta Q = (-152.48 \text{ kJ}).$$

Note that heat is evolved so that the heat absorbed is negative.

W, the work done, will be the mechanical work associated with the change in volume of the system which may be taken as equal to the volume of gas evolved. The volume changes for other components of the system will be so small by comparison that they may be neglected.

The work done therefore will be $P\Delta V$ where P is the pressure and ΔV the volume of gas evolved.

From the gas laws we know PV = nRT and n, the number of moles of gas evolved, is unity.

$$\therefore \ \Delta V = \frac{nRT}{P}$$

hence:

$$P\Delta V = nRT$$

$$= 1 \times 8.314 \times 291 = 2420 \text{ J} = 2.42 \text{ kJ}$$

$$\therefore \ W = -2.42 \text{kJ}$$

Note the value of W is negative as this is work done by the system on the environment (surroundings) when it increases its volume.

Since $\Delta U = \Delta Q + W$:

$$\Delta U = -152.48 + (-2.42) = -154.90 \text{ kJ}$$

2.4 The reaction for the complete oxidation of glucose is:

$$C_6H_{12}O_{6(s)} + 6O_{2(g)} \rightarrow 6CO_{2(g)} + 6H_2O_{(l)}$$

Here there is no net change in volume (6 molecules of O_2 are used and 6 of CO_2 are produced, all at the same temperature).

$$\therefore \ \Delta U = \Delta H = -2810 \text{ kJ mol}^{-1}$$

For the complete oxidation of palmitic acid the reaction is:

$$C_{16}H_{32}O_{2(s)} + 23O_{2(g)} \rightarrow 16CO_{2(g)} + 16 H_2O_{(l)}$$

Here the net volume change is a fall of 7 moles of gas per mole of palmitic acid oxidised ($23O_2$ - 16 CO_2).

$$\therefore \quad \Delta H = \Delta U + nRT$$

$$= \left(-10012 + \frac{(-7)(8.314)(310)}{1000}\right) \text{ kJ mol}^{-1} = -10030 \text{ kJmol}^{-1}$$

(R is 8.314 JK^{-1} mol^{-} so we must divide by 1000 to convert to kJ).

Palmitic acid is clearly a much more energy rich storage compound than glucose.

Even if more refined calculations were to be carried out to take into account the fact that the glucose would probably be stored as a polysaccharide and the palmitic acid as a triglyceride it would still be overwhelmingly richer as an energy source.

2.5

1) $\frac{dQ}{T}$;

2) is zero;

3) an increase;

4) decrease

2.6

1) Decreasing the volume would make the system more ordered in space and increase our information on its exact state so that the entropy would fall.

2) Increasing the temperature would decrease our knowledge of its exact state (eg there would be an increase in movement of molecules) and a decrease in its order so that the entropy would rise.

2.7

G, the Gibbs function, is the most useful because in biology we are concerned mainly with systems at constant temperature and pressure.

2.8

K_{eq}	$\Delta G^{o'}$ kJ mol^{-1}
0.00001	28.5
0.001	17.1
0.1	5.7
1.0	0
10.0	-5.7
1000.0	-17.1
100000.0	-28.5

2.9

1)

$$K_{eq} = \frac{[G-1-P]}{[G-6-P]} = \frac{[0.01]}{[0.19]} = 0.0526$$

2)

$$\Delta G^{o'} = -2.303 \; RT \log_{10} K_{eq}$$

$$= -5.706 \log_{10} 0.0526$$

$$= 7.297 \text{ kJ mol}^{-1}$$

for the reaction: [G-6-P] $\rightarrow$ [G-1-P]

3) For the reverse reaction: [G-1-P] $\rightarrow$ [G-6-P]

$$\Delta G^{o'} = -2.303\ RT \log_{10}\frac{[0.19]}{[0.01]} = -5.706 \log_{10} 19 = -7.297 \text{ kJ mol}^{-1}$$

2.10

since:

$$\frac{\Delta G_2'}{T_2} = \frac{\Delta G_1'}{T_1} - \frac{\Delta H^{o'}(T_2 - T_1)}{T_1 T_2}$$

and:

$$T_1 = 18°C = 291K \quad T_2 = 58°C = 331K$$

$$\Delta H^{o'} = -20\ 100 \text{ J mol}^{-1} \quad \Delta G_1' = -30\ 300 \text{ J mol}^{-1}$$

(Note that it can be useful to use ΔG values in the form of joules (J) rather than kilo joules (kJ) to avoid confusion between k (kilo) and K (kelvin) when specifying units.)

Substituting in the Gibbs-Helmholtz equation gives:

$$\frac{\Delta G_2'}{331} = \frac{(-30\ 300)}{291} - \frac{-20\ 100\ (331 - 291)}{291 \times 331}$$

$$= -104.1 + \frac{804000}{96321} = -104.1 + 8.35 = -95.75$$

$$\therefore\ \Delta G_2' = -95.75 \times 331$$

$$= 31\ 700\ \text{ J mol}^{-} = -\ 31.700 \text{ kJ mol}^{-1}$$

Thus, the change in Gibbs function $\Delta G'$ for a 40 degree rise in temperature will be (-31.7 - (30.3)) = 1.4 kJ mol^{-1} (about 4%). The effects of this small change in $\Delta G'$ for what, in physiological terms, is a large temperature change, means that the energetics and equilibria of biological processes involving ATP will not be unduly affected by physiological temperature changes.

2.11 A large ΔG value means that a reaction is far removed from equilibrium. The reactions catalysed by hexokinase, phosphofructokinase and pyruvate kinase all have large negative ΔG values which implies that the reactions are kinetically irreversible.

These enzymes therefore are catalysing non-equilibrium reactions under physiological conditions and may be identified as potential regulatory sites in glycolysis. The identification of glyceraldehyde-3-phosphate dehydrogenase and phosphoglycerate kinase as regulatory enzymes is more questionable.

It should be noted that identification of regulatory enzymes by this means must be tentative as uncertainties exist in the determination of physiological concentrations.

2.12

$$\Delta G' = \Delta H' - T\Delta S' \qquad T = 303K$$

$$\therefore\ -20 = +52.3 - 303\ \Delta S'$$

$$\therefore\ \Delta S' = \frac{72.3}{303} = +0.24 \text{ kJK}^{-1} \text{ mol}^{-1}$$

$$= +240 \text{ JK}^{-1} \text{ mol}^{-1}$$

It can be seen from a comparison with the trypsin denaturation data and Table 2.10 that this represents a substantial entropy change which indicates that a significant conformational change has resulted from the binding of the nucleotide.

2.13 1) Replacing glu 49 with valine or leucine makes the subunit less likely to unfold.

2) In contrast to the above glutamine substitution produces a form of the enzyme more likely to unfold.

3) From the previous data it is clear that the stability of the α sub-unit is substantially increased by replacement of an amino acid with a strongly polar side chain at position 49 with one with a non polar side chain. The increased stability might lead to a change in sensitivity to allosteric effectors or to a change in the thermal stability of the biologically active protein.

The value of this type of information to the genetic engineer and biotechnologist should be obvious.

2.14 We may represent the reactions in the following way:

$G + Pi \rightarrow G\text{-}6\text{-}P + H_2O$ $\qquad \Delta G^{o'} = + 13.8 \text{ kJ mol}^{-1}$

$\Delta G^{o'}$ is positive because we have reversed the reaction.

$ATP + H_2O \rightarrow ADP + Pi$ $\qquad \Delta G^{o'} = - 32.2 \text{ kJ mol}^{-1}$

therefore:

$G + ATP \rightarrow ADP + G\text{-}6\text{-}P$ $\qquad \Delta G^{o'} = - 18.4 \text{ kJ mol}^{-1}$

The reaction will proceed spontaneously and we may expect the enzyme catalysed reaction to be feasible under physiological conditions. The reaction will be driven by highly exergonic removal of the phosphate group from ATP being coupled to the moderately endergonic phosphorylation of the glucose.

The mechanism of the enzyme catalysed reaction will involve the formation of intermediate enzyme substrate and product complexes. It will not simply be the addition of water to ATP with consequent release of free phosphate which will then react with glucose. This does not however invalidate the analysis of the net ΔG changes.

2.15 1) $-19.15 \text{ kJ mol}^{-1}$.

$$\text{Slope of the graph} = \frac{-\Delta H'}{2.303R} = \frac{0.4}{(3.6-3.2) \times 10^{-3}} = 10^3$$

$$\text{Thus, since: } R = 8.314 \text{ JK}^{-1}\text{ mol}^{-1} : \quad \frac{-\Delta H'}{2.303 \times 8.314} = 10^3$$

$$\text{then: } -\Delta H' = 10^3 \times 19.15 \text{ J mol}^{-1} = 19.14 \text{ kJ mol}^{-1} \text{ so: } \Delta Ho = -19.15 \text{ kJ mol}^{-1}$$

2) -57.5 J mol^{-1}.

$$\text{Since: } \log K_{eq} = \frac{\Delta S'}{2.303R} - \frac{\Delta H'}{2.303RT}$$

$$\text{when: } \log K_{eq} = 0 \text{ then } \frac{\Delta S'}{2.303R} = \frac{\Delta H'}{2.303RT}$$

$$\text{therefore } \Delta S' = \frac{\Delta H'}{T}: \quad \log K_{eq} = 0 \text{ when } \frac{1}{T} = 3.0 \times 10^{-3}$$

$$\text{therefore } \Delta S' = -19.15 \times 3.0 \; 10^{-3} \text{ JK}^{-1}\text{ mol}^{-1} = -0.0575 \text{ JK}^{-1}\text{ mol}^{-1} = -57.5 \text{ JK}^{-1}\text{ mol}^{-1}$$

3) Approximately -2.0 kJ mol^{-1}.

Since: $\Delta G' = \Delta H' - T\Delta S'$

then: $\Delta G' = -19150 - 298\ (-57.5)\ J\ mol^{-1}$

$= -19150 + 17135\ J\ mol^{-1} = -2015\ J\ mol^{-1}$ = approximately $-2.0\ kJ\ mol^{-1}$

2.16

$$\Delta G = \Delta G' + 2.303\ RT \log \frac{[\text{LEU.GLY.}][H_2O]}{[\text{GLY}][\text{LEU}]}$$

By convention $[H_2O] = 1$.

The maximum concentration of dipeptide will correspond to the equilibrium condition at $\Delta G = 0$.

Let the dipeptide concentration equal $10^{-x}\ mol\ l^{-1}$.

$$0 = +14 + 5.7 \log \frac{[10^{-x}][1]}{[10^{-3}]]10^{-3}]}$$

$$-14 = 5.7\ (-x + 6) = -5.7\ x + 34.2$$

$$\therefore\ -5.7\ x = -48.2\ \therefore\ x = 8.46\ \therefore\ [\text{LEU.GLY.}] = 2.9 \times 10^{-8}\ mol\ l^{-1}$$

The very low concentration of the dipeptide form suggests that the direct reaction of amino acids condensing to form peptide links is not likely to be the biosynthetic route within the cell. The concentration of peptide would be 100 000 times lower than that of the individual amino acids. This would imply that very high amino acid concentrations would be needed if peptide and protein synthesis proceeded by this route in the cell. Such concentrations are not found. In biological systems, to get sufficient peptide bond formation, there is a need to couple the reaction to an exergonic reaction to produce a more favourable ΔG value and thus increase the formation of the peptide. We will not go into details of this at this stage but you will learn of many examples in later chapters by which unfavourable equilibria are made favourable by coupling the endergonic reaction to strongly exergonic reactions.

Responses to Chapter 3 SAQs

3.1

Reduction is a gain of electrons so that reactions 1) and 2) as written are reductions. For reaction 3) as written, electrons have been lost and the process is consequently an oxidation although the reverse process 2) is a reduction.

Note that if 1) and 3) are mixed we would have an oxidant and a reductant couple and the subsequent reaction would be a redox reaction:

$$Zn + Cu^{2+} \rightleftarrows Zn^{2+} + Cu$$

The oxidation - reduction (redox) reaction consists of a transfer of electrons from a reducing agent to an oxidising agent. In consequence there is no reduction without an oxidation taking place.

3.2

Red_L	Ox_L	RedR	OxR
Cu(s)	Cu^{2+}(aq)	Zn(s)	Zn^{2+}(aq)

(s) = solid, (aq) = aqueous solution

3.3

1) $Cu\ (s) \mid Cu\ SO_4\ (aq) \parallel Zn\ SO_4\ (aq) \mid Zn\ (S)$

or more simply but perfectly correctly:

$$Cu\ (s) \mid Cu^{2+}\ (aq) \parallel Zn^{2+}\ (aq) \mid Zn\ (s)$$

or $Cu \mid Cu^{2+} \parallel Zn^{2+} \mid Zn$

2) $Pt \mid Fe^{2+}, Fe^{3+}$

3) $Pt, H_2 \mid H^+ \parallel Cl^{-1} \mid AgCl, Ag$

Note the hydrogen gas can be regarded as being absorbed onto the platinum electrode.

3.4

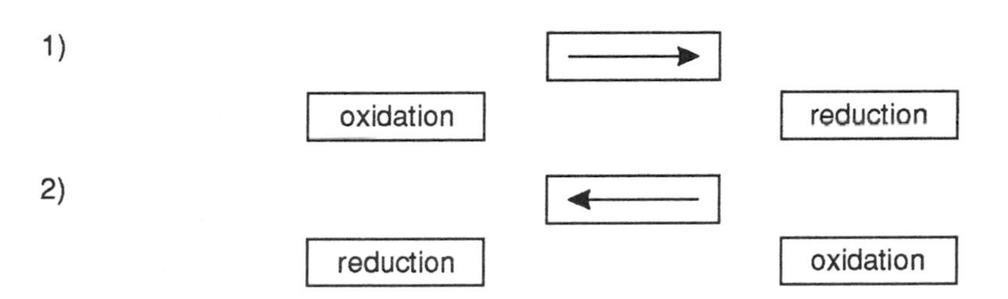

3.5

1) Yes because this would change the molecular environment of the iron atom. In the case of bacterial P_{450} it has been reported that the binding of camphor increased the reduction potential from -0.303 to -0.17OV. In the case of rabbit liver micromal P_{450} however the reduction potential was reported as unchanged at -0.33 V in the presence of typical substrates.

2) The change in E° is only -0.0034V when the $H^+/½\ H_2$ couple is replaced by $D^+/½D_2$. The change in atomic weight is proportionately much greater in this case than would result from any other substitution likely to be made in biochemical work. It is therefore most unlikely that isotopic labelling of biochemically important molecules would have any significant or even measurable effect upon redox characteristics. This is one reason why isotopes are so widely used in metabolic studies.

3.6

Chlorine, hypochlorous acid, ozone and fluorine gas are all stronger oxidants than oxygen and candidates for oxidising germicides.

3.7

The redox reaction involves two oxidation-reduction couples and the half cell reactions are:

$$½\ O_2 + 2H^+ + 2e^- \rightleftarrows H_2O \qquad E_o' = +\ 0.82V$$

and:

$$NAD^+ + 2H^+ + 2e^- \rightleftarrows NADH + H^+ \qquad E_o' = -\ 0.32V$$

Note $\Delta G^{o'}$ refers to pH7.0 unless otherwise specified as does E_o'.

$E_{cell} = E_R - E_L$ placing the couple with the more positive electrode potential on the right hand side of the cell.

$$E_{cell} = (+0.82 - (-0.32))\ V = +1.14V$$

n = 2 for the reaction so as:

$$\Delta G^{o'} = -n\ F\ E_{cell}$$
$$= -2 \times 96\ 487 \times (+1.14)\ J\ mol^{-1}$$
$$= -220000\ J\ mol^{-1} = -220\ kJ\ mol^{-1}$$

The sign of E_{cell} was positive and of $\Delta G^{o'}$ negative so the spontaneous reaction will be from left to right in the reaction equation as written ie in the direction of the reduction of oxygen to water.

Note that the convention for the reaction corresponding to the electrochemical cell

$$\underset{\text{left}}{Ox \mid Red} \parallel \underset{\text{Right}}{Ox \mid Red}$$

is:

$$\text{left (Red)} + \text{right (Ox)} \rightleftarrows \text{left (Ox)} + \text{right (Red)}.$$

3.8

For the reaction in question n = 2 and a = 2, so that on substitution into Equation 3.7 we get:

$$E_o' = E_o + \frac{0.059}{2} \log \frac{[^1/_2O_2]}{[H_2O]} + \frac{0.059}{2} \log [H^+]^2$$

but for standard conditions:

$$\log \frac{[^1/_2O_2]}{[H_2O]} = \log 1 = 0$$

so we can write:

$$E_o' = E^o + \frac{0.059}{2}\ 2 \log [H^+]$$

but at pH7 $[H^+] = 10^{-7}$

$$= E_o + 0.059 \log [0.0000001]$$
$$= +1.23 + 0.059 \times (-7) = +0.82V.$$

Note that this agrees with the figure in Table 3.3.

3.9

1) E_o' for the pyridine nucleotides is in the order of -0.30 V whilst E_o' for the fumarate/succinate couple is slightly positive. Substances such as nicotinamide dinucleotide would thus tend to spontaneously be oxidised not reduced by the succinate/fumarate couple. That is to say they would tend to bring about the reverse reactions (fumarate → succinate). The E_o' values for flavoproteins on the other hand can be as high as + 0.2 V so that they could spontaneously accept electrons from succinate and hence be likely natural cofactors.

2) The direction of electron flow will be in the direction of increasing E_o' so that the likely direction of electron flow is:

$NAD^+ / NADH + H^+$

↓

flavoprotein ox/red

↓

cytochrome C Fe^{3+} / Fe^{2+}

↓

cytochrome A_3 Fe^{3+} / Fe^{2+}

↓

$\frac{1}{2}O_2 / H_2O$

3.10 When E = + 0.217V the ratio of oxidised to reduced ascorbic acid can be obtained by the use of the Nernst equation:

$$E = E_o' + \frac{2.303\ RT}{nF} \log \frac{[Ox]}{[Red]}$$

n = 2 and for ascorbic acid $E_o' = + 0.08$ V;

so at E = +0.217V:

$$+0.217 = +0.08 + \frac{0.059}{2} \log \frac{[Ox]}{[Red]}$$

$$+0.137 = +0.0295 \log \frac{[Ox]}{[Red]}$$

$$4.644 = \log \frac{[Ox]}{[Red]}$$

$$\frac{[Ox]}{[Red]} = 4.4 \times 10^4$$

The percentages of ascorbic acid still in the reduced form is therefore 0.002%.

At a value of E = + 0.217V DCPIP will be 50% reduced as $E_o' = + 0.217$ V for DCPIP. This means that under conditions where enough reduced ascorbic acid has been added to half the concentration of the coloured oxidised form of DCPIP only 0.002% of reduced ascorbic acid will remain. In other words under conditions where the absorbency has been halved the reaction will have gone near enough to completion for all practical purposes. The method could therefore be used for the colourimetric assay of reduced ascorbic acid.

3.11

$$\frac{[H^+]_{out}}{[H^+]_{in}} = \frac{65}{1} = 65$$

$$\therefore \Delta pH = \log 65 = 1.81$$

ie the pH inside the mitochondria is more alkaline by 1.81 pH units.

$$\Delta\Psi = -\frac{2.303RT}{F} \log \frac{[^{86}Rb^+]_{in}}{[^{86}Rb^+]_{out}}$$

$$= -0.059 \log 108 = -0.059 \times 2.033 = -0.12 \text{ V}$$

$$\Delta p = \Delta\Psi - 0.059\ \Delta pH$$

$$\Delta p = -0.12 - 0.059 \times 1.81 = -0.12 - 0.109 = -0.227 \text{ V}$$

From the adenosine phosphate data for the reaction ADP + P_i → ATP.

$$\Delta G = \Delta G^{o'} + RT \ln \frac{[ATP]}{[ADP][P_i]}$$

Therefore:

$$= 30.5 + 2.303 \text{ rt} \log \frac{[1.1 \times 10^{-3}]}{[2.1 \times 10^{-5}][1.01 \times 10^{-3}]}$$

$$= 30.5 + 5.7 \log 51\,862 = 30.5 + 26.9 = 57.4 \text{ kJ mol}^{-1}$$

Now:

$$\Delta G = -n F\Delta p$$

so:

$$\Delta p = \frac{57.4}{n \times 96.487} \text{ V} = \frac{0.595}{n} \text{ V}$$

However from the earlier calculation using the relative concentration data, we have:

$$\Delta p = -0.227 \text{ V}$$

so that:

$$n = 2.62$$

This value for n, which is reasonably typical of what is often obtained in bench experiments, does not fit in very well with any of the proposed values. It also serves to show why the mechanism of oxidative phosphorylation is still a subject of controversy.

Responses to Chapter 4 SAQs

4.1

1) False. The substantial energy barrier between A and B (the activation energy) means that few molecules of A will possess sufficient energy (A + *) to react.

2) False. The first part of this is correct: the activation energy requirement is now less and more molecules will possess at least this level of energy. However, the enzyme does not alter the overall energy yield of the reaction A → B.

3) False. The reaction is spontaneous and may occur without an enzyme, albeit at a very low rate. Enzymic catalysis is based on the general principle that rate increases result from a lowering of activation energy.

4) True. B has a lower energy content than A: conversion of A to B will result in release of energy (amounting to the difference in energy contents of A and B) and the reaction will tend to proceed spontaneously. The high activation energy requirement may make the rate of reaction negligible.

5) True. Re-read the responses to questions 2) and 3) if in doubt.

4.2

1) This is false. Enzymes certainly increase the rate of a reaction but do not alter the equilibrium position.

2) True. Enzymatic catalysis is achieved by lowering the activation energy of a reaction; this is accomplished by providing an alternative reaction route. Substrates must bind to their enzyme and form an ES complex for catalysis to occur.

3) False. Look again at Figure 4.3 b) and the comments in the paragraph which follows it. Binding of 2 groups of the substrate may not be sufficient to guarantee distinction between 2 stereoisomers. With only 2 recognition sites, both stereoisomers could bind.

4) False: at least as a generalisation. Enzymes may be highly specific and only act on a single compound. Others, however, show rather loose specificity (such as hydrolytic enzymes involved in breaking foodstuffs into smaller fragments) and act on a number of related compounds. Thus wide variations in the specificity displayed occur.

5) Correct. A substrate only binds reasonably tightly and correctly positioned for catalysis to occur if there is a good match between the shape and charge distribution of active site and substrate. This is how enzymes discriminate between different molecules.

4.3

1) We hope you resisted any temptation to plot v_0 against [S], then attempting to extrapolate the line to get V_{max} (along the lines of Figure 4.6). Instead, the data should be transformed such that a straight line will be formed (providing Michaelis Menten kinetics apply; assume so at this stage). Whilst there are various ways to do this, we have described the Lineweaver-Burk plot. Reciprocals of v_0 and [S] have to be calculated, giving 1/v and 1/[S], respectively:

	[S] mmol l^{-1}	1/[S] l $mmol^{-1}$	v_0 µmol min^{-1}	$1/v_0$ min µmol^{-1}
A	0.6	1.67	1.14	0.88
B	1.0	1.0	1.7	0.59
C	1.5	0.67	2.36	0.42
D	3.0	0.33	3.64	0.27
E	6.0	0.17	5.0	0.2
F	15.0	0.067	6.46	0.15

These are then plotted as follows:

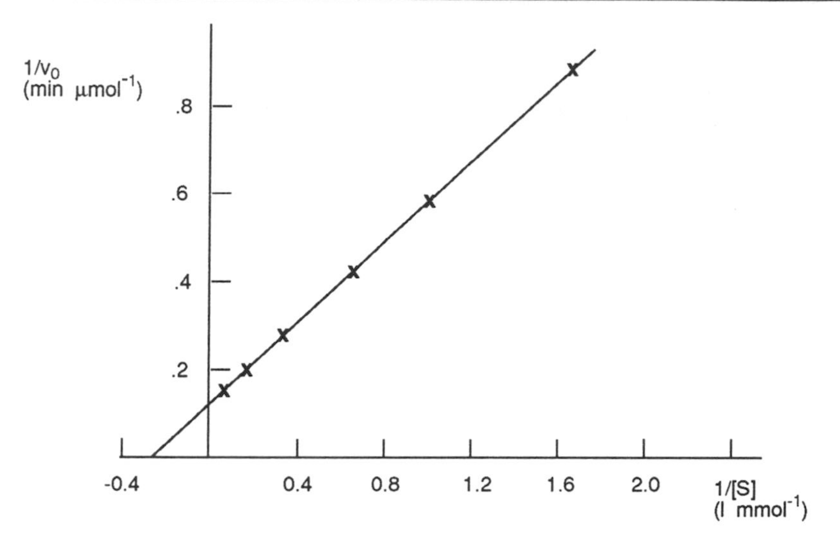

The line of best fit is determined (by eye; better by regression analysis) and K_M and V_{max} obtained as:

a) The intercept on the $1/v_0$ axis = $1/V_{max}$; in this case it is 0.12; thus:

$$1/V_{max} = 0.12 \text{ and } V_{max} = 8.33 \ \mu\text{mol min}^{-1}$$

b) K_M is determined from the intercept on the 1/[S] axis; this is -0.255 in this case; thus:

$$-1/K_M = -0.255 \quad K_M = 1/0.255 = 3.9 \text{ mmol l}^{-1}$$

2) The crucial data are obtained from substrate concentrations around the K_M value, particularly in the range 0.2-5.0x K_M. Inspection of Figure 4.6 shows that significant change in the rate occurs over this range. Assays at very low [S] may cause problems, because they tend to be more prone to error; additionally, when expressed as 1/[S], they give very high values on the x axis. Within the range given above, try to arrange substrate concentrations which will give a good spread of points when plotted as reciprocals. Note how this was done with the numerical data above; whilst it was a highly non-linear distribution in mol 1^{-1} terms, they were more evenly spaced as reciprocals.

Suitable values would be:

[S], mmol l⁻¹	giving	1/[S], l mmol⁻¹
0.8		1.25
1.2		0.83
2.0		0.5
4.0		0.25
10.0		0.1
20.0		0.05

4.4 The correct answer is 4). Non-competitive inhibition means that less active enzyme is available when inhibitor is bound to any enzyme molecules; thus V_{max} is lowered. Since I can bind to either E or ES, there is no effect on substrate binding ie K_M is not affected.

1) would be expected for mixed inhibition, where adverse effects on both substrate binding and catalysis are seen;

2) is an interesting case, in which the inhibitor makes it easier for substrate to bind, whilst none-the-less lowering enzyme activity. This occurs if inhibitor can bind to ES but not to E. This is a form of inhibition which is known as uncompetitive inhibition. It is rare with single substrate enzymes but can occur when more than one substrate is involved;

3) is the result expected for competitive inhibition. In this case, since S can displace the inhibitor more substrate will be needed to achieve V_{max}, hence the increase in K_M;

5) is consistent with extreme damage to the enzyme, such as denaturation or perhaps irreversible inhibition.

4.5 1) Whilst plotting v against [S] may give some indication of the type of inhibition, it is unlikely to distinguish between the possibilities. The data needs to be linearised, as in the Lineweaver-Burk plot. Thus 1/v and 1/[S] for both inhibited and non-inhibited assays must be calculated, giving:

(S)	**1/(S)**	**A (- inhibition)**		**B (+ inhibition)**	
mmol l^{-1}	l $mmol^{-1}$	v_0	$1/v_0$	v_0	$1/v_0$
.02	50.0	31.3	.032	16.1	.062
.03	33.3	41.7	.024	22.2	.045
.04	25.0	47.6	.021	27.8	.036
.06	16.7	57.1	.0175	35.7	.028
.12	8.33	71.4	.014	52.6	.019

These are then plotted as 1/v against 1/[S], and straight lines can be fitted, as shown on the next page:

It is apparent that the intercept on the $1/v_0$ axis is unchanged in the presence of inhibitor: within experimental error; V_{max} has not been affected (intercept on $1/v_0 = 1/V_{max}$). The intercept on the negative side of the 1/[S] axis is changed: K_M has been increased (since this intercept = $-1/K_M$). This pattern is characteristic of competitive inhibition. As substrate can displace the inhibitor, the original V_{max} can still be obtained. However, a raised [S] will be required to achieve this, hence the increase in K_M.

2) All experiments should be repeated several times before conclusions are made. In this case, if reversible competitive inhibition is suspected, then regeneration of activity after careful removal of the inhibitor must be demonstrated. If we cannot regenerate activity, then irreversible inhibition must be suspected. Additional experiments which address this question should be conducted.

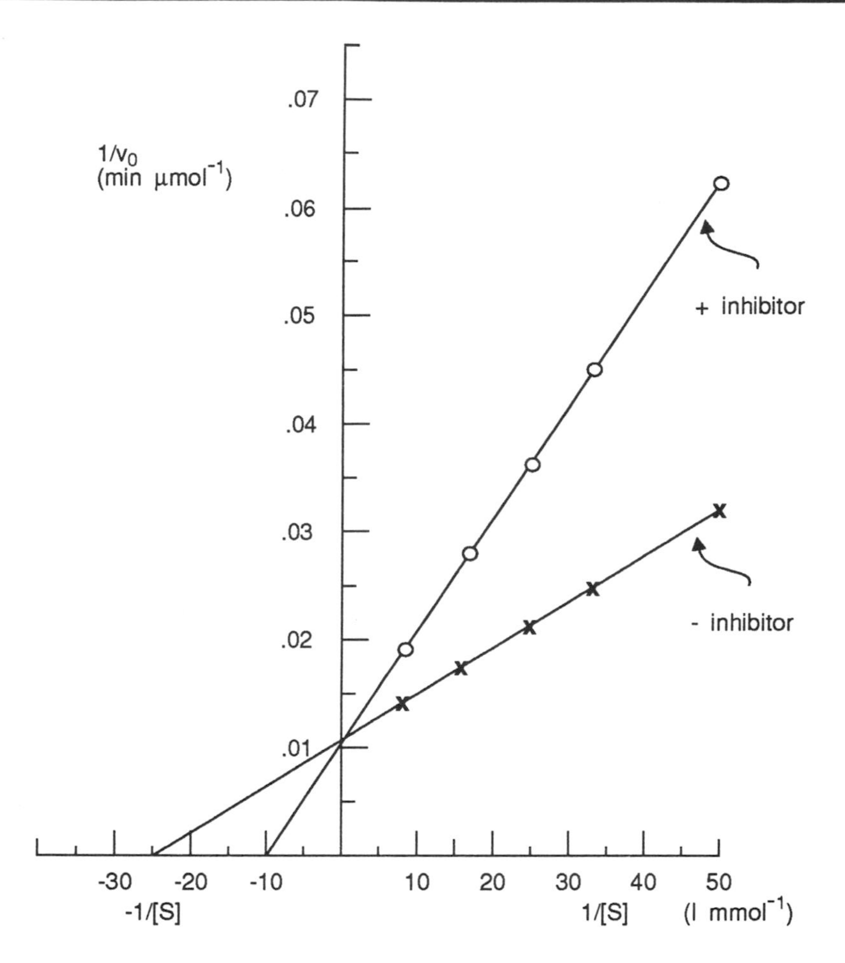

4.6

1) Not true: since metabolism totally depends on enzymes, they must be able to work at whatever temperatures organisms can live at: even at very high temperatures (100°C) in hot springs (so-called thermophilic bacteria)! The enzymes of different organisms tend to show characteristic activity ranges - which simply represent adaptation at the molecular level complementing that of physiological and anatomical adaptation. Increasing temperature routinely has two contrasting effects on enzyme activity: it increases the reaction velocity but, at elevated temperatures, it also causes denaturation. Both these effects have to be borne in mind.

2) Usually true, although change in charge of the substrate may also be important and could be the crucial effect. The requirement for substrate binding and/or catalysis is that appropriate amino acid residues are in the 'correct' ionic form (protonated or deprotonated) to fulfil their role. If they are not, the rate will be less. Thus if a carboxyl group must be deprotonated (and hence negatively charged) in order to interact with a positively charged group on the substrate, lowering the pH in the region of the pKa of the carboxyl group will lead to its protonation thereby leading to a lessening (or complete loss) of activity.

3) False. We have already noted (in response 1) above) that denaturation occurs as the temperature is raised.

4) False. Of course you do! When assaying enzymes you should try to measure the initial rate, before the reaction rate changes. It will frequently slow down: one cause

of this is denaturation of the enzyme, whereby, with time, some of the enzyme molecules are denatured; as a consequence the rate slows (since there are fewer active enzyme molecules and rate α[E] (or should be!). Denaturation is most likely to be a problem at elevated temperature.

5) True. This is why reactions (both enzyme-catalysed and non-enzymic) proceed faster at higher temperatures.

4.7

1) Whilst all enzymes are probably best thought of as being flexible, allosteric enzymes undergo conformational changes as a consequence of substrate binding, which is not necessarily the case with non-allosteric enzymes. This statement is thus not true.

2) If you wrote 'true', re-read the previous section - the text states the answer! Think back to Section 4.3 and Michaelis Menten kinetics: if an enzyme giving a hyperbolic rate against [S] plot and gives a straight line when transformed to a double reciprocal plot, it is impossible for a sigmoid plot to also give a straight line after a similar transformation - at least over a wide [S] range. Further, K_M is not very helpful with allosteric enzymes, since the rate against [S] relationship changes depending on the circumstances.

3) Yes, because allosteric enzymes show a greater change in activity over a given [S] range, through co-operative binding of substrate. They are also amenable to activation and inhibition by compounds other than the substrate.

4) True. The requirement for allosteric behaviour is that multiple active sites (for substrate and perhaps also for regulatory molecules) are present. This enables binding of substrate at one site to influence substrate binding at other sites (co-operativity). This is routinely achieved by the possession of several subunits (often 4); however, recent research has identified allosteric behaviour in a single polypeptide.

4.8

1) This is true, at least in some cases, but is not the answer. The existence of allosteric inhibition means that activity at a given [S] can be lessened - but this does not cause the sigmoidal curve (it could, in principle occur with a hyperbolic curve, with the line in the presence of inhibitor simply being lower ie non-competitive inhibition)!

2) This is again true, but this is only a partial explanation. Some enzymes, such as malate dehydrogenase, have more than one subunit but do not display a sigmoid rate curve. Thus presence of subunits does not, on its own, result in allosteric effects.

3) Although dissimilar subunits can occur, this does not always result in allosteric behaviour.

4) This is the real reason for allosteric phenomena and a sigmoid rate against [S] plot. Several binding sites are required, which can be switched between low- and high-affinity forms. Binding of substrate to one subunit 'helps' binding of substrate to other subunits in the group (if we think of the 4 subunits of a tetramer as a group). This communication or 'co-operativity' is vital.

4.9

1) What is required here is that the system adjusts to changes in concentration by diverting the compound. This is what will happen if there are two isoenzymes, of differing Michaelis constants, such that one only becomes significantly active at high [S]. If this relatively high K_M isoenzyme is confined to a particular tissue, at high [S] material will be diverted in this tissue (assuming enzymes for subsequent steps are present). Thus choice c) is preferred.

2) Choice a) looks most suitable. Co-ordination between the pathways is required. This is most effectively accomplished by allosteric regulation, involving activation and inhibition.

3) Choice d) *De novo* synthesis of an enzyme, only required when the new substrate is present, is the most likely strategy. Thus induction of the enzyme, such that it is only synthesised when the nutrient is present, solves this problem in the most economical way (in energy terms).

4) Choice b) Reversible phosphorylation is frequently used to switch enzymes on and off, where a rapid and dramatic change in activity is needed. This strategy enables an instantaneous change in activity to be made.

Responses to Chapter 5 SAQs

5.1

Responses to 1) and 5) apply to fat-soluble vitamins. Responses 2), 3) and 4) apply to water-soluble vitamins.

5.2

The main purpose of this exercise was to get you to review the preceding material. This should have enabled you to make the following identifications.

1) Several of the compounds we have discussed contained sugars (for example, NAD^+, $NADP^+$, FAD, FMN, coenzyme A, cobalamin). NAD^+, $NADP^+$, coenzyme A and cobalamin are used as coenzymes, not cofactors, and the sugar involved (ribose) is not a sugar alcohol. The only cofactors which contain a sugar alcohol are FAD and FMN, which contain ribitol (Section 5.3.2).

2) The answer is coenzyme A, re-read Section 5.3.6 if in doubt.

3) The only cofactor which forms a Schiff's base with anything is PLP (Section 5.3.4). The reversible formation of a Schiff's base in this way (with enzyme or amino acid) is essential for its role in transamination reactions.

4) This is the third activity in this chapter involving this aspect of the organisation of metabolism - the answer is NADPH. Re-examine the relevant section (5.3.1) if in doubt.

5) The only cofactor (or vitamin) about which any reference to colour has been made was riboflavin and its derivatives, which are yellow (in fact PLP is also yellow but this was not described earlier). As noted in Section 5.3.2, FAD and FMN are bleached upon reduction.

6) Thiamine pyraphosphate (TPP); the one specific example given for the role of TPP (Section 5.3.5) was in reactions catalysed by transketolase. This enzyme is able to catalyse interconversions of this type.

If you feel that you struggled to make these identifications, re-read the preceding sections. The objective of this section is that you should gain an appreciation of what these cofactors and coenzymes are (in general terms) and of their typical roles or actions.

5.3

The common feature of NAD^+, $NADP^+$, FMN and FAD is that they are acceptors and carriers of reducing equivalents. Thus NADH and NADPH can be thought of as mobile carriers of reducing equivalents, whilst $FMNH_2$ and $FADH_2$ are cofactors, acting in more fixed locations. To these can be added coenzyme Q as another mobile carrier, and haem (or Fe^{2+}) as a further cofactor accepting reducing equivalents. The fact that Fe^{3+} only accepts an electron, NAD^+ accepts H^- and FAD accepts 2H is not relevant. They are all reducing equivalent acceptors.

Responses to Chapter 6 SAQs

6.1

1) The structure drawn is a homopolysaccharide since it only contains one type of monosaccharide unit (fructose). It is similar to inulin produced by Jerusalem artichokes.

2) The linkages shown are β1-2. They join fructose residues together at C1 of one residue to C2 of the other and the linkages are in the β configuration. Although most β-linked polysaccharides (especially polymers of glucose) are resistant to hydrolysis, not all β-linked polysaccharides are difficult to hydrolyse. In this instance the β-linked fructose units are easily hydrolysed.

3) The first stages in the catabolism of this polysaccharide is to convert it into a form which could be transported into the cells. This almost certainly means it will be hydrolysed to monosaccharides or disaccharides which can either diffuse into the cell or be actively transported across the cell membrane. If the polysaccharide had been produced within the cells, the first product of polysaccharide breakdown would probably have been the phosphate derivation of the monosaccharide units.

6.2

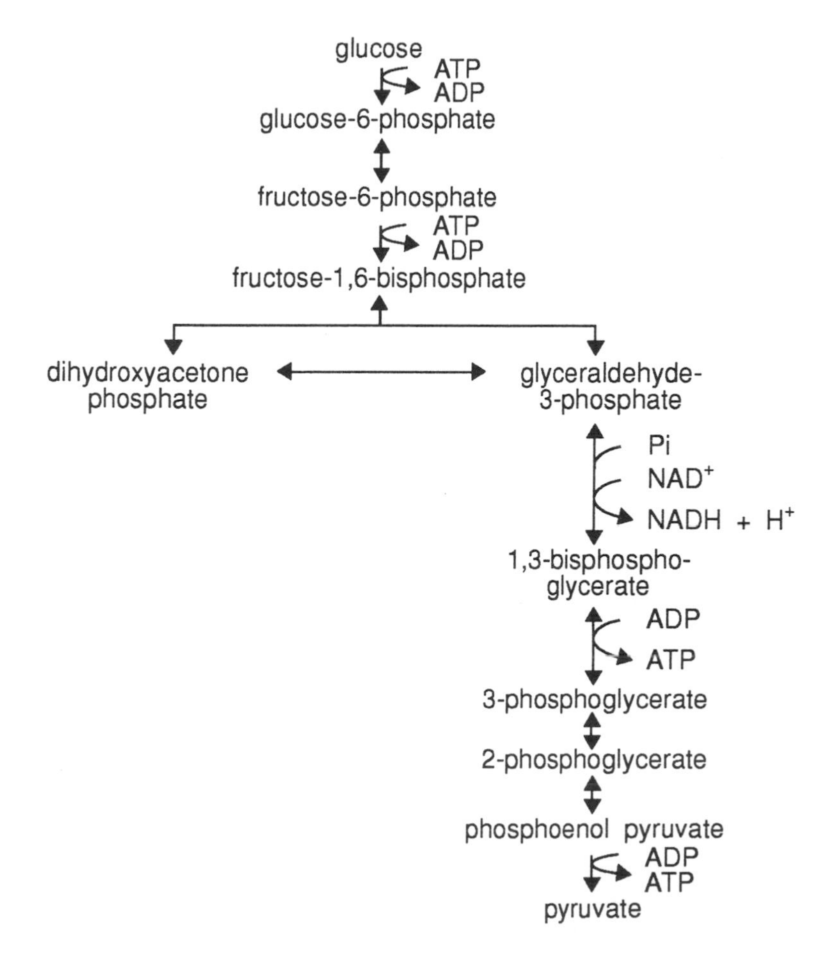

Check your diagram for completeness and check also that you have indicated correctly the one-way and the reversible reactions. The importance of irreversible reactions will be discussed later.

6.3 The overall equation is:

$$\text{glucose} + 2\text{ADP} + 2\text{Pi} + 2\text{NAD}^+ \rightarrow 2\text{ pyruvate} + 2\text{ATP} + 2\text{NADH} + 2\text{H}^+ \qquad (\text{E - 6.1})$$

You might have finished with the following equation:

$$\text{glucose} + \text{NAD}^+ \rightarrow 2\text{ pyruvate} + \text{NADH} + \text{H}^+$$

This latter equation, though not correct, shows that you have been working on the right lines but you have forgotten to take into account the fact that the dihydroxyacetone phosphate is converted to a second molecule of glyceraldehyde-3-phosphate.

The series of reactions can be written as follows and Equation (E - 6.1) obtained after cancelling out compounds which occur on both sides of the equation.

$$\text{glucose} + 2\text{ATP} \rightarrow \text{gly-3-P} + \text{DHAP} + 2\text{ADP}$$

$$\text{gly-3-P} + 2\text{ADP} + \text{NAD}^+ + \text{Pi} \rightarrow \text{pyruvate} + 2\text{ATP} + \text{NADH} + \text{H}^+$$

$$\text{DHAP} \rightarrow \text{gly-3-P}$$

$$\text{gly-3-P} + 2\text{ADP} + \text{NAD}^+ + \text{Pi} \rightarrow \text{pyruvate} + 2\text{ATP} + \text{NADH} + \text{H}^+$$

$$\text{glucose} + 2\text{ADP} + 2\text{Pi} + 2\text{NAD}^+ \rightarrow 2\text{ pyruvate} + 2\text{ATP} + 2\text{NADH} + 2\text{H}^+ \qquad (\text{E - 6.1})$$

Note: in these equations gly-3-P is glyceraldehyde-3-phosphate and DHAP is dihydroxyacetone phosphate.

6.4 We can represent the flow of carbon atoms in the Embden Meyerhof pathway in the following way:

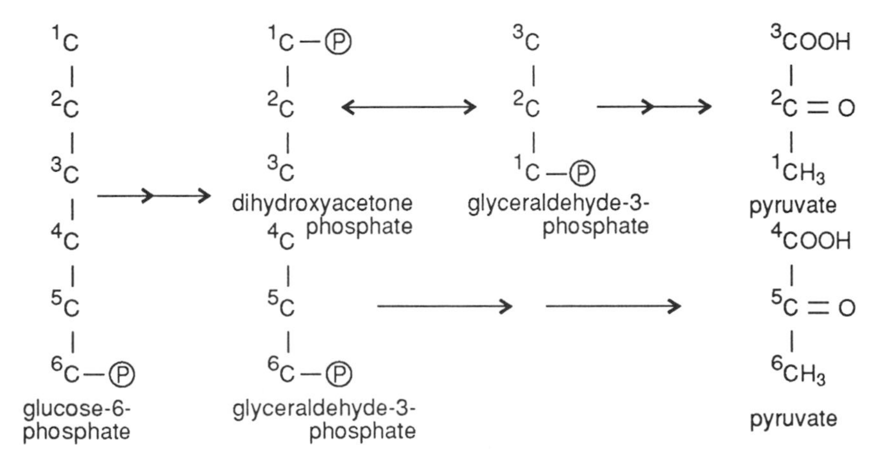

(Note we have numbered the carbon atoms so that we can follow them through the pathway).

But carbon dioxide is produced from the carboxyl (COOH) group of pyruvate. Thus metabolism of glucose by the Embden Meyerhof pathway leads to the release of carbons 3 and 4 of glucose as carbon dioxide.

6.5 No - the removal of erythrose-4-phosphate would leave fructose-6-phosphate, but, as we have seen, this features in the Embden Meyerhof pathway and would not, therefore, be wasted. The tetrose is often siphoned off from this pathway.

6.6 If NADPH + H^+ is not required then it is obvious that the oxidative part of hexose monophosphate pathway is not needed (Equation (E 6.2)). Thus the only way that pentoses can be produced is by operating the second half of the hexose monophosphate pathway in reverse (Equation (E 6.3)).

Let us write down the Equation (E 6.3), this time in reverse and see if the hypothesis above is reasonable.

2 fructose-6-phosphate + glyceraldehyde-3-phosphate ⇄ 3 ribulose-5-phosphate

Thus as well as the two substrates we shall require the two enzymes transaldolase and transketolase. We know that the substrates are available from glycolysis and that the two enzymes are ubiquitous, therefore our hypothesis is feasible.

6.7 Carbon dioxide is produced from carbons 1 and 4 of glucose when glucose is converted to pyruvate via the Entner Doudoroff pathway and the pyruvate is converted to CO_2 and acetyl CoA. This contrasts with the Embden Meyerhof pathway; when glucose is converted to acetyl CoA by this pathway, the carbon dioxide is produced from carbons 3 and 4 of glucose. If you have any doubts about this, re-examine SAQ 6.4 and draw out the two pathways yourself and check your answer.

6.8

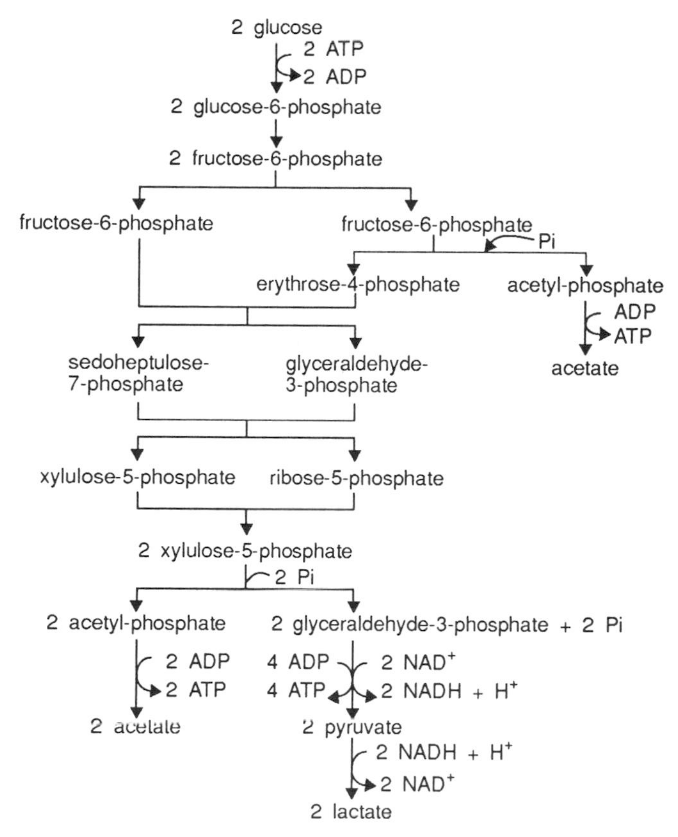

Overall reaction: 2 glucose + 5ADP + 5Pi → 2 lactate + 3 acetate + 5ATP

Net yield: 5ATP from 2 glucose molecules = 5/2 = 2.5 ATP from each glucose molecule.

Possible source of error - remember that because you need a fructose-6-phosphate and an erythrose-4-phosphate for transketolase activity you must start with two glucose molecules. Thus many compounds in the flow diagram need to be doubled.

6.9

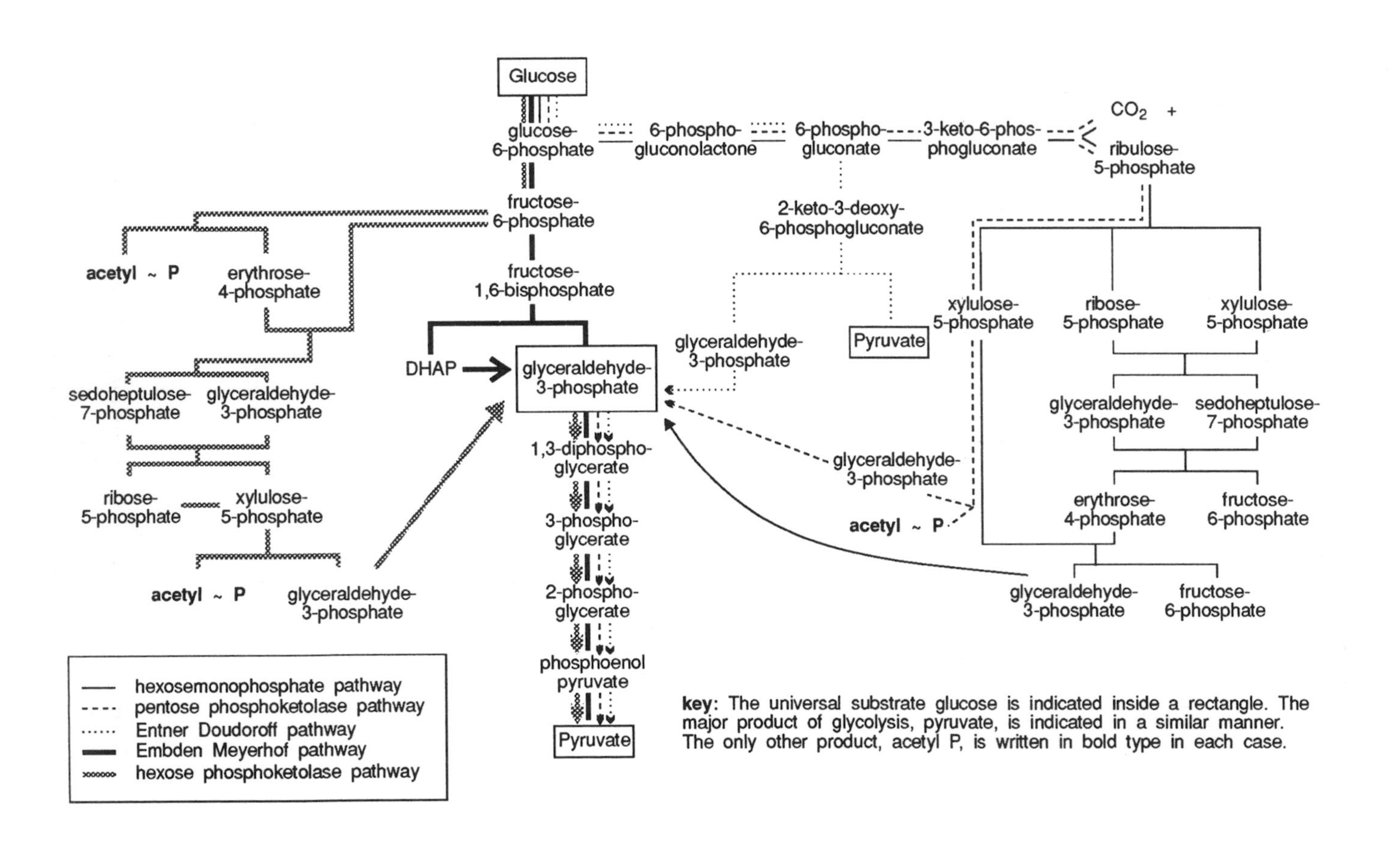

key: The universal substrate glucose is indicated inside a rectangle. The major product of glycolysis, pyruvate, is indicated in a similar manner. The only other product, acetyl P, is written in bold type in each case.

Responses to Chapter 7 SAQs

7.1 The diagram below shows the sequence of events which occur in the first turn of the TCA cycle using labelled pyruvate. From the diagram we can see that:

a) the carboxyl group of pyruvate is lost immediately as CO_2;

b) the keto group of the pyruvate eventually becomes one of the carboxyl groups of oxaloacetate;

c) the methyl group of pyruvate becomes the keto or methylene group of oxaloacetate.

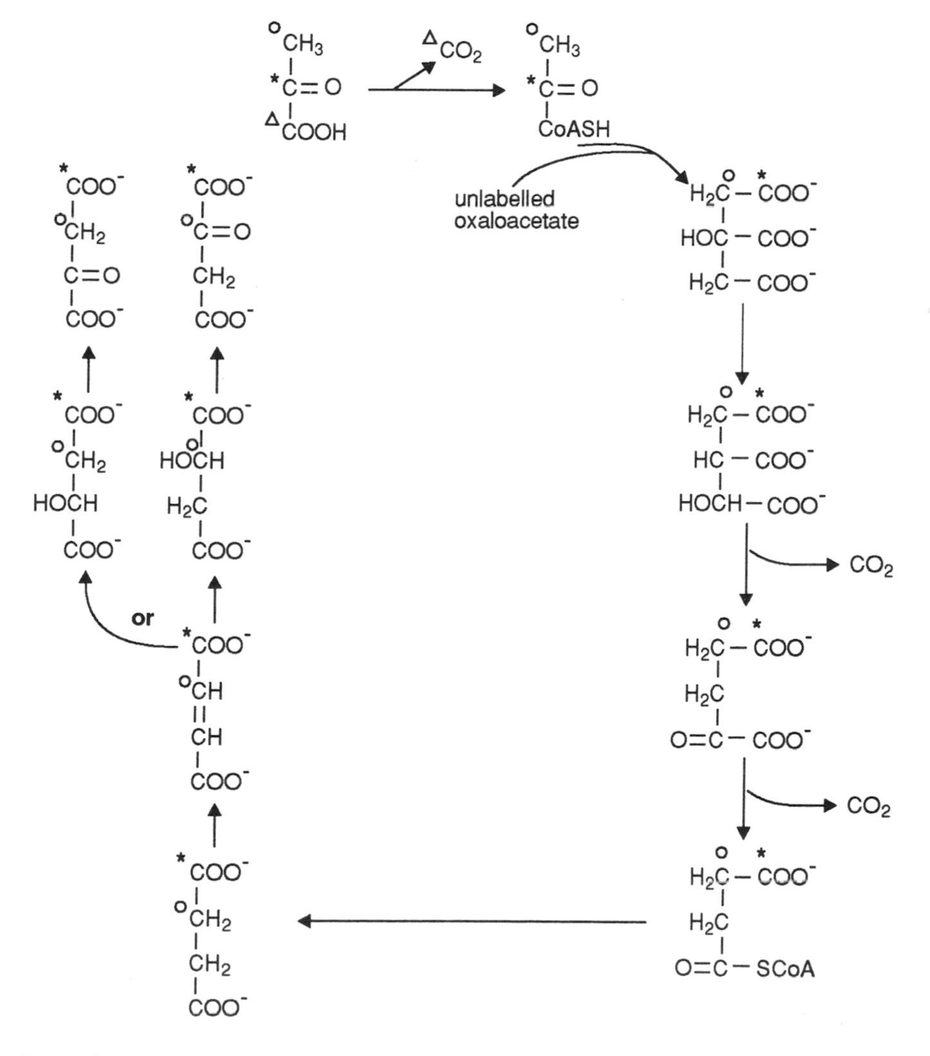

Therefore after one turn of the cycle all of the labelled carbon from the carboxyl of pyruvate has been removed but the carbon of the other two labelled carbons remains in oxaloacetate in the positions indicated.

Two points however require clarification. First let us consider the conversion of citrate to isocitrate. Citrate is a symmetrical molecule and you might have been puzzled as to whether the hydroxyl becomes attached to either or both of the two carbons when it forms isocitrate.

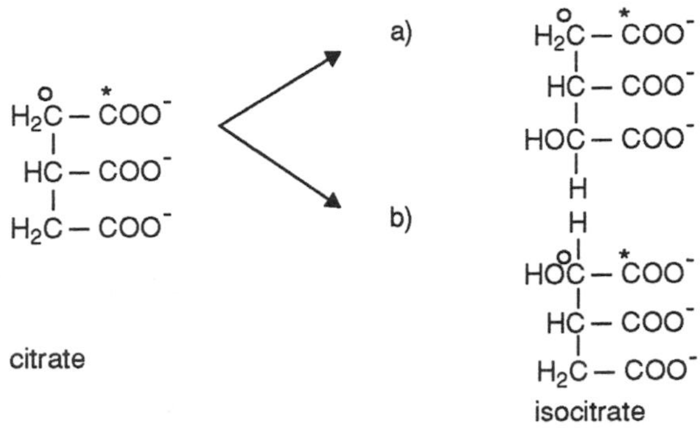

citrate: $H_2\overset{o}{C}-\overset{*}{C}OO^-$ / $HC-COO^-$ / H_2C-COO^-

a) $H_2\overset{o}{C}-\overset{*}{C}OO^-$ / $HC-COO^-$ / $HOC-COO^-$ / H

b) H / $HO\overset{o}{C}-\overset{*}{C}OO^-$ / $HC-COO^-$ / H_2C-COO^-

isocitrate

In fact the OH with respect to our labelling always moves as indicated on the previous page and as in option a) above. Compounds which have an asymmetric carbon can exist in one of two forms and are called chiral molecules. Many enzymes are specific for only one form and will thus only catalyse reactions involving that form. However we have noted that citrate does not have an asymmetric carbon; in fact it is an example of a prochiral molecule, that is it is potentially capable of reacting asymmetrically having two halves (in three dimensions) to the molecule which are different to each other. Aconitase recognises these different halves and always puts the hydroxyl group into one particular place.

The second point requiring clarification is the formation of malate from the symmetrical molecule fumarate. The latter is not a prochiral molecule and hence two possible forms of malate with respect to our labelling are possible, as shown on the previous page.

The following table summarises the fate of these two remaining carbons after three further turns of the cycle and shows the position of the label in oxaloacetate, the position of the label in isocitrate and also underlines the carbons removed as CO_2 by isocitrate dehydrogenase and α-oxoglutarate dehydrogenase.

oxaloacetate — isocitrate

start of 2nd cycle: $O=\overset{o}{C}-\overset{*}{C}OO^-$ / H_2C-COO^- + acetyl CoA → H_2C-COO^- / $H\overset{o}{C}-\overset{*}{\mathbf{C}}\mathbf{OO}^-$ / $HOC-COO^-$ / H

lost by isocitrate dehydrogenase stage

or

$O=C-COO^-$ / $H_2\overset{o}{C}-\overset{*}{C}OO^-$ + acetyl CoA → H_2C-COO^- / $HC-\mathbf{COO}^-$ / $HO\overset{o}{C}-\overset{*}{C}OO^-$ / H

lost at the α-oxoglutarate dehydrogenase stage

start of 3rd cycle: $\overset{o}{C}$ can be present in any position of oxaloacetate

$O=\overset{o}{C}-\overset{o}{C}OO^-$ / $H_2\overset{o}{C}-\overset{o}{C}OO^-$

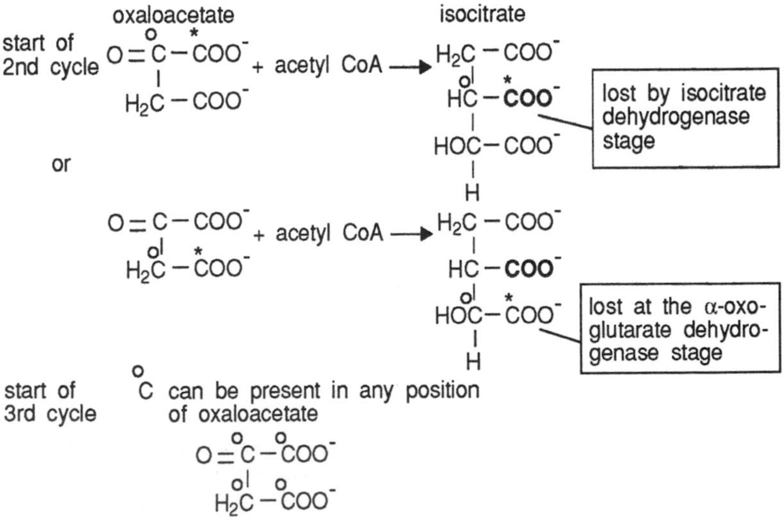

Note that whichever of the two possibilities of oxaloacetate is formed the carbon labelled C is lost on turn two of the cycle.

In summary

1) Carbon 1 of pyruvate (COOH) is lost in the bridging reaction.

2) Carbon 2 of pyruvate (C = 0) is lost at the second turn of the cycle by the action of isocitrate dehydrogenase.

3) Carbon 3 of pyruvate (CH_3) will be distributed amongst the four carbons of oxaloacetate. If we followed this around a third cycle, we would show that 50% of this label would be lost since 50% of the carbons of oxaloacetate are released as CO_2 by isocitrate dehydrogenase and α-oxaglutarate dehydrogenase.

7.2 The table below shows the areas of ATP production by aerobic organisms using glycolysis and the TCA cycle when glucose is completely oxidised to $6CO_2$. Remember that one molecule of glucose gives rise to two molecules of pyruvate and thus two molecules of acetyl CoA. Remember also that oxidation of each NADH produced in the cytophasm gives rise to net yield of 2ATPs whilst NADH produced by the TCA cycle in the mitochondrion gives rise to 3ATPs by oxidative phosphorylation.

ATP production by:-

Pathway	substrate level phosphorylation	reoxidation of $FADH_2$ by ETC and Ox Phos	reoxidation of NADH + H^+ by ETC and Ox Phos
glycolysis	2	-	4
bridging reaction	-	-	6
TCA cycle	2	4	18
subtotal	4	4	28
		32	
final total	**36**		

ETC = electron transport chain; Ox Phos = oxidative phosphorylation.

Thus 4ATP are produced by substrate level phosphorylation and 32 by the electron transport chain and oxidative phosphorylation system (ETC and Ox Phos); of the latter, 28 are produced from the reoxidation of NADH + H^+.

7.3 The critical point is that the TCA cycle is a cycle. If oxaloacetate is not available to react with acetyl CoA the cycle stops. If any of the intermediates are diverted less oxaloacetate will be produced and the cycle will slow down or even stop. This point of principle applies to all cycles.

7.4 Your scheme should have the following reactions:

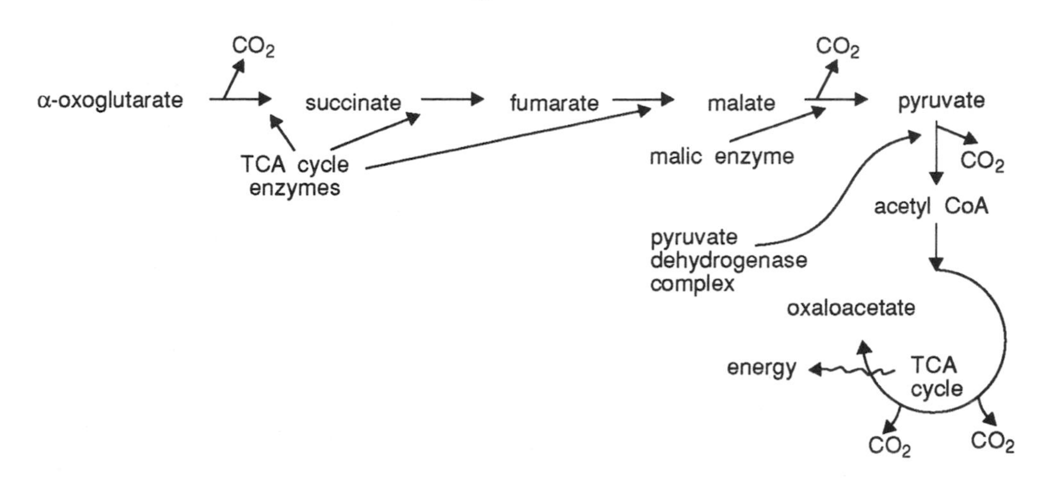

Thus C_5 (oxoglutarate) and C_4 (succinate, fumarate and malate) can be oxidised via the TCA cycle provided the malic enzyme is present.

7.5

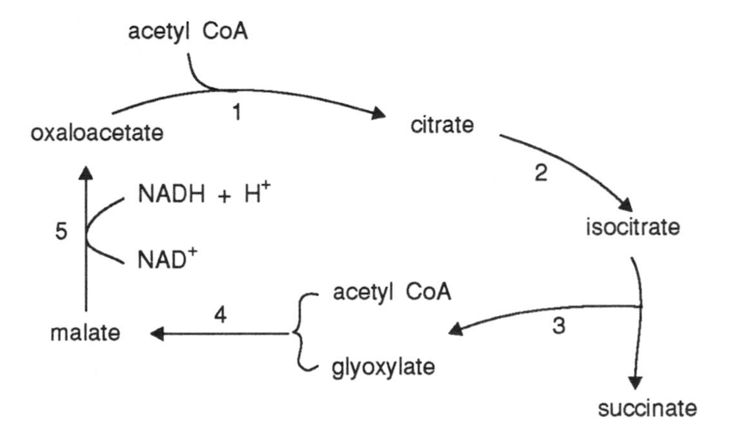

Isocitrate produces glyoxylate which is immediately utilised by malate synthase to produce malate. The task then is to regenerate isocitrate. The three TCA cycle enzymes necessary are enzymes 1, 2 and 5 above. The cycle drawn above is the glyoxylate cycle.

The overall reaction is:

$$2 \text{ acetyl CoA} + NAD^+ \rightarrow \text{succinate} + NADH + H^+$$

The end-product is therefore succinate.

If the derivation of an overall equation is difficult try listing the individual reactions and cancelling as for a series of simultaneous equations. This technique as demonstrated in Chapter 6 is useful, particularly for the elucidation of complicated pathways.

$$\text{isocitrate} \rightarrow \text{glyoxylate} + \text{succinate}$$

$$\text{glycoxylate} + \text{acetyl CoA} \rightarrow \text{malate} + \text{CoASH}$$

$$\text{malate} + \text{NAD}^+ \rightarrow \text{oxaloacetate} + \text{NADH} + \text{H}^+$$

$$\text{oxaloacetate} + \text{acetyl CoA} \rightarrow \text{citrate} + \text{CoASH}$$

$$\text{citrate} \rightarrow \text{isocitrate}$$

$$2\ \text{acetyl CoA} + \text{NAD}^+ \rightarrow \text{succinate} + \text{NADH} + \text{H}^+ + 2\text{CoASH}$$

7.6 Pathways 1), 2), 3) and 7) occur in the cytosol in all living systems. We have to be careful when writing down the location of 4), 5) and 6). If your answer put them in the mitochondrion, you are correct but you should have added the words 'in eukaryotic cells'. Remember that prokaryotic cells do not have mitochondria. In many cases they have organised but relatively non-membraneous cytosolic interiors. The possession of mitochondria in eukaryotic cells can be beneficial in that they can act as a control barrier and compartmentalise pathways. However problems can arise because sophisticated mechanisms of transport across the membrane(s) may be necessary.

7.7 Inhibition is brought about by raised levels of GTP, ATP, NADH + H^+ and acetyl CoA, in other words when the energy charge of the cell is high and intermediates are plentiful. Relief of inhibition is brought about by higher concentrations of factors such as NAD^+ and AMP, those expected to be in excess when the energy charge is low. These facts are logical in that they slow down production of an energy-yielding intermediate which would enter an energy-yielding cycle.

Responses to Chapter 8 SAQs

8.1 Calculating the outside relative to the inside:

1)

$$\Delta\ \text{potential} = 0 - (-.12) = 0.12\ \text{V}$$

$$\Delta\text{pH} = 6\text{-}8 = -2$$

$$\therefore\ \text{pmf} = 0.12\ \text{V} + (2 \times 0.06) = 0.24\ \text{V}$$

2)

$$\Delta\ \text{potential} = -0.06 - (+0.06) = -0.12$$

$$\Delta\text{pH} = 7\text{-}9 = -2$$

$$\therefore\ \text{pmf} = -0.12 + (2 \times 0.06) = 0\text{V}$$

8.2

1) False. The change in Gibbs function (free-energy released) is the same, the respiratory chain merely releases the energy in small steps.

2) True.

3) False. The voltage gradient is created by moving protons across the inner mitochondrial membrane, the electrons are passed to oxygen.

4) True.

5) False. Although protons can pass through the membrane via ATP synthetase and there is some leakage of protons across the membrane, the membrane should be regarded as being essentially impermeable to protons and is certainly not freely permeable.

6) False. The energy for pyruvate import into the matrix comes directly from the pmf.

7) False. Synthesis of ATP by ATP synthetase requires the import of protons into the matrix.

8.3

1) True. There is no membrane involved so the F_1 will work in one direction only, hydrolysing all the added ATP. (Actually it will hydrolyse the ATP until the thermodynamic equilibrium position is reached. Since K_{eq} for ATP hydrolysis is high, more-or-less all of the ATP will be hydrolysed before equilibrium is reached).

2) True. F_1 is on the outside of the vesicles and will use the energy of ATP hydrolysis to drive H^+ through F_o into the inside. You should note that in intact mitochondria the F_1 is located on the matrix side of the membrane.

3) False. Although F_1 has six ATP binding sites, ATP hydrolysis occurs in only one of these sites at a time.

4) False. ADP weakens the binding of the inhibitor to F_1, thus allowing ATP synthesis to proceed at a favourable rate.

5) True. Oligomycin blocks the movement of protons through F_o.

8.4

1) $\frac{1}{2}O_2 + 2H+ + 2e^+ \rightarrow H_2O$

2) $NAD^+ + H^+ + 2e^- \rightarrow NADH$

3) Yes, because ΔE_o for the reaction is positive ($\Delta E_o = +0.29 - (+0.22)$).

4) Fumarate + $FADH_2 \rightarrow$ succinate + FAD^+.

5) $\Delta E_o = 0.03 - (-0.06) = 0.09V$

8.5

The missing words are:

1) impermeable;

2) protons, matrix, electrons;

3) ATP hydrolysis, ATP synthesis.

8.6

At equilibrium $\qquad \Delta G_{H^+} = -0.096$ pmf

and $\qquad \Delta G_{H^+} = 210 \times (-0.096) = 20.16$ kJ mol^{-1}

$\therefore$ number of protons = 220/20.16 = 10.9 mol

This is the maximum value if an equilibrium was attained.

8.7

At equilibrium: ΔG_{ATP} synthesis $+ n\Delta G_{H+} = 0$

where n = number of protons pumped.

$$pmf = (0.5 \times 60) + 180 = 210mV$$

Thus: $\Delta G_{H+} = -0.096 \times 210 = -20.16 \text{ kJ mol}^{-1}$

So, at equilibrium we have: $-60 + n\,(-20.16) = 0$

$$n = -60/-20.16 = 2.97$$

8.8

We first calculate the free-energy of the NADH half reaction:

$$\Delta G^{\circ} = -2 \times 96.5 \times 1.14 = -220 \text{ kJ mol}^{-1} \quad (\text{since } \Delta G^{\circ} = -nF\Delta E_o)$$

Note 2 electrons are transfered hence n = 2.

For calculation of the value of ΔG° of the coupled reaction described we should remember that it contains the ATP half reaction in reverse order, and three times. Thus for the coupled reaction we have:

$$\Delta G^{\circ} = -220 + (3 \times 30.5) = -128.5 \text{ kJ mol}^{-1}$$

From an energetic point of view this reaction therefore strongly tends to move to the right.

8.9

1) Uncoupling electron flow from ATP generation removes respiratory control and leads to the speeding up of electron transport and substrate utilisation. So body fuel (fat) and carbohydrate is used more rapidly and weight loss occurs. (For reasons which should be obvious, this method of slimming is NOT recommended: fatalities have been recorded).

2) When the energy released during electron transport is not coupled to ATP synthesis, it is dissipated as heat - hence the rise in body temperature.

3) The clinical symptoms are all consistent with the patient having a poorly coupled electron transport chain. This is confirmed by the P:O data since a tightly coupled system has a ratio of three.

4) The inner mitochondrial membrane is not entirely impermeable to protons and there is some leakage from the outside into the matrix which is not associated with ATP synthesis. This allows the passage of some electrons along the respiratory chain and the consumption of small amounts of oxygen.

5) One possible explanation is the sudden consumption of a great deal of ATP by energy-requiring reactions. This would suddenly stimulate ATP synthesis via ATP synthetase resulting in an increased flow of protons through the membrane and thus a sudden dissipation of the pmf. The lowered pmf allows more electrons to pass down the respiratory chain (since respiratory control is slackened) and more protons are pumped out of the matrix, thus re-establishing the pmf at a constant value.

8.10

	Oxidation of		
	isocitrate	succinate	cytochrome C
rotenone	+	-	-
antimycin	+	+	-
cyanide	+	+	+

Note: + = inhibition - = no inhibition

8.11

1) Decrease. Oligomycin blocks proton movement by F_o and then the pH gradient is maintained. This restricts the flow of further electrons as it reduces further proton pumping.

2) Increase. Dinitrophenol uncouples oxidative phosphorylation from electron transport thus releasing the system from respiratory control. There is now unrestricted flow of electrons to oxygen.

3) Increase. Dinitrophenol bypasses F_o and, therefore, overrides the effect of oligomycin. However, oligomycin still inhibits the action of ATP synthetase in the presence of dinitrophenol because the protons cannot leave the pump.

4) Increase. An increased concentration of ADP would increase the rate of ATP synthesis by decreasing the affinity of binding of the F_1 inhibitor. This reduces the electrochemical proton gradient and respiratory control would be slackened.

8.12

The net result of the tricarboxylic acid cycle is the complete oxidation of pyruvate to CO_2 with the production of four molecules of NADH. Each of the NADH molecules can be oxidised to NAD^+ through the electron transport chain, producing three ATP molecules per NADH oxidised. In addition, two molecule(s) of ATP are formed during the flavin-linked oxidation of succinate, and one molecule(s) of ATP is formed by the substrate level phosphorylation at the expense of succinyl CoA. Thus a total of 15 ATP molecules are synthesised for each turn of the cycle. If we suppose that the glucose is being metabolised via glycolysis then two molecules of pyruvate are formed per molecule of glucose. This means that there are two turns of the TCA cycle which produces 30 molecules of ATP. In addition, glycolysis generates 2ATPs by substrate level phosphorylation and a further two NADH molecules. Oxidation of each of these extramitochondrial NADH molecules, via the electron transport chain, may each generate two or three molecules of ATP depending on the shuttle used to transfer reducing equivalent into the mitochondrial matrix. If we assume that they glycerol phosphate shuttle is operating then four ATPs are generated from the extramitochondrial NADH. This means that aerobic respiration produces 36 ATP molecules per molecule of glucose oxidised.

8.13

		Moles ATP equivalent
stearic acid to acetyl-CoA		
ATP	(activation)	-2
8 x $FADH_2$		16
8 x NADH		24
tricarboxylic acid cycle and oxidative phosphorylation		
9 x GTP		9
9 x $FADH_2$		18
27 x NADH		81
	Total moles ATP per mole of stearic acid:	146

Note 9GTPs produced by substrate level phosphorylation between succinyl CoA and succinate in the TCA cycle.

Note all of the remainder of the ATP is produced by oxidative phosphorylation.

8.14

$$\Delta G^\circ \text{ for the ATP synthesis } = +30.5 \times 130 = +3965 \text{ kJ mol}^{-1}.$$

Energy capturing efficiency = (3965/9790) x 100 = 40% of the standard free-energy of oxidation of palmitic acid is recovered as phosphate bond energy. As you can see glucose and palmitate oxidation are carried out with similar efficiencies.

Responses to Chapter 9 SAQs

9.1

1) The redox potential for nitrate reduction (+0.4V) is lower than that for reduction of oxygen (+0.8V). This explains why two ATP molecules rather than three, are generated by oxidative phosphorylation when nitrate is used as the terminal electron acceptor.

2) Inhibition of nitrate reductase by O_2 would be advantageous to a denitrifying bacterium because aerobic respiration generates more ATP than respiration using nitrate.

3) Four ATP. Four electrons are required to reduce nitrate to nitrous oxide and 2 ATP are generated per pair of electrons transferred. This should be regarded as being less energy efficient than the reduction of one atom of oxygen because the amount of ATP generated per pair of electrons transferred is lower.

4) Condition a) *Paracoccus denitrificans* would not grow because chlorate inhibits nitrate reductase and the absence of O_2 prevents aerobic respiration.

Condition b) *Paracoccus denitrificans* would not grow because antimycin prevents aerobic respiration and oxygen inhibits nitrate reductase.

9.2

1) True. Sulphite is an intermediate in the reduction sequence of sulphate to elemental sulphur. *Desulphovibrio* can use any of the intermediates as a terminal electron acceptor.

2) True. Two high-energy bonds are broken in the reduction of sulphate to sulphite.

3) True. Fumarate reduction can be coupled to NADH oxidation. This can pump protons through the NADH dehydrogenase complex (site 1 ATP synthesis) of the respiratory chain with the generation of 1 ATP per pair of electrons transferred to fumarate.

4) False. Methanogenic bacteria are killed in the presence of oxygen.

5) True. Iron II is formed by reduction of the terminal electron acceptor iron III.

9.3

Statement 1) is applicable to fermentation, c).

Statement 2) is applicable to anaerobic respiration and fermentation, b) and c). You will recall that fumarate can serve as a terminal electron acceptor for anaerobic respiration.

Statement 3) is applicable to all three modes of metabolism.

Statement 4) is applicable to fermentation, c).

Statement 5) is applicable to all three modes of metabolism.

Statement 6) is applicable to all three modes of metabolism.

9.4

1) Homolactic fermentation:

$$\text{glucose} + 2\ \text{ADP} + 2\ \text{Pi} \rightarrow 2\ \text{lactate} + 2\text{ATP}$$

Heterolactic fermentation:

$$\text{glucose} + \text{ADP} + \text{Pi} \rightarrow \text{lactate} + \text{ethanol} + CO_2 + \text{ATP}$$

2) For homolactic fermentation:

$$\Delta G^{o'} = -137 + (2 \times 49) = -39 \text{ kJ mol}^{-1} \text{ glucose.}$$

Energy capturing efficiency = 98/137 x 100 = 72%.

3) We can see from Figure 9.4 that pyridine nucleotides produced in the pentose phosphate pathway are reoxidised by the formation of ethanol. The formation of lactate from glyceraldehyde-3-P is also in oxidation-reduction balance with respect to NADH.

9.5

We can see from Figure 9.5 that the formation of lactate from pyruvate oxidises 1 NADH. The formation of acetate and formate does not involve oxidation of NADH and therefore leaves 1 NADH unoxidised. Other products must therefore be formed to maintain the oxidation-reduction balance. These products could be ethanol and/or succinate since for each pyruvate consumed in the formation of one of these compounds, 2 NADH + H^+ are oxidised.

9.6

$$\Delta G' = -306 + 5.7 \log \frac{(0.001).(0.001).(0.001)^2}{(0.001)}$$

$$= -357.3 \text{ kJ mol}^{-1}$$

In the fermentation; 3 ATP are produced (2 in glycolysis and one in the formation of acetate). So the amount of energy conserved = 3 x (+49) = +147 kJ mol^{-1}.

The efficiency is thus: 147/357 = 41%

9.7

From the range of fermentation end-products we can conclude that the organism is a mixed acid fermenter. We know that, in order to maintain oxidation-reduction balance all the NADH generated in glycolysis must be reoxidised in reactions leading from pyruvate to end-product. From Figure 9.5 we can deduce that:

1) The formation of 10 moles of lactate requires 5 moles of glucose and all the NADH produced is reoxidised.
2) The formation of 10 moles of hydrogen and carbon dioxide is equivalent to the formation of 10 moles of formate. This, in turn, requires 5 moles of glucose and 10 NADH are produced. 10 moles of acetyl CoA are also generated at this stage.
3) The 5 moles of acetate are derived from 5 moles of acetyl CoA; no additional glucose is required and NADH is not generated. 5 moles of acetyl CoA remain.

The formation of the end-products shown in the list, therefore, utilises 10 moles of glucose and generates 10 NADH which are so far not reoxidised. If the 5 moles of acetyl CoA left in 3) produce 5 moles of ethanol, this would reoxidise the 10 NADH. Thus, 10 moles of glucose were supplied and 5 moles of ethanol produced.

9.8

1) Organism b) should be regarded as a butanediol fermenter because i) butanediol is an end-product and ii) the ratio CO_2:H_2 is greatly in favour of CO_2. In comparison, a mixed acid fermenter, such as organism a), does not produce butanediol and produces CO_2 and H_2 in a 1:1 ratio, via formate.
2) Organisms a, b and d. In organisms a and b there is much more H_2 released than formate, therefore, the organism must have utilised formate hydrogenlyase to break down the formic acid thereby increasing the pH. Organism b and d produce significant quantities of 2,3 - butanediol thereby switching metabolism away from the production of acid. This is corroborated by the high levels of CO_2 released in comparison with H_2.

9.9

1) a) False. Yeasts decarboxylate pyruvate directly to acetaldehyde whereas solvent fermenters produce acetyl CoA as an intermediate.

   b) False. Basic butyrate fermenters produce acetate, not acetone.

   c) True. See Figure 9.6.

   d)True. The H_2 is required to reduce NAD^+ and thus maintain the oxidation-reduction balance.

9.10 a) is lactic acid; b) is formic acid; c) is acetone; d) is succinic acid; e) is glucose; f) is pyruvic acid; g) is acetic acid; h) is butyric acid; i) is ethanol; j) is butanol.

9.11 1) The end-products are:

1) lactate; 2) 2, 3-butanediol; 3) formate; 4) H_2 (or CO_2); 5) CO_2 (or H_2); 6) ethanol; 7) acetic acid; 8) acetone; 9) isopropanol; 10) butyric acid; 11) butanol; 12) succinate; 13) propionate; 14) CO_2 (or H_2); 15) H_2 (or CO_2).

2) Reaction a) is used by homolactic, heterolactic and mixed acid fermenters.

Reaction b) is used by butanediol fermenters.

Reaction c) is used by mixed acid fermenters.

Reaction d) is used by basic butyrate and solvent fermenters.

Reaction e) is used by mixed acid and butanediol fermenters.

Reaction f) is used by mixed acid and propionate fermenters.

Reaction g) is used by mixed acid and solvent fermenters

Reaction h) is used by mixed acid and basic butyrate fermenters.

9.12 1) a) is homolactic fermentation: thus 2 mol of ATP are produced per mole of glucose metabolised; similarly b) is ethanol fermentation; 2ATP mol^{-1} glucose; c) is mixed acid fermentation; 2ATP from glycolysis and some extra ATP from acetate formation, depending on how much is produced. Therefore, the yield here is more than 2 mols of ATP per mole of glucose fermented.

Since Y_{ATP} = constant, bacterium c) would have the highest growth yield; bacteria a) and b) would be equal to each other but less than c).

2) Since:

$$(\text{ATP mol}^{-1}\text{ glucose}) \times Y_{ATP} = \text{molar growth yield and } Y_{ATP} = 10.5\text{g mol}^{-1}\text{ ATP}$$

thus for A):

$$\text{ATP mol}^{-1}\text{ glucose} = \frac{22}{10.5} = 2$$

and for B):

$$\text{ATP mol}^{-1}\text{ glucose} = \frac{33}{10.5} = 3$$

Therefore A) is homolactic fermentation and B) is heterolactic fermentation.

Appendix 1

Units of measurement

For historical reasons a number of different units of measurement have evolved. The literature reflects these different systems. In the 1960s many international scientific bodies recommended the standardisation of names and symbols and a universally accepted set of units. These units, SI units (Systeme Internationale de Unites) were based on the definition of: metre (m), kilogram (kg); second (s); ampare (A); mole (mol) and candela (cd). Although, in the intervening period, these units have been widely adopted, their adoption has not been universal. This is especially true in the biological sciences.

It is, therefore, necessary to know both the SI units and the older systems and to be able to interconvert between both sets.

The BIOTOL series of texts predominantly uses SI units. However, in areas of activity where their use is not common, other units have been used. Tables 1 and 2 below provides some alternative methods of expressing various physical quantities. Table 3 provides prefixes which are commonly used.

Mass (SI unit: kg)	Length (SI unit: m)	Volume (SI unit: m^3)	Energy (SI unit: $J = kg\ m^2\ s^{-2}$)
$g = 10^{-3}$ kg	$cm = 10^{-2}$ m	$l = dm^3 = 10^{-3}\ m^3$	cal = 4.184 J
$mg = 10^{-3}\ g = 10^{-6}$ kg	$Å = 10^{-10}$ m	$dl = 100\ ml = 100\ cm^3$	$erg = 10^{-7}$ J
$\mu g = 10^{-6}\ g = 10^{-9}$ kg	$nm = 10^{-9}\ m = 10Å$	$ml = cm^3 = 10^{-6}\ m^3$	$eV = 1.602 \times 10^{-19}$ J
	$pm = 10^{-12}\ m = 10^{-2}\ Å$	$\mu l = 10^{-3}\ cm^3$	

Table 1 Units for physical quantities

Concentration (SI units: mol m^{-3})

a) M = mol l^{-1} = mol dm^{-3} = 10^3 mol m^{-3}

b) mg1^{-1} = μg cm^{-3} = ppm = 10^{-3} g dm^{-3}

c) μg g^{-1} = ppm = 10^{-6} g g^{-1}

d) ng cm^{-3} = 10^{-6} g dm^{-3}

e) ng dm^{-3} = pg cm^{-3}

f) pg g^{-1} = ppb = 10^{-12} g g^{-1}

g) mg% = 10^{-2} g dm^{-3}

h) μg% = 10^{-5} g dm^{-3}

Table 2 Units for concentration

Fraction	Prefix	Symbol	Multiple	Prefix	Symbol
10^{-1}	deci	d	10	deka	da
10^{-2}	centi	c	10^2	hecto	h
10^{-3}	milli	m	10^3	kilo	k
10^{-6}	micro	μ	10^6	mega	M
10^{-9}	nano	n	10^9	giga	G
10^{-12}	pico	p	10^{12}	tera	T
10^{-15}	femto	f	10^{15}	peta	P
10^{-18}	atto	a	10^{18}	exa	E

Table 3 Prefixes for SI units

Appendix 2

Chemical Nomenclature

Chemical nomenclature is quite a difficult issue especially in dealing with the complex chemicals of biological systems. To rigidly adhere to a strict systematic naming of compounds such as that of the International Union of Pure and Applied Chemistry (IUPAC) would lead to a cumbersome and overly complex text. BIOTOL has adopted a pragmatic approach by predominantly using the names or acronyms of chemicals most widely used in biologically-based activities. It is recognised however that there remains some potential for confusion amongst readers of different background. For example the simple structure CH_3COOH can be described as ethanoic acid or acetic acid depending on the environment or industry in which the compound is produced or used. To reduce such confusion, the BIOTOL series makes every effort to provide synonyms for compounds when they are first mentioned and to provide chemical structures where clarity and context demand.

Appendix 3

Abbreviations used for the common amino acids

Amino acid	Three-letter abbreviation	One-letter symbol
Alanine	Ala	A
Arginine	Arg	R
Asparagine	Asn	N
Aspartic acid	Asp	D
Asparagine or aspartic acid	Asx	B
Cysteine	Cys	C
Glutamine	Gln	Q
Glutamic acid	Glu	E
Glutamine or glutamic acid	Glx	Z
Glycine	Gly	G
Histidine	His	H
Isoleucine	Ile	I
Leucine	Leu	L
Lsyine	Lys	K
Methionine	Met	M
Phenylalanine	Phe	F
Proline	Pro	P
Serine	Ser	S
Threonine	Thr	T
Tryptophan	Trp	W
Tyrosine	Tyr	Y
Valine	Val	V

Index

A

B